BIOSEPARATIONS SCIENCE AND ENGINEERING

TOPICS IN CHEMICAL ENGINEERING: A SERIES OF TEXTBOOKS AND MONOGRAPHS

SERIES EDITOR
Keith E. Gubbins, North Carolina State University

ASSOCIATE EDITORS
Mark A. Barteau, University of Delaware
Douglas A. Lauffenburger, MIT
Manfred Morari, ETH
W. Harmon Ray, University of Wisconsin
William B. Russel, Princeton University

Deen, *Analysis of Transport Phenomena*
Doraiswamy, *Organic Synthesis Engineering*
Floudas, *Nonlinear and Mixed Integer Optimization: Fundamentals and Applications*
Friedlander, *Smoke, Dust, and Haze, Second Edition*
Fuller, *Optical Rheometry of Complex Fluids*
Harrison, Todd, Rudge, and Petrides, *Bioseparations Science and Engineering*
Larson, *The Structure and Rheology of Complex Fluids*
Lauffenburger and Linderman, *Receptors: Models for Binding, Trafficking, and Signalling*
Morrison, *Understanding Rheology*
Ogunnaike and Ray, *Process Dynamics, Modeling, and Control*
Pearson, *Discrete-Time Dynamic Models*
Phan-Thien and Kim, *Microstructures in Elastic Media*
Pozrikidis, *An Introduction to Theoretical and Computational Fluid Dynamics*
Pozrikidis, *Numerical Computation in Science and Engineering*
Schmidt, *The Engineering of Chemical Reactions*
Varma and Morbidelli, *Mathematical Methods in Chemical Engineering*

BIOSEPARATIONS SCIENCE
AND ENGINEERING

Roger G. Harrison
*University of Oklahoma,
Norman, Oklahoma*

Paul Todd
*Space Hardware Optimization Technology, Inc.
Greenville, Indiana*

Scott R. Rudge
*FeRx, Inc.
Aurora, Colorado*

Demetri P. Petrides
*Intelligen, Inc.
Scotch Plains, New Jersey*

New York   Oxford
OXFORD UNIVERSITY PRESS
2003

ty Press

York
ngkok   Buenos Aires   Cape Town   Chennai
Delhi   Hong Kong   Istanbul   Karachi   Kolkata
r   Madrid   Melbourne   Mexico City   Mumbai   Nairobi
Shanghai   Singapore   Taipei   Tokyo   Toronto

© 2003 by Oxford University Press, Inc.

d by Oxford University Press, Inc.
dison Avenue, New York, New York, 10016
www.oup-usa.org

ord is a registered trademark of Oxford University Press

Library of Congress Cataloging-in-Publication Data

Bioseparations science and engineering / Roger G. Harrison . . . [et al.].
    p. cm — (Topics in chemical engineering)
  Includes bibliographical references and index.
  ISBN 0-19-512340-9 (acid-free paper)
    1. Biomolecules—Separation. 2. Biochemical engineering. I. Harrison, Roger G., 1944-
  II. Topics in chemical engineering (Oxford University Press)

TP248.25.S47 B55 2002
660.6—dc21

                                                                2002019609

Printing number: 9 8 7 6 5 4 3 2

Printed in the United States of America
on acid-free paper

We dedicate this book to our families.

Contents

CHAPTER 5
Sedimentation   142

CHAPTER 9
Crystallization   272

CHAPTER 10
Drying   290

CHAPTER 11

Bioprocess Design   319

CHAPTER 12
Laboratory Exercises in Bioseparations   373

APPENDIX
Table of Units and Constants   392

Preface

Bioseparations has developed as a significant, separate discipline within the general field of biochemical engineering. Bioseparations involves the separation and purification of compounds of biological origin, which are derived from cells grown in bioreactors or from cells contained in animal or plant tissue. The biotechnology industry, which originated in the late 1970s, gave added importance to bioseparations because many of the products of biotechnology are proteins that often are difficult to purify and frequently must be purified to homogeneity or near homogeneity, leading to high costs. Practitioners of bioseparations include biochemical engineers and biochemists in the pharmaceutical, biotechnology, food, and chemical industries. This new text focuses on the science and engineering aspects of bioseparations and is designed for juniors, seniors, and graduate students. The book is also intended to be useful for practitioners in industry.

The first chapter contains introductory material in two parts, the first of which is basic information about biochemistry. Bioprocess engineers need to know about the physical properties of the materials they must purify, which include antibiotics, vitamins, and vaccines, and, in the field of biotechnology, proteins, nucleic acids, antibodies, subcellular particles, and whole cells. An understanding of cell structure and function is necessary for the intelligent choice of early postfermentation processes. The second part of this chapter is a brief review of engineering analysis, which is used in various forms throughout the book.

Because we believe that good analytical methods are the foundation of any effort to develop, optimize, and then operate and troubleshoot any bioprocess, the introductory chapter is followed by a chapter on analytical methods. A good understanding of analytical methods is important for anyone involved in bioseparations process development or production.

The important unit operations are thoroughly covered in separate chapters that follow (Chapters 3–10), and the order of the chapters is similar to the order in which the operations are used in a typical bioseparations process. The approach in each of the chapters on unit operations is to start with a qualitative description indicating the importance and general application of the unit operation, describe the scientific foundation of the operation, develop the necessary mathematical theory, and finally describe the applications of the theory in engineering practice with an emphasis on design and scaleup.

In deciding on the subject matter in each of the chapters on unit operations, we have been guided by a desire to emphasize the aspects of unit operations that are most important as applied to bioseparations at the commercial scale. For example, we have not discussed all methods of industrial drying, just those that are the most important in general for drying bioproducts; nor have we included details concerning unit operations currently used primarily in the laboratory, such as preparative electrophoresis.

A comprehensive treatment of bioseparations process design (Chapter 11) follows the unit operations chapters and focuses on how to integrate the individual unit operations in developing a process design that is the "best" among several plausible alternatives. The use of a process simulator (SuperPro Designer; Intelligen, Inc., Scotch Plains, NJ) is illustrated to analyze and evaluate the production of three biological products—citric acid, recombinant human insulin, and monoclonal antibodies. All the problems in this chapter are intended as open-ended assignments, which can be worked either with or without a biochemical process simulator. The use of a bioprocess simulator, however, will greatly facilitate the process analysis, including generating flowsheets, carrying out material balances, and analyzing costs. It is recommended that students have 6 to 8 weeks to complete their problems.

SuperPro Designer is our strong preference for a biochemical process simulator that can handle batch as well as continuous processes. A functional evaluation version of SuperPro Designer can be accessed at the website www.intelligen.com. A site license (one time fee) of SuperPro Designer can be obtained at a discount for courses at universities adopting *Bioseparations Science and Engineering*.

Illustrative example problems are included within the text throughout Chapters 1 to 11. The purpose of these examples is often to show how to apply previously presented theory; in some cases, the theory is extended to other situations. One way that these examples can be effectively used in the classroom is by using the "thinking aloud pair problem-solving" (TAPPS) method, where one student is the problem solver and talks through the example and the other student is the listener who questions and prompts the solver to keep talking, giving clues if necessary.

Numerous problems developed for this book are given at the ends of Chapters 1-11. These problems are intended to help the student both understand and be able to apply the material presented. A few of the problems are deliberately underspecified or of an open-ended nature. The problems can be assigned as out-of-class homework or can be tackled in class by students working in groups (of two, three, or four, depending on the complexity of the problem). Groups in class respond to the task in the following way: (1) each student formulates his or her answer, (2) students share their answers with their partner or partners, (3) students listen carefully to partner's or partners' answers, and (4) groups create a new answer that is superior to each member's initial formulation through the process of association, building on each other's thoughts, and synthesizing.

The practicum (Chapter 12) describes a set of bioseparations experiments that has been thoroughly tested by students (University of Colorado). A few selected experiments, or all of them, could be used in a course on bioseparations.

Additional information supporting this textbook and bioseparations in general can be found at the website www.biosep.ou.edu. Material at this website includes new problems and examples, which are added periodically; links to useful databases (such as for proteins); and links to manufacturers of bioseparations equipment and supplies.

Acknowledgments

We are grateful for the contributions that many people made, either directly or indirectly, to this book. The comments we received from the external reviewers of the book—Larry Erickson, Juan Hong, Harold Monbouquette, and Todd Przybycien—were very helpful. Ed Cussler, Antonio Garcia, and Richard Willson gave us encouragement and feedback in their reviews of the prospectus of the textbook. Jean-Bernard Gros read the final draft and checked the solutions of the problems giving us helpful feedback. Eric Dunlop deserves special recognition for his extensive direct contributions to several chapters and for helping establish the book's didactic paradigm of applying basic principles of physical chemistry during the earliest years of text development; with the assistance of Harold Null of Monsanto, he created a set of lecture notes on bioseparations well before the participation of the authors. The selection of content, sequencing, and language of this textbook were driven in part by the early efforts of Robert Davis, Dale Gyure, Subhas Sikdar, and Geoffrey Slaff while team teaching and developing a bioseparations course at the University of Colorado. We appreciate the feedback we have received over several years from students taking courses and from the teaching assistants in these courses at the Universities of Colorado and Oklahoma, where material on bioseparations was presented.

Finally, we are thankful for the patience and support from our families during the writing of this book.

Roger G. Harrison
Paul Todd
Scott R. Rudge
Demetri P. Petrides

Introduction to Bioproducts and Bioseparations

Bioproducts—chemical substances or combinations of chemical substances that are made by living things—range from methanol to whole cells. They are derived by extraction from whole plants and animals or by synthesis in bioreactors containing cells or enzymes. Bioproducts are sold for their chemical *activity:* methanol for solvent activity, ethanol for its neurological activity or as a fuel, penicillin for its antibacterial activity, taxol for its anticancer activity, streptokinase (an enzyme) for its blood clot dissolving activity, hexose isomerase for its sugar-converting activity, and whole *Bacillus thuringiensis* cells for their insecticide activity, to name a few very different examples. The wide variety represented by this tiny list makes it clear that bioseparations must encompass a correspondingly wide variety of methods. The choice of separation method depends on the nature of the product, remembering that purity, yield, and activity are the goals, and the most important of these is activity.

This first chapter therefore reviews the chemical properties of bioproducts with themes and examples chosen to heighten awareness of those properties that must be recognized in the selection of downstream processes that result in acceptably high final purity while preserving activity. The final part of this chapter is an introduction to the field of bioseparations, which includes a discussion of the stages of downstream processing, the basic principles of engineering analysis as applied to bioseparations, and the various factors involved in developing a bioproduct for the marketplace.

The pharmaceutical, agrichemical, and biotechnology bioproduct industries account for several billion dollars in annual sales—neglecting, of course, commodity foods and beverages. By "bioproduct" we mean chemical substances that must be extracted and purified to some degree before being suitable for market. Figure 1.1 indicates a clear inverse relationship between bioproduct market size and cost.

Owing to intense competition, cost, price, and value are very closely related, except in the case of completely new products that are thoroughly protected by patents, difficult to copy, and of added value to the end user. Products with these characteristics—"biotechnology products"—have typically been developed at considerable cost (over $100 million for R&D alone in the case of recombinant therapeutic proteins requiring clinical trials) and marketed at prices that allow for the recovery of the development, production, and marketing costs. Their total contribution to the bioproduct market is also in the billions of dollars

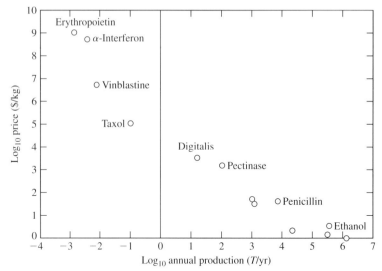

Figure 1.1 World production levels and prices of bioproducts, showing the inverse relationship between price and production. (Data from sources in the early 1990s time frame.)

annually (see Table 1.1) [1]. Such products must recover their development costs within a few years of initial sales, owing to potential competing products, expiration of patents, and economic pressures. Eventually such products move "down" the curve of Figure 1.1, but they stay on the curve. In the future, movement down the curve will require increased scale and reduced cost of production. Therefore, while existing processes meet many of the requirements of the current market, innovation will eventually be required for the economic large-scale production of biotechnology products. Thus, the following chapters are a practical guide to biochemical separations as currently practiced and as might be practiced in the future.

TABLE 1.1
Summary of U.S. Biotechnology Product Sales Forecasts (Millions of 2000 Dollars) by Key Market Sectors: 2000–2010[a]

Key Sectors	Sales Forecasts ($ million)			Annual Growth, 2000–2010 (%)
	2000	2005	2010	
Human therapeutics	$11,700	20,600	36,300	12
Human diagnostics	2,500	3,700	5,400	8
Agriculture	980	2,250	5,100	18
Specialty chemicals	550	1,170	2,400	16
Nonmedical diagnostics	320	480	700	8
Totals	16,050	28,200	49,900	12

[a]Data from reference 1.

1.1 Instructional Objectives

After completing this chapter, the reader should be able to do the following:

- Broadly classify bioproducts as small molecules, macromolecules, and particulate products including cells.
- Explain the differences between the structures of the various bioproducts.
- Explain the difference between a primary metabolite and a secondary metabolite.
- Outline the structure of proteins at four levels and their stability and functions.
- Explain the structures of other macromolecular substances that are commercial bioproducts, such as nucleic acids and polysaccharides.
- Outline the four stages of downstream processing, the objectives of each stage, and typical unit operations for each stage.
- Explain the engineering analysis concepts of material balance, equilibria, and transport phenomena.
- Calculate purity, specific activity, and yield as quality indicators in purification.

1.2 Broad Classification of Bioproducts

Bioproducts can be broadly classified into the following categories: small molecules, consisting of fine chemicals, antibiotics, hormones, amino acids, and vitamins; large molecules, consisting of proteins, polysaccharides, and nucleic acids; and particulate products, consisting of cells, spores, liposomes, and subcellular particles or organelles. Table 1.2

TABLE 1.2
Broad Categories of Bioproducts and Their Sizes

Bioproduct	Examples	Molecular Weight (Da)	Typical Radius
Small molecules	Sugars	200–600	0.5 nm
	Amino acids	60–200	0.5 nm
	Vitamins	300–600	1–2 nm
	Organic acids	30–300	0.5 nm
Large molecules	Proteins	10^3–10^6	3–10 nm
	Polysaccharides	10^4–10^7	4–20 nm
	Nucleic acids	10^3–10^{10}	2–1,000 nm[a]
Particles	Ribosomes		25 nm
	Viruses		100 nm
	Bacteria		1 μm
	Organelles		1 μm
	Yeast cells		4 μm
	Animal cells		10 μm

[a]Single pieces of genetic DNA can have end-to-end lengths up to millimeters, depending on method of isolation; the diameter of the DNA double-stranded α helix is 2.5 nm.

indicates the sizes and masses of each category of product. These three broad categories correspond to three broad categories of separation procedures for purifying them. As we shall see in this book, small molecules cannot be sedimented, but they can be separated by extraction; many large molecules cannot withstand the conditions of solvent extraction, but they are highly adsorptive; and particulate products can be collected by sedimentation or filtration. In addition, separation processes depend on property differences. Therefore it is important to learn as much as we can about the physical properties and chemical characteristics of the categories of products for which we must design separation procedures. The following subsections summarize the bioproduct types in order of increasing molecular size and complexity, which is also increasing order of complexity of processing.

1.3 Small Biomolecules

Small biomolecules include naturally occurring compounds and metabolites such as citric acid, vitamins, amino acids, and antibiotics. Besides their importance in fermentation processes, many of these compounds are important commercial products, some of them at very large production rates. Small biomolecules can be divided into two categories—primary metabolites and secondary metabolites.

1.3.1 PRIMARY METABOLITES

A primary metabolite is one that is formed during the primary growth phase of the organism. Figure 1.2 shows the key central metabolic intermediates of biosynthetic pathways in heterotrophs, organisms that use organic compounds as carbon sources. The intermediates

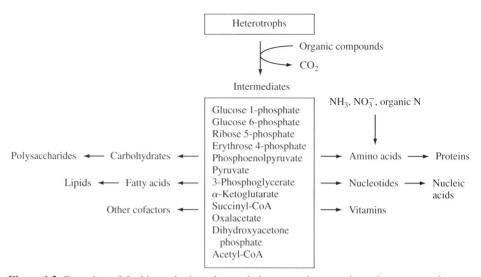

Figure 1.2 Overview of the biosynthetic pathways in heterotrophs, organisms that use organic compounds as carbon sources, showing the key central metabolic intermediates.

shown in this overview are used in catabolism, the processes by which microorganisms obtain energy from organic compounds, and well as in biosynthesis, also called anabolism.

Sugars

Naturally occurring sugars are found as monosaccharides, disaccharides, or polysaccharides. One of the world's highest tonnage bioproducts is one of these, sucrose, a disaccharide, whose structure is shown in Figure 1.3. In the food industry sucrose has been refined from sugarcane for centuries by using an aqueous solution and crystallization process. Mono- and disaccharides are also used in the biochemical process industry as solutes for the control of osmotic pressure and, most importantly, as carbon sources for the growth of organisms in fermenters. Glucose, a monosaccharide (Figure 1.3), is the most common fermentation nutrient, and lactose (milk sugar) and maltose are popular commercial disaccharides.

Sugars can also be obtained as products in bioprocesses. For example, the acetic acid bacteria can partially oxidize mannitol to fructose (Figure 1.4), and the enzyme glucose isomerase can convert glucose to fructose. This latter process is being used commercially to produce large amounts of high fructose corn syrups, which are used as sweeteners in soft drinks.

Organic Alcohols, Acids, and Ketones

Organic alcohols, acids, and ketones can be produced by the anaerobic fermentation of microorganisms. Many of these organisms proceed from glucose to pyruvic acid and then on to the final bioproduct. Ethanol, isopropanol, acetone, acetic acid, lactic acid, and propionic acid are some of the organic alcohols, acids, and ketones that have been produced in anaerobic fermentations [2]. The most economically important of these processes is the production of ethanol. Another way of producing organic alcohols, acids, and ketones is by using the acetic acid bacteria to perform partial oxidation. For example, the acetic acid product produced by *Acetobacter* bacteria with ethanol as a substrate is vinegar. Other end products obtained from acetic acid bacteria are shown in Figure 1.4.

Vitamins

The name for vitamins was originally derived from "vital amines," but we have known for decades that not all vitamins are amines, and some are not even vital. Generally, animals do not synthesize vitamins, but generalizations have exceptions, and humans and guinea pigs do not synthesize vitamin C, ascorbic acid, while nearly all other mammals do. Vitamin C has many more functions than originally thought. While it is required in catalytic quantities to support the hydroxylation of proline in collagen, which gives collagen its strength and

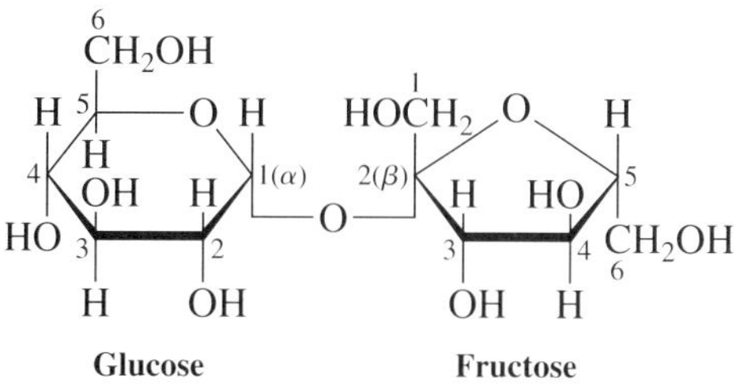

Figure 1.3 Sucrose, or table sugar, is a disaccharide composed of the monosaccharides glucose and fructose.

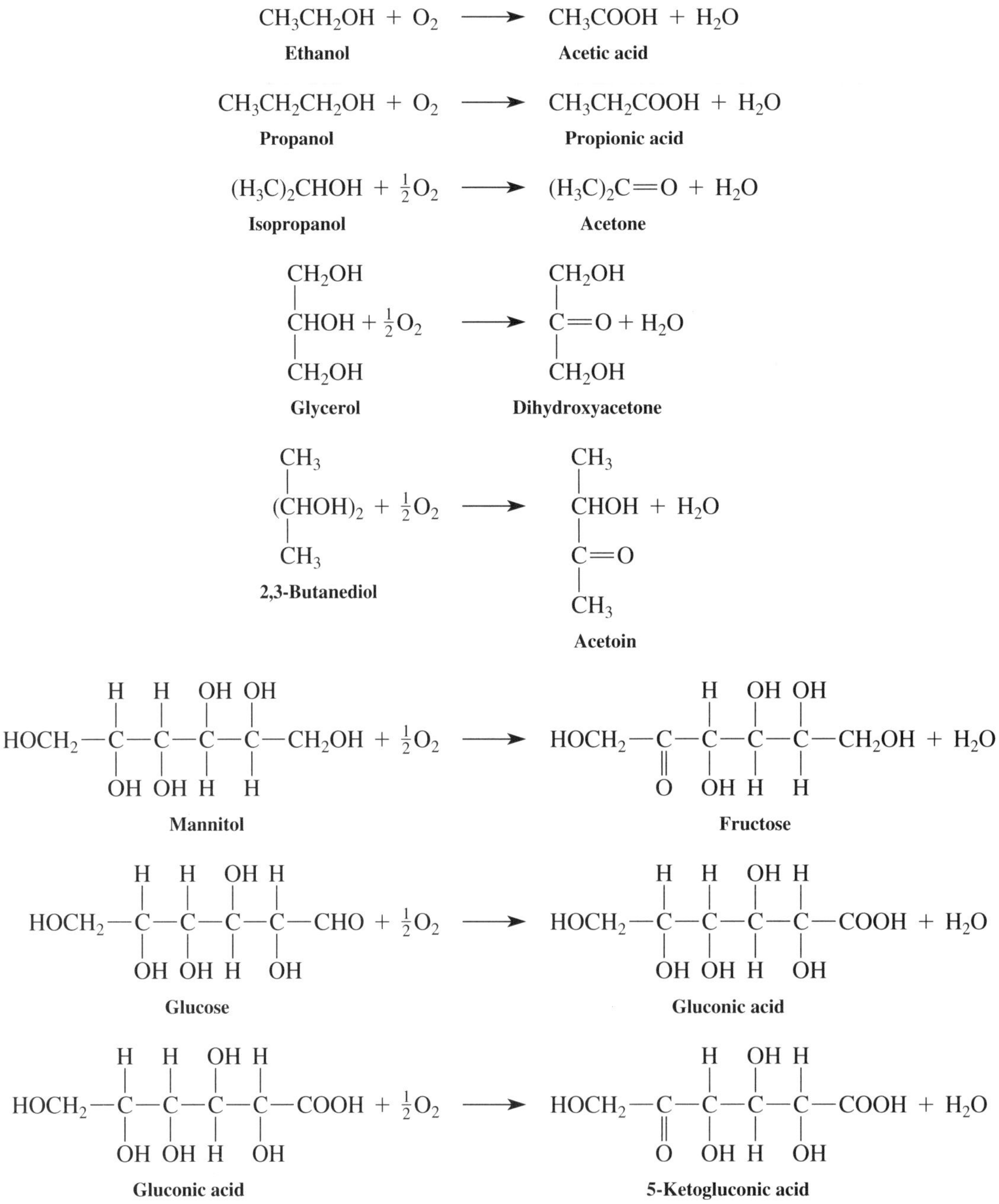

$$CH_3CH_2OH + O_2 \longrightarrow CH_3COOH + H_2O$$

Ethanol   Acetic acid

$$CH_3CH_2CH_2OH + O_2 \longrightarrow CH_3CH_2COOH + H_2O$$

Propanol   Propionic acid

$$(H_3C)_2CHOH + \tfrac{1}{2}O_2 \longrightarrow (H_3C)_2C{=}O + H_2O$$

Isopropanol   Acetone

$$\begin{array}{c} CH_2OH \\ | \\ CHOH \\ | \\ CH_2OH \end{array} + \tfrac{1}{2}O_2 \longrightarrow \begin{array}{c} CH_2OH \\ | \\ C{=}O \\ | \\ CH_2OH \end{array} + H_2O$$

Glycerol   Dihydroxyacetone

$$\begin{array}{c} CH_3 \\ | \\ (CHOH)_2 \\ | \\ CH_3 \end{array} + \tfrac{1}{2}O_2 \longrightarrow \begin{array}{c} CH_3 \\ | \\ CHOH \\ | \\ C{=}O \\ | \\ CH_3 \end{array} + H_2O$$

2,3-Butanediol   Acetoin

Mannitol $+ \tfrac{1}{2}O_2 \longrightarrow$ Fructose $+ H_2O$

Glucose $+ \tfrac{1}{2}O_2 \longrightarrow$ Gluconic acid $+ H_2O$

Gluconic acid $+ \tfrac{1}{2}O_2 \longrightarrow$ 5-Ketogluconic acid $+ H_2O$

Figure 1.4 Examples of useful end products obtained by partial oxidations in acetic acid bacteria.

low solubility, its presence at high concentrations supports antioxidant functions and the sulfation of cholesterol. Biological antioxidants are thought to prolong life and promote health, so vitamin C and its fat-soluble counterparts, vitamins A and E, have become popular consumer products. Both natural and synthetic forms are produced and marketed.

The B vitamins are water soluble and are metabolic precursors of cofactors involved in enzymatically catalyzed reactions. For example, vitamin B_6, pyridoxine, becomes phosphorylated inside the cell and serves as a nitrogen atom shuttle in transamination reactions, which convert keto acids into amino acids and vice versa. Niacin becomes nicotinamide adenine dinucleotide (NAD), the cell's principal redox compound and carrier of hydrogen atoms. The water-soluble vitamins are not phosphorylated because cell membranes cannot transmit organic phosphates; instead, the cell phosphorylates them after they enter. After phosphorylation (and in some cases additional modifications) they become cofactors, or "coenzymes." Folic acid, also water soluble and named for its origin from leaves, is required in minute quantities for hydrogen atom transfer in very special chemical reactions, such as methyl group transfer in the synthesis of the important amino acid methionine and the important DNA base thymine. Folic acid is present in plant tissues in extremely low quantities and was among the earliest challenges in bioprocessing in terms of purifying usable amounts of a product initially present at minuscule concentrations. Folic acid is now synthesized in bacterial or fungal fermentations and is sold as a common ingredient in daily vitamin tablets.

Vitamin A, a carotenoid, becomes the energy transducer of the pigment proteins of the retina. Vitamin D, perhaps more appropriately called hormone D, is a steroid that regulates the passage of calcium ions in and out of cells. Vitamins A and D are derived from carrots and milk, respectively, and are sold after extraction and purification (or after laboratory synthesis) as over-the-counter consumer products.

Vitamins are an excellent example of a present-day market tension: natural vs synthetic. Most vitamins can be synthesized in organic chemical reactions, but some argue that unnatural solvents and reagents used in these syntheses could be present as impurities, possibly harming the consumer. Extraction from plants and fermentation produces vitamins naturally—at higher cost. This issue is important in bioseparation technology, because the downstream processes are very different in the two approaches. In the latter case, hundreds of unknown contaminating solutes are present, and the product itself is at a low concentration. In the former case (synthesis), fewer impurities are present, but in some cases these may be more objectionable. The chapters that follow will give much more attention to natural products.

Amino Acids

Amino acids are the building blocks of proteins; understanding their chemistry is critical to an understanding of the stability of protein bioproducts. The general structure of an α-amino acid is given in Figure 1.5, where it is seen that every such amino acid has a side

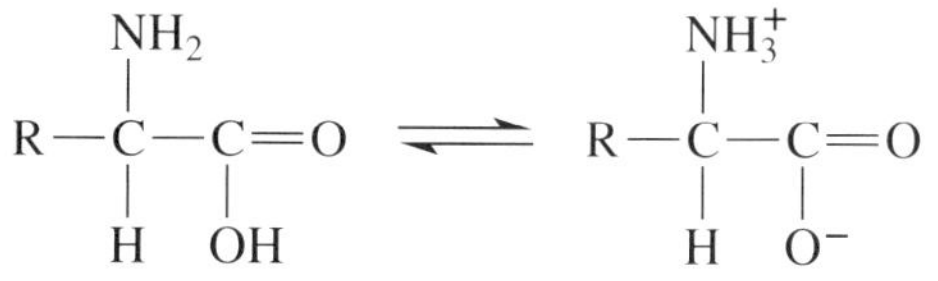

Figure 1.5 α-Amino acid structure showing zwitterionic equilibrium at neutral pH.

TABLE 1.3

List of α-Amino Acids Indicating R Group (Side Chain) Structure, Abbreviations, pK of R Group,[a] and Class: For General Structure, See Figure 1.5

R Group	Name	Abbreviations	pK of R group	Class
—H	Glycine	Gly, G		Aliphatic
—CH$_3$	Alanine	Ala, A		
—CH(CH$_3$)$_2$	Valine	Val, V		
—CH$_2$CH(CH$_3$)$_2$	Leucine	Leu, L		
—CHCH$_3$CH$_2$CH$_3$	Isoleucine	Iso, I		
—CH$_2$OH	Serine	Ser, S		Hydroxyl or sulfur containing
—CHOHCH$_3$	Threonine	Thr, T		
—CH$_2$SH	Cysteine	Cys, C	8.3	
—(CH$_2$)$_2$SCH$_3$	Methionine	Met, M		
—CH$_2$COOH	Aspartic acid	Asp, D	3.9	Acids and corresponding amides
—CH$_2$CONH$_2$	Asparagine	Asn, N		
—(CH$_2$)$_2$COOH	Glutamic acid	Glu, E	4.2	
—(CH$_2$)$_2$CONH$_2$	Glutamine	Glu, Q		
—(CH$_2$)$_3$CH$_2$NH$_2$	Lysine	Lys, K	10.5	Basic
—(CH$_2$)$_3$NHCNHNH$_2$	Arginine	Arg, R	12.5	
	Histidine	His, H	6.0	
	Phenylalanine	Phe, F		Aromatic
	Tyrosine	Tyr, Y	10.1	
	Tryptophan	Try, W		
	Proline	Pro, P		Imino acid
—CH$_2$—S—S—CH$_2$—	Cystine	—		Disulfide

[a]See reference 3.

chain, or "R group," a negatively charged carboxyl group, and a protonated (positively charged) amino group bonded to a carbon atom, which is called the α carbon because it is adjacent to the carboxyl group. At neutral pH all amino acids are zwitterionic; that is, they carry positive and negative charges simultaneously.

Table 1.3 lists the amino acids, the three-letter and one-letter abbreviations of each, and their respective side chain structures. This table is grouped according to side chain class. The single-letter code allows easy representation of the primary structure of proteins.

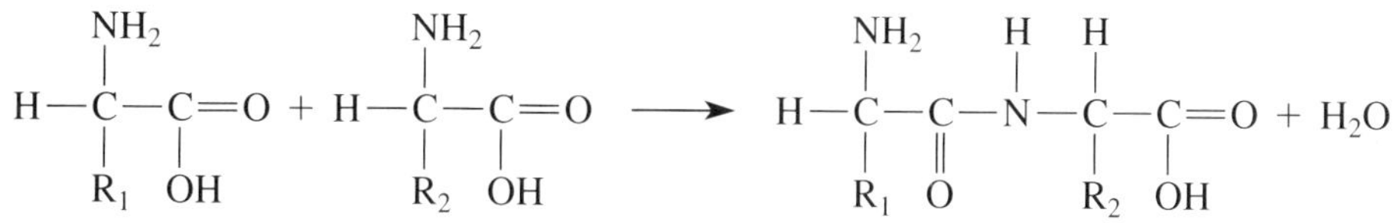

$$H-\underset{\underset{R_1}{|}}{\overset{\overset{NH_2}{|}}{C}}-\underset{\underset{OH}{|}}{\overset{}{C}}=O \ + \ H-\underset{\underset{R_2}{|}}{\overset{\overset{NH_2}{|}}{C}}-\underset{\underset{OH}{|}}{\overset{}{C}}=O \ \longrightarrow \ H-\underset{\underset{R_1}{|}}{\overset{\overset{NH_2}{|}}{C}}-\underset{\underset{O}{\|}}{\overset{\overset{H}{|}}{C}}-\underset{}{\overset{\overset{H}{|}}{N}}-\underset{\underset{R_2}{|}}{\overset{\overset{H}{|}}{C}}-\underset{\underset{OH}{|}}{\overset{}{C}}=O \ + \ H_2O$$

Figure 1.6 The formation of a peptide bond (amide linkage) between two amino acids and the resulting peptide backbone of proteins that results from polymerization.

It can be seen that all the α-amino acids except glycine contain asymmetric carbon atoms, indicating that amino acids, like sugars, should be optically active. With almost no exceptions, the amino acids synthesized by living things are levo rotatory and not racemic, so proteins can be studied on the basis of optical rotary dispersion (ORD); each protein, made of a different combination of amino acids, will have a unique ORD spectrum.

When amino acids are polymerized to form proteins, their carboxyl carbons are linked to their neighbors' amino nitrogens in a classical amide linkage called the *peptide bond*. This occurs through a dehydration condensation, and equilibrium laws dictate that this reaction is highly unlikely in aqueous solutions. However, living cells solve this problem with a catalytic unit known as a *ribosome*. This reaction and a resulting peptide backbone are depicted in Figure 1.6.

Specific properties of amino acid side groups can be exploited in purification methods. A protein rich in acidic or basic amino acids on its surface can be adsorbed by ion exchange or separated by electrophoresis. Many aliphatic side chains can result in preferential adsorption onto or extraction into nonpolar separation media. The free —SH (sulfhydryl) group of cysteine can be used to bind proteins to immobilized mercury. Histidine forms coordination complexes with metals, a fact that is being heavily exploited in protein purification by adsorption methods, as we shall see in Chapter 7.

Lipids

The natural fats consist of fatty acids, lipids, steroids, and steroid precursors. This family of bioproducts is highly extractable into nonpolar solvents. Fatty acids are synthesized by cells by building up two-carbon fragments contributed by a precursor compound known as acetyl–coenzyme A (acetyl-CoA), the cell's principal mechanism of acetyl group transfer. Fatty acids are usually esterified to glycerol to form di- and triglycerides, and diglycerides are usually esterified to a phosphate group (hence "phospholipids"), which may in turn be esterified to ethanolamine or choline, rendering the phosphate "head" zwitterionic or amphoteric (having both charges). Fatty acids and phospholipids are *amphiphilic,* having a strongly polar "head" and a strongly nonpolar tail. Such molecules form layers at liquid–liquid interfaces, and just such layers form the membranes that surround living cells and are also found inside eukaryotic cells.

Steroids are cyclical compounds, and the most common is cholesterol. Most other steroids have hormone activity, partly because they are able to penetrate the nonpolar cell membrane and get inside cells, where they bind and modify the activity of an intracellular protein. It is no surprise that steroids are important bioproducts. Similarly, a family of very potent lipids can profoundly affect the activities of animal cells, namely prostaglandins, leukotrienes, and thromboxane; these are also significant commercial bioproducts.

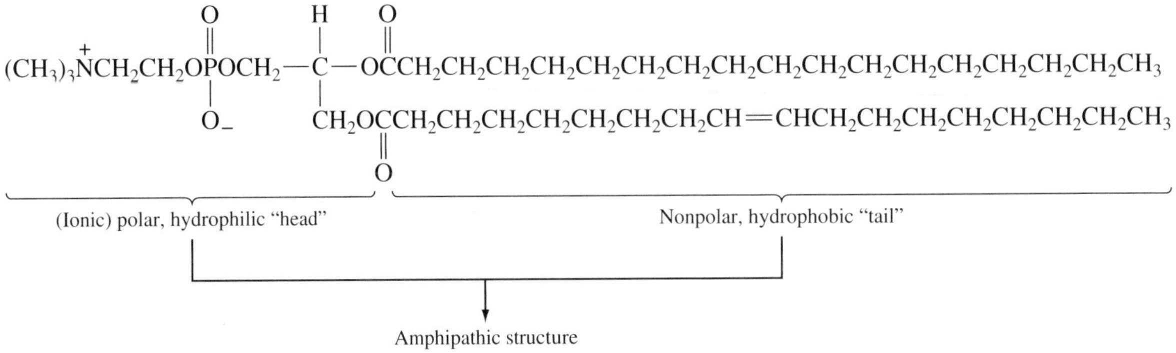

Phosphatidylcholine

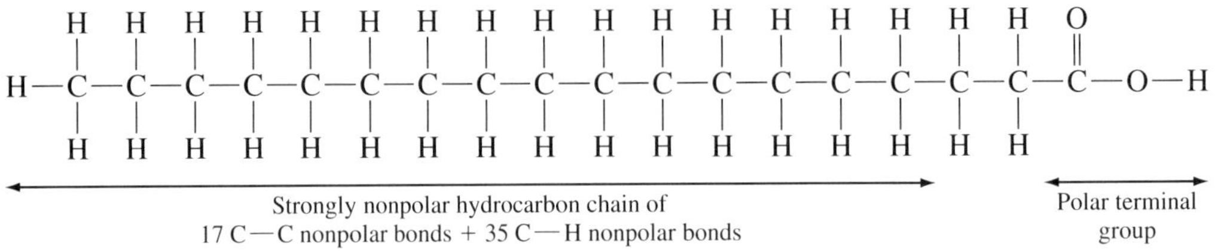

Stearic acid

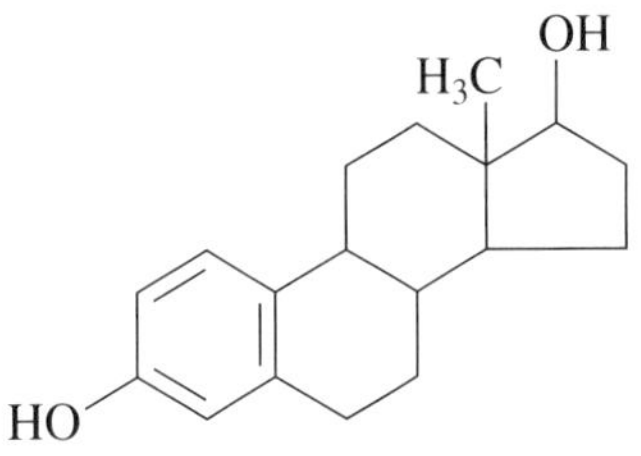

Estradiol 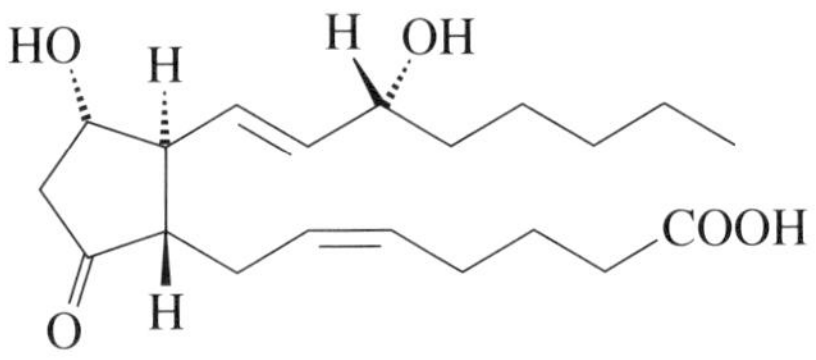   **PGE$_2$**

Figure 1.7 Structures of typical compounds: a phospholipid (phosphatidylcholine), a fatty acid (stearic acid), a steroid (estradiol), and a prostaglandin (PGE$_2$).

Figure 1.7 gives typical structures of four classes of compounds: fatty acid, phospholipid, steroid, and prostaglandin.

Commercial Uses

Many primary metabolites are important commercially; some important examples are given in Table 1.4. The products listed in Table 1.4 are sold in the fermentation, food, and biochemical research marketplaces, and several fine-chemical houses have adopted the sale of primary metabolites as a specialty or as a product line in a warehouse of general chemicals. The biochemical research market demands high variety and low volume. This market is served by a small number of large retail firms who purchase from specialty producers, some of whom also sell directly to customers.

TABLE 1.4
Some Examples of Primary Metabolites That Are Marketed and Their Uses

Primary Metabolite	Commercial Uses
Citric acid	Beverages
Acetic acid	Food (vinegar) and fine chemicals
Glutamic acid	Food flavoring (MSG)
Lactic acid	Biodegradable polymers
Glycerol	Solvent
Butanol	Solvent
Fructose	Food, fermentation
Formic acid	Fine chemicals

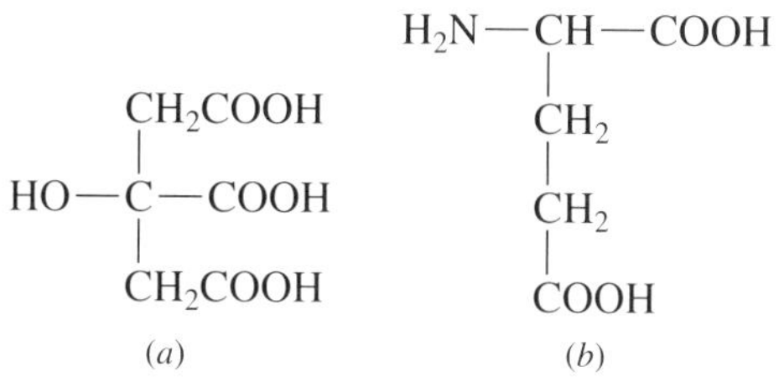

$$H_2N-CH-COOH$$

CH$_2$COOH

HO-C-COOH   CH$_2$

CH$_2$COOH   CH$_2$

(a)   COOH

(b)

Figure 1.8 Structures of (*a*) citric acid and (*b*) glutamic acid (an amino acid).

Exemplified by ethanol at the bottom of Figure 1.1, these low molecular weight compounds are the easiest products to purify, mainly owing to their thermal stability. Not surprisingly, they are produced in highest quantity and are used in the beverage, food, feed, solvent, and specialty chemicals industries. Some of the alcohols are traditionally purified using distillation. This is true of none of the other substances discussed in the paragraphs that follow. The highest volume fermentation products in this category include ethanol, acetic acid, butyric acid, lactic acid, citric acid, glutamic acid ("MSG"), tryptophan, and glycine. Two examples are given in Figure 1.8.

1.3.2 SECONDARY METABOLITES

Secondary metabolites are not produced during the primary growth phase of a microorganism, but at or near the beginning of the stationary phase. Antibiotics are the best known and most extensively studied secondary metabolites. They are synthesized by fungi as a successful means of competing with bacteria (and sometimes other fungi) for unimpeded growth in natural (dirty) environments. Most primitive eukaryotes are heavily endowed with mechanisms of defense against prokaryotic intrusion, and humans have learned to discover and use these mechanisms. While some antibiotics can be prepared by chemical synthesis, Figure 1.9 illustrates, on the basis of molecular complexity, that their synthesis is more profitably performed by fungal cultures in fermentation vessels. Indeed, large-scale fermentation of any product other than ethanol began with the penicillin industry. Fungi are not the only sources of antibiotics, and bacteria, higher plants, and even animal tissues that produce them can be found.

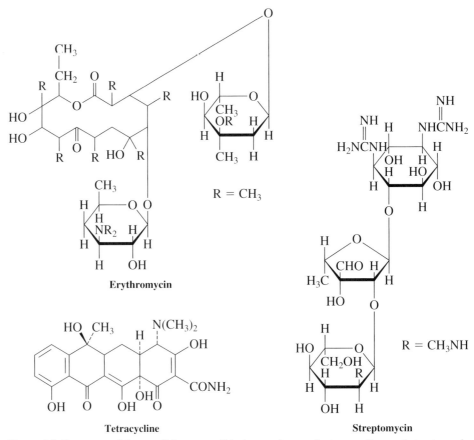

Figure 1.9 Structures of three well-known antibiotics: erythromycin, tetracycline, and streptomycin. The choice of purification method is influenced by the profound differences in the chemical structure of these compounds, which have, respectively, macrocycle, polycyclic aromatic, and oligosaccharide structures.

A number of different secondary metabolites and their general routes of synthesis are shown in Figure 1.10. Clearly, primary metabolites are the raw material for the synthesis of secondary metabolites. The primary metabolites, the level of phosphate in the culture medium, and the cell's energy charge are important modulators of the pathways leading to secondary metabolites [2].

Secondary metabolites made by plants and fungi have found many roles in human culture. Apart from steroid hormones, hallucinogens, and other psychoactive drugs, there are numerous cytotoxic secondary metabolites that have found their way into cancer therapy, cell and biochemical research, and physiological applications.

1.3.3 SUMMARY OF SMALL BIOMOLECULES

Any small-molecule bioproduct could, given suitable conditions, be synthesized by a chemist from petroleum-derived shelf reagents. Where there is large bulk demand, as in the

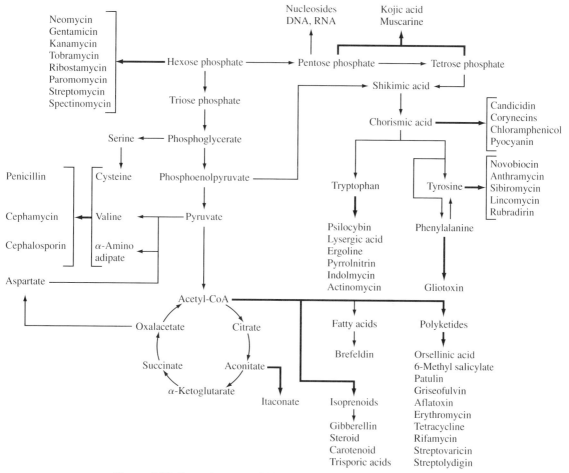

Figure 1.10 General routes of biosynthesis of secondary metabolites. The heavy arrows signify enzyme-catalyzed reaction sequences, and the products at the end of these arrows are secondary metabolites.

cases of methanol, acetone, and aspirin, chemical synthesis has displaced extraction of natural products as the primary source. Many steroids, especially those popular in preventive gynecology, are synthesized, whereas they were once extracted in large quantities from the tubers of a variety of beets.

In many cases, bioprocess technology is still economically superior. Examine, for example, the structure of taxol in Figure 1.11. Its complete synthesis is extremely difficult, and laboratories are actively pursuing an economical synthetic route owing to its potential importance in cancer therapy and its limited supply from trees. All three upstream methods are used to produce small-molecule bioproducts: extraction from plant and animal tissues, synthesis by fermentation, and conversion in enzyme-based catalytic reactors. Concerning the last, enantioselective synthetic methods are important, and enantioselective separation methods have been developed for bioproducts with a specific active enantiomer.

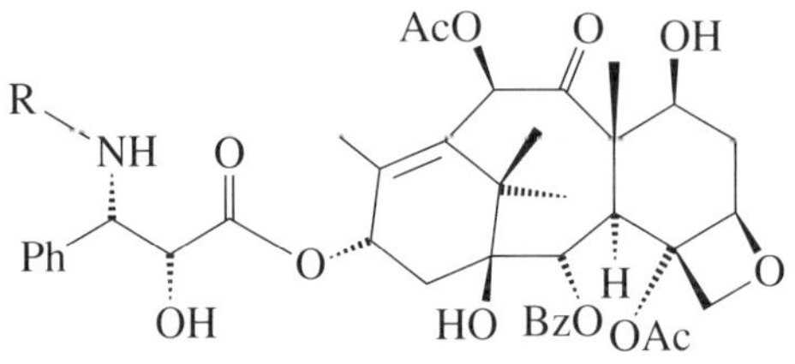

Figure 1.11   Structure of taxol, an anticancer compound extracted from the bark of *Taxus brevifolia,* the Pacific yew tree, and certain other natural sources. It functions by immobilizing microtubules in dividing cells and preventing the redissolution of the mitotic spindle.

1.4 Macromolecules: Proteins

The principal source of excitement in the biotechnology industry during the past decade has been the production of biological macromolecules. From presoak laundry enzymes to therapeutic peptides for the treatment of rare diseases, the world of protein chemistry has undergone remarkable metamorphosis. Protein purification is now approaching billion-dollar-industry status on its own. Vendors of adsorption equipment, ultrafiltration technology, and large-scale fermentors are moving rapidly to capture the "picks-and-shovels" market of biotechnology.

Nearly all biological macromolecules are characterized at four levels of structure: primary (the covalent amino acid sequence), secondary (the hydrogen-bonded structures), tertiary (the folding pattern of hydrogen-bonded and disulfide-bonded structures), and quaternary (the formation of multimeric complexes by individual protein molecules).

1.4.1 PRIMARY STRUCTURE

Primary structure is the sequence of amino acids established by the sequence of nucleotide codons of the messenger RNA for that protein. The amino acids are held together in strictly linear sequence (just as mRNA is strictly linear), and all backbone bonds are covalent. The peptide bond was illustrated in Figure 1.6. Any polymer of α-amino acids is called a *polypeptide*. When peptides are synthesized by cells, using ribosomes and several enzymes as catalysts, all polymer chains are linear; there are no branched peptides, despite the existence of amino and carboxyl side chains capable of forming amide linkages. Even after translation, branching amide linkages are not formed. In nearly all protein molecules having more than about 70 amino acids, the —SH groups of cysteines *might* react with each other to form disulfide bonds and produce peptide rings; such rings can become branches if the cell deletes a portion of the peptide sequence, as in the case of insulin, shown in Figure 1.12.

1.4.2 SECONDARY STRUCTURE

There are two main forms of secondary structure, both due to the formation of hydrogen bonds between oxygen and nitrogen atoms in the peptide backbone. Since these are regularly spaced, a repeating pattern of interaction is expected. Indeed, the two structures that form are the α helix and the β sheet (or "pleated sheet"), shown schematically in Figures 1.13 and 1.14, respectively. In the α helix, each NH group is connected to a CO group by a hydrogen bond at a distance equivalent to three amino acid residues [4]. For the β sheet,

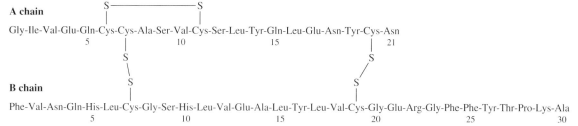

A chain

Gly-Ile-Val-Glu-Gln-Cys-Cys-Ala-Ser-Val-Cys-Ser-Leu-Tyr-Gln-Leu-Glu-Asn-Tyr-Cys-Asn

B chain

Phe-Val-Asn-Gln-His-Leu-Cys-Gly-Ser-His-Leu-Val-Glu-Ala-Leu-Tyr-Leu-Val-Cys-Gly-Glu-Arg-Gly-Phe-Phe-Tyr-Thr-Pro-Lys-Ala

Figure 1.12 Amino acid sequence of bovine insulin. The two polypeptide chains are held together by disulfide bridges, but they originated from a single chain, a portion of which was excised by the cell. The A and B chains are connected by interchain disulfide bonds. The chains are numbered, by convention, from their N-terminal ends.

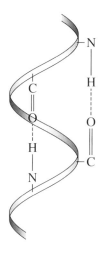

Figure 1.13   Schematic representation of a peptide chain in an α-helix configuration, showing hydrogen bonding (dashed lines) that stabilizes the structure.

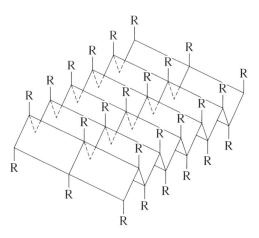

Figure 1.14 Representation of three parallel peptide chains in a β-sheet structure. All the R groups (side chains) project above or below the β-sheet structure.

the side-by-side polypeptide chains are cross-linked by interchain hydrogen bonds; all the peptide linkages participate in this cross-linking, thus giving the structure great stability [3]. Another secondary structural element that occurs in proteins is the β turn, which is a hairpin turn of the polypeptide chain. This turn is stabilized by the CO group of residue n of the polypeptide being hydrogen bonded to the NH group of residue $(n + 3)$ [ref. 5]. This kind of turn allows proteins to have compact three-dimensional structures. β turns often connect antiparallel strands in β-sheet structures.

A commonly used means of following changes in the secondary structure of a protein is the circular dichroism (CD) method, which measures the difference in the absorbances of left circularly polarized light and right circularly polarized light. This difference occurs because secondary structures, such as an α helix, interact one way with a left circularly polarized beam and differently with a right circularly polarized beam. CD can be used to estimate the fraction of a protein in α-helix, β-sheet, and random coil configurations [6].

1.4.3 TERTIARY STRUCTURE

A complete protein molecule will consist of various combinations of various secondary structures, with each helix or each sheet called a *domain*. The three-dimensional folding of the coiled or pleated polypeptide chain establishes the following:

- Surface properties
- Catalytic activity
- Stability
- Mechanical strength
- Shape

In protein synthesis, once secondary structures have been established, in most cases the hydrophobic amino acids fold inward, owing to entropic organization of the surrounding polar water molecules. The outcome of the folding step may be constrained by the strong covalent disulfide bonds between cysteine amino acid residues; some proteins, like trypsin inhibitor, can still be hydrophobic on the outside, owing to a large number of covalent disulfide bonds per mole, which stabilize the molecule against folding according to polarity. Finally, the polar and charged amino acids fold toward the surface and confer a net charge on the protein molecular surface that depends on the numerical combination of Asp, Glu, Lys, and Arg on the surface, as well as on the pH.

In globular proteins in aqueous solution, for example, many of the amino acids with nonpolar side chains are forced to the inside of the ball-shaped molecule owing to the affinity of the other, polar amino acids for the aqueous environment. Temperature, pH, surfactants, and organic solvents can reverse this picture, leading to *denaturation* of globular protein molecules, a subject discussed in greater detail in Section 1.4.7. Once formed, a tertiary structure is often stabilized by the formation of disulfide bridges between cysteine residues that have folded to positions near one another. A ribbon model of the three-dimensional structure of ribonuclease, determined by x-ray diffraction analysis, is shown in Figure 1.15.

Beyond these simple thermodynamic rules, the tertiary structure of a protein depends highly upon the way in which it is synthesized. For example, the eukaryotic cell takes great

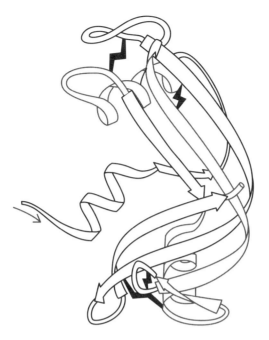

Figure 1.15 Ribbon model of ribonuclease. Sections of α helix are shown as helices, β sheets as parallel ribbons, and disulfide bridges in solid black. (Reproduced by permission of Professor Jane Richardson.)

care to fold a protein properly during synthesis. In many cases, as the growing peptide chain rolls off the polyribosome translation assembly, an early-synthesized part of the chain is extracted into the adjacent lipid phase membrane. This "signal sequence" holds the N-terminus of the peptide in place while the rest of the molecule is synthesized, glycosylated, and folded. If the same gene is expressed in a prokaryotic cell, which has no internal lipid membranes, however, these things do not happen, and the growing peptide chain folds according to simpler, more "thermodynamic" rules that might apply in free solutions. Heterologous proteins expressed in prokaryotic cells often do not fold correctly, leading to a precipitate called an "inclusion body." Synthesis in other environments can lead to yet other results. The stability of the tertiary structure is further discussed later, in Section 1.4.7.

EXAMPLE 1.1

Effect of a Reducing Agent on Protein Structure and Mobility   Upon the addition of 1 mM dithiothreitol (a reducing agent), the relative mobility of a protein decreases by 10% during the migration of the protein through a polyacrylamide gel under the influence of an electric potential. What is the likely reason for this shift?

SOLUTION

Dithiothreitol is a reducing agent that causes the disulfide bonds between cysteines in the protein to be broken. The breaking of the disulfide bonds causes the protein to unfold and the three-dimensional structure to become less compact, which causes the protein to move more slowly in the polyacrylamide gel. The analysis of proteins by electrophoresis is discussed further in Chapter 2 on analytical methods.

1.4.4 QUATERNARY STRUCTURE

Quaternary structure specifies the way in which the individual, folded peptide chains interact with one another in the native conformation of an oligomeric protein. Thus, a complete hemoglobin molecule contains two α-globin and two β-globin peptide chains (total of four), and bacterial RNA polymerases contain several subunits, one of which confers specificity based on the gene being transcribed. Cytochrome oxidase, located in mitochondria, can have up to 13 peptide chains. Quaternary structure is maintained by intermolecular bonds, including ionic and covalent linkages. In the latter case, cysteine residues form disulfide bridges; in the former case, opposite charges attract (e.g., Lys and Asp).

1.4.5 PROSTHETIC GROUPS AND HYBRID MOLECULES

Not all proteins are pure peptides, and the majority of them are *conjugated proteins* that contain not only amino acids, but other organic and inorganic compounds. The non–amino acid portion of a conjugated protein is called its *prosthetic group*. Conjugated proteins are classified on the basis of the chemical nature of their prosthetic groups (Table 1.5). Proteins containing lipids or sugar moieties have also been called *hybrid* compounds.

As the cell synthesizes each new peptide chain by the process of translation, highly specific chemical steps occur that determine whether chemical derivatives of the peptide will be made. Any changes to protein molecules other than the addition of amino acids via peptide bond formation are called *posttranslational modifications*.

During the past decade a great deal of attention was given to glycosylation, in which oligosaccharides (usually branched) are added to active side chains of specific amino acids (serine, asparagine, or threonine) while the nascent peptide is in the Golgi apparatus and being prepared for secretion. Also, phosphorylation of membrane-associated proteins occurs during intracellular signal transduction. The immunogenicity of certain peptides is established by their oligosaccharide content.

Side chain oligosaccharides added during glycosylation serve extremely important functions, such as stability, catalytic function, binding specificity, solubility, targeting, and transport. Glycosylation is extremely important in bioprocessing, since many commercial proteins must be properly glycosylated to function and be marketable. An excellent example is human erythropoietin, which is normally made in cells of the kidney and transported via the blood to the bone marrow, where the target cells reside. Since bacteria do not glycosylate peptides, this valuable product cannot be manufactured by recombinant bacteria; rather, it is

TABLE 1.5
Classification of Proteins According to Prosthetic Group

Class	Prosthetic Group	Example
Lipoproteins	Lipids	β_1-lipoprotein of blood
Glycoproteins	Oligosaccharides	Gamma globulin
Phosphoproteins	Phosphate	Protein kinase C
Hemoproteins	Iron porphyrin	Hemoglobin
Flavoproteins	Flavin nucleotides	Succinate dehydrogenase
Metalloproteins	Iron, zinc	Ferritin, alcohol dehydrogenase

made in large cultures of animal cells. Animal or yeast cells are chosen to perform syntheses when the product is a glycosylated protein.

Antibodies are hybrid molecules that have become a major product of biotechnology companies, and many start-up biotechnology companies have initiated their revenue streams by selling antibodies, mainly to be used in diagnostic kits. There is also hope that the "seek and destroy" function of antibodies can someday be exploited in therapeutic medicine, and clinical trials of antibodies that can carry poisonous or radioactive compounds to tumors, for example, have been under way for some years.

Antibodies are globular proteins synthesized by B lymphocytes in animals. All B lymphocytes that are members of a "clone" produce identical antibody molecules, but different clones produce different antibodies, with a combinatorial number of possibilities, perhaps more than 10^8 different possible peptide sequences. How do they do that? Genes that code for proteins are supposed to be constant and stable, but antibody genes are an exception; they are susceptible to internal rearrangements of DNA base sequences that encode the "hypervariable" region of antibodies, the part that binds antigens. Once the immune system has recognized a substance as "foreign," the B lymphocytes that make antibodies against it expand their clone at the request of specific T lymphocytes, and the blood is flooded with one kind of antibody. If they are truly all identical, they are called "monoclonal antibodies," and a trick has been devised for producing them in bioreactors. This trick consists of hybridizing one B lymphocyte with a lymphoma (blood cell tumor) cell. The hybrid, called a hybridoma, was developed by Jerne, Koehler, and Milstein, who won the Nobel Prize for their innovative work. Hybridomas produce progeny indefinitely *in vitro,* and the progeny make monoclonal antibodies, which can be purified and sold as diagnostic reagents or as affinity reagents for downstream bioprocessing, about which we shall learn more detail later (Chapter 7).

The structure of antibodies is dictated by their function. They are Y-shaped protein molecules with an invariant stem (Figure 1.16). The stems are not essential to antigen binding,

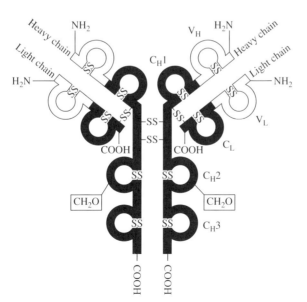

Figure 1.16 Structural features of an antibody (immunoglobulin G, or IgG). Each pair of N-terminal residues constitutes an antigen binding site. The variable and constant regions are shown in white and black, respectively. V_H, variable region of the heavy chain; V_L, variable region of the light chain; C_H2 and C_H3, constant regions of the heavy chain; C_L, constant region of the light chain. CH_2O denotes carbohydrate, and SS denotes a disulfide bond.

so sometimes reagents are made without them. Peptide bonds that are sensitive to hydrolysis and disulfide bonds that are sensitive to reducing agents can be broken by chemists who want to separate parts of antibody molecules.

1.4.6 FUNCTIONS AND COMMERCIAL USES OF PROTEINS

Proteins are of interest to biotechnology only on the basis of their function. Table 1.6 lists some functions of proteins, with examples. The 10 proteins produced commercially by recombinant DNA technology that had the highest worldwide sales in 1997 are shown in Table 1.7, along with the activity or use of each protein.

Enzymes are catalysts of reactions, usually involving organic substrates. Enzymes are highly specific both in the reaction catalyzed by each and in the substrates utilized. The most important feature of every enzyme is an *active site* that binds substrates and contains the amino acid residues that are directly involved in the making and breaking of bonds. The active site of an enzyme is typically a cleft or crevice in its three-dimensional structure [5].

TABLE 1.6
Classification of Proteins According to Function

Class	Examples
Enzymes	Ribonuclease
	Trypsin
	Urokinase
	DNA polymerase
	Cellulase
Transport proteins	Hemoglobin
	Serum albumin
	Myoglobin
	β_1-Lipoprotein
Nutrient, storage protein	Ovalbumin
	Casein
Contractile or motile	Actin
	Myosin
	Tubulin
Structural	Keratin (skin)
	Fibroin (silk)
	Collagen (tendons, ligaments)
	Elastin (joints)
	Proteoglycans (cell walls)
Defense	Immunoglobulin (antibodies)
	Venom proteins
	Ricin
	Interferon
Regulatory	Insulin
	Growth hormone
	Lymphokines/cytokines
	Protein kinases (signal transduction)
	DNA binding repressors and activators
Inhibitors	Soybean trypsin inhibitor
	Plasminogen activator inhibitor
	HIV protease inhibitor

TABLE 1.7
The 10 Leading Biotechnology Proteins, in Order
of Worldwide Sales in 1997[a]

Protein	Activity/Use
Erythropoietin	Red blood cell growth
Colony-stimulating factor	White blood cell growth
Insulin	Diabetes
α-Interferon	Anticancer, infections
Vaccine against hepatitis B	Hepatitis B
Glucocerebrosidase	Genetic deficiency
Tissue plasminogen activator	Heart attack/stroke
Growth hormone (somatotropin)	Growth deficiencies
GPIIb/IIIa antibody	Prevents blood clots
Interferon β-1a	Multiple sclerosis

[a]See reference 7.

From the biotechnology point of view, it is crucial that a protein product be purified without destroying any of its functions. Therefore, tertiary and quaternary structures must be preserved. Recombinant proteins are often produced in bacteria in a nonnative state, and this problem is solved by "refolding" certain recombinant proteins made in bacteria. Refolding is a reversible denaturing–renaturing process, about which more is explained in the Section 1.4.8.

Human growth hormone (hGH) is an excellent example of a commercial *polypeptide hormone*. For several decades it was prepared in the United States from hemipituitaries extracted from human cadavers (the other half of the anterior pituitary lobe was used for research) under auspices of the National Pituitary Agency. A few cases of fatal virus infection discouraged the medical profession from continuing to use this source, and recombinant sources were sought. As a nonglycosylated peptide, hGH is ideally suited for production by recombinant fermentation. It should be added that the first recombinant human peptide produced in bacteria was another peptide hormone, somatostatin, for which the gene was synthesized chemically and transfected into laboratory bacterial strains. In addition to endocrine peptide hormones and their releasing factors, and in addition to tissue-targeted monoclonal antibodies, there exists a broad category of "biological response modifiers" (BRMs), proteins that cause a biological response to a particular environmental insult, such as infection, wounding, clotting, or cancer.

1.4.7 STABILITY OF PROTEINS

One important aspect of quality control in bioprocessing is the production and monitoring of active proteins. Since the active function of proteins is dependent on tertiary and quaternary structures (which in turn depend on primary and secondary structures), the three-dimensional configuration of each molecule must be maintained: denaturation is to be prevented if possible, but if it occurs, it must be detected. Proteins degrade via several pathways: deamidation of asparagine and glutamine, oxidation of methionine, oligomerization, aggregation, fragmentation (depolymerization), cross-linking, inter- and intramolecular rearrangements involving cysteine disulfide exchanges, and denaturation (breakdown of hydrogen bonds and ionic bonds that determine tertiary structure). In the United States, the legal evaluation of recombinant and other protein products is the responsibility of subdivisions of the U.S. Food and Drug Administration (FDA), including the Center for Biologics Evaluation and Research (CBER), located in Rockville, Maryland. This agency is responsible for, among other things, authorizing processes used for testing the stability of biopharmaceutical protein preparations.

It is possible to test samples for each of these chemical modifications by fairly routine methods of analytical biochemistry. For example, gel electrophoresis can detect changes in molecular weight and hence aggregation, oligomerization, or fragmentation; isoelectric focusing, which depends on charge, can detect deamidation on the basis of the appearance of dissociating amino groups [8]. High performance liquid chromatography (HPLC), using several adsorption chemistries, provides information on both purity and structure. Thus, methods used to purify proteins are also used as analytical methods for quality control and for off-line product analysis following purification steps. For this reason, some of our chapters include a discussion of the analytical applications of each separation method, and Chapter 2 emphasizes these applications.

It is important to be aware of the variables that have effects on the stability of proteins. Variables that most commonly affect protein stability are temperature, pH, mechanical shear, chemical agents, and irradiation.

Temperature   As temperature is increased, denaturation of a protein will start to occur at some point. Thermal denaturation for many proteins begins to occur at 45 to 50°C [2]. An example of the thermal denaturation of the protein ribonuclease is shown in Figures 1.17 and 1.18. When the temperature is high enough, the hydrogen bonds (dotted lines in Figure 1.17) are broken, and the β-pleated sheet structure of the protein is disrupted. Heating,

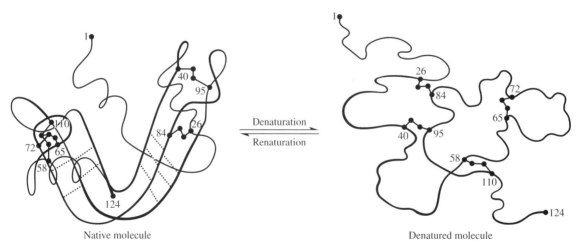

Native molecule   Denatured molecule

Figure 1.17 Schematic drawing showing the conformational change that occurs when ribonuclease is heated above its thermal denaturation temperature. Dotted lines indicate hydrogen bonds, and numbers indicate the positions of the amino acid cysteine, which forms disulfide bonds between chains. (Copyright © 1996 by the Benjamin/Cummings Publishing Company, Inc. Reprinted by permission of Pearson Education, Inc.)

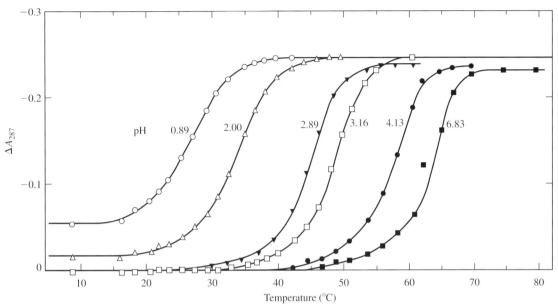

Figure 1.18 Thermal denaturation of ribonuclease as measured by ΔA_{287}, the change in absorbance at 287 nm, compared with a reference solution at zero absorbance. The pH values of the solutions (ribonuclease concentration 1.90 mg/ml, and ionic strength 0.16 M) are indicated. (Data from J. Hermans and H. A. Scheraga, "Structural studies of ribonuclease. V. Reversible change of configuration," *J. Am. Chem. Soc.,* vol. 83, p. 3283, 1961.)

however, still leaves the disulfide bonds intact. As observed in Figure 1.18, the denaturation of ribonuclease (as monitored by the change in absorbance) is an S-shaped function of temperature and depends strongly on the solution pH. Chromophores buried deep in the interior of the protein become exposed to the solvent when the protein is denatured and cause the absorbance to change. Some enzymes are still active at temperatures much higher than 45 to 50°C. For example, enzymes isolated from thermophilic bacteria inhabiting hot springs are active at temperatures above 85°C.

pH   Solution pH is a parameter that is often varied during the downstream processing of proteins. The impact of pH on the protein ranges from mild denaturation that is reversible to complete hydrolysis of the protein. In mild denaturation, protonation or deprotonation of the protein leads to a change in the structure of the protein. Figure 1.19 illustrates the phenomenon for four enzymes for which the effect of pH on the relative activity of the enzymes has been determined. In these cases the pH changes are leading to a change in the ionization state and structure of the active site. Changing the pH, however, can lead to more than just a perturbation of the active site of an enzyme. Studies with serum albumin, for example, have shown that this protein undergoes several major reversible conformational changes when the pH is varied [9]. Albumin, which usually exists in a compact globular form, undergoes an abrupt opening of the molecule at pH 4.0 that involves unfolding of one of the three molecular domains (domain III) from the rest of the molecule. Below pH 4.0, there is a significant expansion of the molecule to a form that resembles a series of balls and strings.

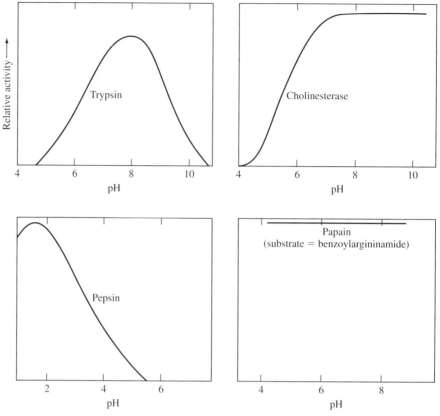

Figure 1.19 The effect of pH on the relative activity of the enzymes trypsin, cholinestrase, pepsin, and papain. (Data from A. L. Lehninger, *Biochemistry,* 2nd ed., p. 196, Worth, New York, 1975.)

At extremes of pH, proteins are not only denatured but can be completely hydrolyzed to their constituent amino acids. Typically, 6 M hydrochloric acid having a pH of −0.8 is used for acid hydrolysis, while 4 M sodium hydroxide having pH 14.6 is used for alkaline hydrolysis [10]. To obtain complete hydrolysis of proteins, temperature is usually elevated to 110°C. If strong acids or bases are used in adjusting the pH of protein solutions during a processing step, transient high concentrations of acid or base are present; and since some dipeptides are more readily hydrolyzed than others, the possibility of at least limited hydrolysis of protein exists. Thus, strong acids or bases should be avoided when even a slight amount of protein hydrolysis is undesirable. Strategies that have been developed to avoid extremes of pH during pH adjustment of protein solutions include the following: (1) using dilute weak acids or bases (1 M acetic acid or Tris) in the pH range of 5 to 8, and (2) for pH values below 5 or above 8, using 5 M acetic acid at pH 4.0 (NaOH) to lower the pH, and 5 M ammonia at pH 9.5 (HCl) to raise the pH [ref. 11].

Shear   The effect of mechanical shear on proteins has been extensively studied, and the general conclusion is that shear effects are not significant except when gas–liquid interfaces are present [12, 13]. In one study with alcohol dehydrogenase, no loss of enzyme activity was observed even at shear rates up to 26,000 s^{-1} using a completely filled concentric cylinder (Couette) shear device. Significant activity losses were, however, observed in a rotating disk reactor that had a gas phase caused by incomplete filling of the reactor [13]. This loss of activity was attributed to the proteins being adsorbed to the gas–liquid interfaces in an unfolded partially active or inactive state; if the turbulence in the solution is sufficiently high, the interface with protein adsorbed breaks, causing unfolded protein to be entrained in the bulk solution. As this process continues, more and more protein becomes inactivated. Therefore, it is important that gas–liquid interfaces be avoided or minimized in general, particularly when mechanical shear is present. Also, it should be recognized that shear rates experienced by proteins in process-scale separation and purification processes are much less than in the alcohol dehydrogenase study, typically no more that 1000 s^{-1}.

1.4.8 Recombinant Protein Expression

Recombinant proteins can be synthesized in a variety of organisms, which can vary from bacteria to mammalian cells. Typically, the host organism is transfected with a plasmid (circular DNA) that contains an efficient promoter sequence for expression of the desired protein at high levels. Two issues that can be encountered early in the development of a purification process for a recombinant protein are whether the protein should be genetically engineered to aid in the purification and whether the protein must be solubilized and refolded from "inclusion bodies," insoluble and inactive aggregates of protein.

Genetic Engineering to Improve Protein Purification

Various approaches can be taken to genetically engineer proteins to aid the purification process. Affinity peptide or protein "tags" can be genetically attached to recombinant protein molecules by splicing a gene that creates a "fusion protein" consisting of the product and its tag. One attractive tag is a histidine oligomer, the most commonly used of which has six histidines. Histidine tags have been added to either terminus of the protein or in the linker region between two segments of a fusion protein. Histidine binds transition metals such as nickel, zinc, and copper. The histidine tag will therefore bind to metal ions immobilized in a purification medium such as a chromatography column, membrane, filter, precipitant, or extraction solvent [14].

Genetic engineering has been applied to avoid the expression of insoluble inclusion bodies through the expression of the target insoluble protein as part of a fusion protein. The target heterologous protein is expressed as a C-terminal fusion with another protein called a carrier protein, which has a higher solubility than the target protein. Carrier proteins that have been used to solubilize target heterologous proteins as part of fusion proteins are glutathione-*S*-transferase (GST) [15], maltose binding protein (MBP) [16], NusA protein [17], and thioredoxin [18]. GST and MBP are themselves affinity tags that have been used to facilitate purification of fusion proteins containing GST and MBP. Histidine tails have been used to purify NusA and thioredoxin fusion proteins. The target protein can be removed from the fusion protein during purification either by proteases such as bovine enterokinase that specifically cleave a short peptide sequence inserted at the N-terminus of the target protein, or by self-cleaving proteins called inteins that are also inserted at the N-terminus of the target protein [19].

Solubilization and Refolding of Proteins in Inclusion Bodies

Inclusion bodies have been commonly observed in the expression of heterologous proteins in bacteria and yeast cells. Although initial purification of inclusion body material by cell lysis and centrifugation is relatively simple, the agglomerated protein must be resolubilized by strong denaturants; and the protein must be refolded to its correct three-dimensional conformation. The success in refolding is protein dependent, and for some proteins relatively high refolding yields can be obtained.

Solubilization   Inclusion bodies are found in the cell sediment after the cells have been lysed and centrifuged. The proteins in these inclusion bodies are usually solubilized by the addition of buffer containing a chaotropic agent such as guanidinium chloride or urea. One protocol calls for incubating the inclusion bodies in 6 M guanidium chloride or 8 M urea with 0.1 M phosphate buffer at pH 7 for 0.1 to 1 h at 20°C (1 ml of denaturing solution per 5 mg of protein) [20]. For proteins that contain cysteine, it is advisable to also add a reducing agent such as β-mercaptoethanol, dithiothreitol (DTT), dithioerythritol, or cysteine, to allow reduction of the interchain disulfide bonds by thiol–disulfide exchange.

Refolding   The refolding of proteins that have been solubilized from inclusion bodies is usually performed by dialysis or dilution [20]. Dialysis is a laboratory procedure that involves the use of ultrafiltration membranes. The denatured protein solution is placed in a bag made from an ultrafiltration membrane, and the bag is placed in a large volume of renaturation buffer. This allows diffusion of the denaturant out of the protein solution. Dialysis can be scaled up by means of diafiltration, in which case the denatured protein solution would flow over an ultrafiltration membrane that allows fluid permeate to flow through, but not the proteins, while at the same time adding renaturation buffer to the denatured protein solution to keep the volume constant (see Chapter 4, Filtration, for a more complete description of these filtration operations). Thus, for both dialysis and diafiltration, the protein being renatured is exposed to intermediate concentrations of denaturant for an extended period of time, which can have either a positive or negative effect on the yield of correctly folded protein. In some cases, it is better to add the denatured protein solution directly into the refolding buffer, which avoids the intermediate concentrations of denaturant. In this situation, it is necessary to mix the renaturation buffer strongly enough to bring the concentration of denatured protein rapidly to a low value [20]. Whether it is better to do the refolding by dialysis/diafiltration or by dilution must be determined separately for any given protein.

During refolding to the native state N, there is a kinetic competition of folding and aggregation of denatured state D according to the following sequence of reactions [20]:

$$N \rightleftharpoons D \longrightarrow \text{aggregates}$$

To minimize aggregation during the refolding process, the refolding usually must be performed at low protein concentrations (<100 mg/liter) [21]. An example is shown for the refolding of lactate dehydrogenase in Figure 1.20. For this enzyme, it is seen that the enzyme concentration should be 1 to 3 mg/liter for reasonably good refolding yields. If it is desired to refold a large quantity of protein, it may not be reasonable to refold the entire batch at such a low concentration. This can be avoided by doing successive additions of the denatured

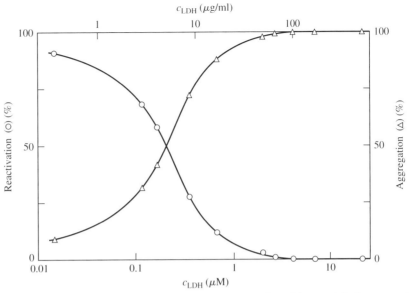

Figure 1.20 Effect of protein concentration c on the renaturation of lactate dehydrogenase (LDH) from pig muscle. Acid-denatured lactate dehydrogenase was renatured in 0.1 M phosphate buffer at pH 7 and 20°C with 1 mM EDTA and 1 mM dithioerythritol. After 192 h, reactivation (circles) and aggregation (triangles) were quantified. (Data from A. Mukhopadhyay, "Inclusion bodies and purification of proteins in biologically active forms." *Adv. Biochem. Eng. Biotechnol.,* vol. 56, p. 61, 1997.)

protein solution to the renaturation solution [20]. Using lactate dehydrogenase as an example, the protein concentration in the renaturation buffer should not increase by more than 1 to 3 mg/liter upon each addition to achieve good refolding yields. The time interval between successive additions should be the time required for renaturation of 90% of the final value.

For proteins that contain disulfide bonds, the renaturation buffer needs to be supplemented with a redox system. Adding a mixture of the reduced and oxidized forms of low molecular weight thiol reagents, such as glutathione, cysteine, or cysteamine, usually provides the appropriate redox potential to allow the formation and reshuffling of disulfides [21]. The concentration of reduced and oxidized reagents are commonly in the ranges of 1 to 5 mM and 0.01 to 0.5 mM, respectively, and optimal regeneration usually occurs with a tenfold molar excess of the reduced component [20].

1.5 Macromolecules: Nucleic Acids and Oligonucleotides

Nucleic acids are polynucleotides whose primary structure consists of repeating units of nucleotides that are joined by phosphodiester linkage between the 3′-hydroxyl group of one ribose (or deoxyribose) and the 5′-hydroxyl group of an adjacent ribose (or deoxyribose). A nucleotide consists of a nitrogenous base, a ribose or deoxyribose sugar, and one

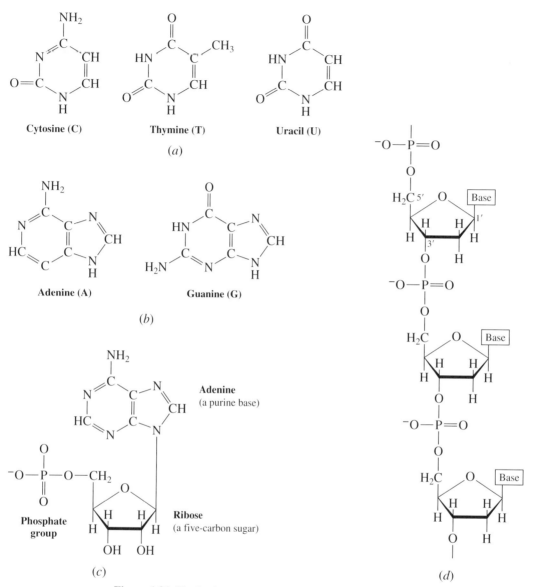

Figure 1.21 The basic structures present in nucleic acids: (*a*) pyrimidine bases, (*b*) purine bases, (*c*) the nucleotide adenosine monophospate (AMP), and (*d*) a section of a DNA chain.

or more phosphate groups (Figure 1.21). Ribonucleic acid (RNA) is a polynucleotide containing only ribose sugar, while deoxyribose (DNA) contains only deoxyriboses (lacking a hydroxyl group a the 2′ position). The structure of a DNA chain is shown in Figure 1.21. The structure of a DNA molecule—two antiparallel polynucleotide chains coiled into a double helix—was deduced by James Watson and Francis Crick in 1953. The nitrogenous bases are on the interior of the helix, with one base on one strand being hydrogen-bonded to an adjacent base on the other strand. Adenine–thymine and guanine–cytosine are the possible base pairs in DNA, while in RNA thymine is replaced by uracil. In shorthand notation, a DNA oligonucleotide can be written as follows:

5′ ATCGGCATCGCCGAT 3′

3′ TAGCCGTAGCGGCTA 5′

The use of DNA in tiny quantities is common in the genetic engineering industry and the gene therapy and DNA vaccines fields. To date, large-scale purification of DNA has never been necessary. However, a new family of nucleic acid products is currently arising, and some of these may be needed in large quantities for diagnostic and therapeutic use. Antisense sequences, ribozymes, and aptamers represent the three major categories of oligo- and polynucleotides that have commercial potential. Antisense molecules can bind to mRNA and inactivate it; ribozymes can react with RNA and catalyze its cleavage at sequence-specific sites, and aptamers are oligonucleotides that bind with high affinity to chosen ligands because they have undergone *in vitro* evolution under selective pressure during several rounds of enzymatic replication. Aptamers are thus considered combinatorial products. These three categories of nucleic acid products are targeted for the therapeutics and diagnostics fields. Upstream processes are currently undergoing major improvements in these cases, and the synthesis of these products is not, in general, achieved by fermentation. The field is dominated by automated polynucleotide synthesizers.

The genetic engineering industry, on the other hand, consists of patented DNA sequences that are sold to customers or used as vectors for recombinant genes in a company's own process. Vectors (plasmids) containing genes for human insulin, human α-interferon, and human erythropoietin are well-known examples of the latter category. Some plasmids are extremely sophisticated and contain the structural gene for the protein product, one or more promoter sequences recognized by the transcription enzyme complex, enhancer elements, origins of replication for bacteria and animal cells, and operator sequences for switching on and off transcription of RNA for the protein product. All these products are generated in recombinant bacteria, characterized by gel electrophoresis (see Chapter 2), and partially purified for transfection into bacterial, yeast, or animal cells that will generate the product. Products used for gene therapy are also produced in this fashion. The most popular means of purifying plasmid DNA and portions thereof is gel electrophoresis, a very low capacity method that is useful in the gene industry because only very small amounts (nanograms) of material are needed to transform cells.

Nucleic acids are chemically more robust than proteins. Their secondary structures are highly stable, being based on base-pairing hydrogen bonds. A word of caution however: nucleic acids are hydrolyzed by enzymes that are ubiquitous in nature, and RNA in particular is susceptible to ribonuclease molecules that seem to be everywhere. In the absence of significant need for scaleup of downstream processing in the nucleic acid industry, this subject will be given less attention than the processing of proteins in this book.

1.6 Macromolecules: Polysaccharides

Of all bioproducts in use, the polysaccharides have the highest molecular weight and the longest end-to-end polymer length. Most naturally derived polysaccharides are branched to various degrees—they are not necessarily linear like peptides and nucleic acids.

The most familiar polysaccharides are starch, glycogen, and cellulose. These are very high molecular weight polymers of a small number of different sugars. Owing to the numerous hydroxyl groups attached to sugar carbons, some polysaccharides can be highly branched via typical glycosidic linkages. Polysaccharides find uses in industrial chemistry (e.g., agar) and foods (e.g., carrageenan). Figure 1.22 depicts two more polysaccharides, amylose (a linear form of corn starch) and chitin (a mucopolysaccharide containing *N*-acetylamino sugars found in fungal cell walls and invertebrate exoskeletons). Bacterial polysaccharides are the active component of a number of vaccines.

Although most polysaccharides are extracted from plants, several are produced microbially, the most important of which is xanthan gum. Other important microbial

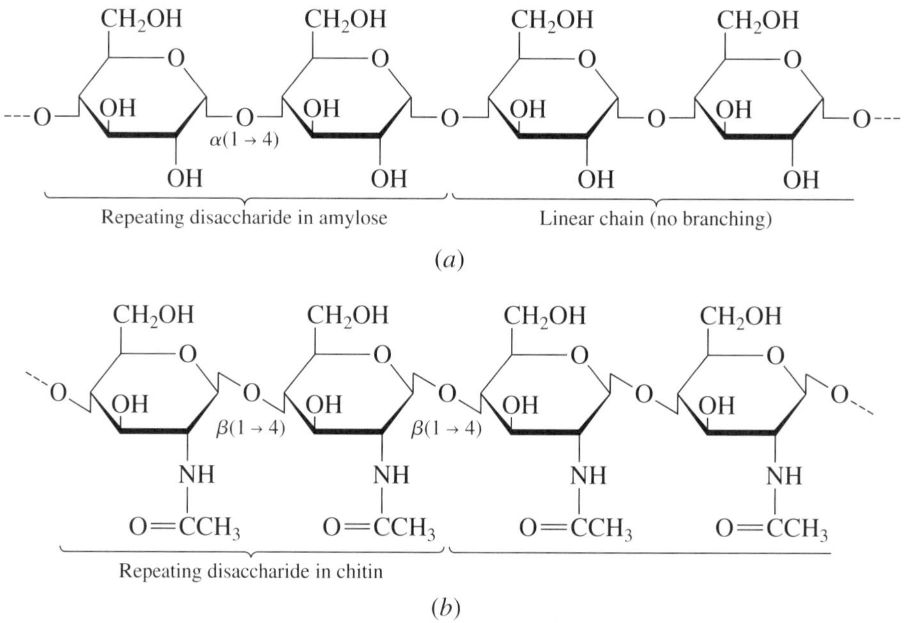

Figure 1.22 Two polysaccharides. (*a*) Amylose, which is poly[$\alpha(1 \rightarrow 4)$ glucose] (straight chain); it has polarity, and the free C_1 end is the "reducing sugar." (*b*) Chitin, which is poly-*N*-acetylglucosamine, a mucopolysaccharide.

TABLE 1.8
Some Polysaccharides, Their Sources and Uses

Polysaccharide Name	Polysaccharide Source	Uses
Starch	Potatoes, corn	Food, clothing
Cellulose	Wood, cotton	Paper, clothing
Carrageenan	Mycophyta (algae)	Food
Agar	Mycophyta	Microbiology, food
Dextran	Corn	Food, medicine
Agarose	Mycophyta	Biochemistry
Xanthan gum	*X. campestris*	Food, industrial chemical, oil field drilling

polysaccharides include dextran, alginate, and pullulan [22]. Except for applications in biochemical research and in the bioprocessing industry itself, highly purified polysaccharides are seldom sold. Sources and uses of selected polysaccharides are listed in Table 1.8. It is of some interest to the field of this textbook that cellulose, agarose, and dextran are used in media for bioseparations, including electrophoresis, chromatography, and extraction (see Chapters 2, 6, and 7).

1.7 Particulate Products

Novel products of biotechnology include particulate materials. These take the form of subcellular particles, bacterial inclusion bodies, whole cells, and insoluble macromolecular aggregates. Theoretically, the logical way to purify these products is by centrifugation. In an increasing number of cases of subcellular particles, this approach fails. Some very small particles have very low sedimentation rates, and therefore require ultracentrifugation—a satisfactory method only at bench scale. Some particle suspensions are contaminated with unwanted particles having identical sedimentation velocities. The reasons for purifying suspended animal and plant cells are numerous, including obtaining pure populations for biochemical study and for food or feed applications, purified living material for transplantation, and starting material for pure bioproducts.

Numerous activities require or involve the separation or purification of cells: using purified cells for relating cell structure to cell function, obtaining clonogenic (multiplying) but rare plant and animal cell subpopulations, isolating hybrid (fused) cells for genetic engineering, separating nonclonogenic cells, obtaining cell subsets for transplantation, separating productive from nonproductive bioreactor cells, and removing cells that produce unwanted products from fermentor harvests.

Microbial cells grown for food or feed applications are known as "single-cell proteins" (SCP). Microorganisms that have been grown for SCP include algae, bacteria, yeast, and filamentous fungi. Raw materials for the production of SCP by fermentation have varied from hydrocarbons, cellulosic materials, and saccharides to carbon-containing wastes [22].

Subcellular components that have been of recent interest include bacterial inclusion bodies (100–1000 nm), ribosomes (25 nm), liposomes or natural vesicles (100 nm), and natural hormone granules (200 nm). Sometimes these products differ in subtle ways from their contaminants, and a pure separation could be challenging. At some point in the life of a number of fermentation products, the product is a component of a particle, such as a cell or organelle. The cell or organelle may not itself be the product, but when it houses the product it must be successfully isolated from the impurity-laden fluid that surrounds it. As will be seen in Chapters 3 to 5, the early bioseparation steps can include flocculation, sedimentation, and filtration—methods designed to isolate particulate separands.

1.8 Introduction to Bioseparations: Engineering Analysis

Modern biotechnology is built on the genetic manipulation of organisms, *genetic engineering,* to produce commercial products or processes. The implementation of these products or processes is dependent upon *biochemical engineering,* which is in turn divided into two disciplines, upstream engineering (fermentation) and downstream engineering (purification, or bioseparations). Purification of bioproducts typically involves a long sequence of steps, and each step requires the use of one or more *unit operations,* such as sedimentation, adsorption, and drying.

Bioproducts have unique properties. The earlier part of this chapter listed several bioproducts and their properties. For bioseparation purposes, important properties data include thermal stability, solubilities, diffusivities, charge, isoelectric pH, reaction rate constants, and separation thermodynamics. A considerable amount of downstream process planning is based on the lability of most bioproducts. Temperature, pH, and concentration must be maintained within specific ranges of values to assure product stability.

1.8.1 STAGES OF DOWNSTREAM PROCESSING

Removal of solids (or recovery), isolation of product, purification, and polishing constitute a sequence of events applied to nearly every product as it is prepared. These four stages make use of various unit operations. A "unit operation" is a single process and the equipment used therefor. These operations are applied in varying combinations to the four stages, approximately as indicated in Table 1.9.

In the unit operations approach used in this book, a separate chapter will be devoted to the analysis and applications of each unit operation. The unit operations can be divided by function into:

- Liquid–solids separations (dewatering, concentration, particle removal)
- Solute–solute separations (isolation, purification)
- Solute–liquid separations (polishing).

TABLE 1.9
Objectives and Typical Unit Operations of the Four Stages in Bioseparations

Stage	Objective(s)	Typical Unit Operations
Separation of insolubles	Remove or collect cells, cell debris, or other particulates Reduce volume (depends on unit operation)	Filtration, sedimentation, extraction, adsorption
Isolation of product	Remove materials having properties widely different from those desired in product Reduce volume (depends on unit operation)	Extraction, adsorption, ultrafiltration, precipitation
Purification	Remove remaining impurities, which typically are similar to the desired product in chemical functionality and physical properties	Chromatography, affinity methods, crystallization, fractional precipitation
Polishing	Remove liquids Convert the product to crystalline form (not always possible)	Drying, crystallization

EXAMPLE 1.2

Initial Selection of Purification Steps   You have been given the assignment to purify the antibiotic erythromycin. You have at your disposal the *Merck Index,* which has the information on erythromycin shown in Figure E1.2. What do you think are the most likely unit operations that should be used for the isolation and purification of erythromycin?

SOLUTION

Erythromycin has limited solubility in water but is soluble in several solvents, including amyl acetate. Since the solubility of amyl acetate in water is low, isolation could be performed by a liquid–liquid extraction of erythromycin using a water–amyl acetate system. For the extraction, it would be desirable to raise the pH of the aqueous phase above the pK_a of erythromycin of 8.8, so that the secondary amino group is converted from the positively

3720. Erythromycin. Erythromycin A; Abomacetin; Ak-Mycin; Aknin; E-Base; EMU; E-Mycin; Eritrocina; Ery Derm; Erymax; Ery-Tab; Erythromast 36; Erythromid; ERYC; Erycen; Erycin; Erycinum; Ermysin; Ilotycin; Inderm; Retcin; Staticin; Stiemycin; Torlamicina. $C_{37}H_{67}NO_{13}$; mol wt 733.94. C 60.55%, H 9.20%, N 1.91%, O 28.34%. Antibiotic substance produced by a strain of *Streptomyces erythreus* (Waksman) Waksman & Henrici, found in a soil sample from the Philippine Archipelago. Isoln: McGuire *et al., Antibiot. & Chemother.* **2,** 281 (1952); Bunch, McGuire, U.S. pat. **2,653,899** (1953 to Lilly); Clark, Jr., U.S. pat. **2,823,203** (1958 to Abbott). Properties: Flynn *et al., J. Am. Chem. Soc.* **76,** 3121 (1954). Solubility data: Weiss *et al., Antibiot. & Chemother.* **7,** 374 (1957). Structure: Wiley *et al., J. Am. Chem. Soc.* **79,** 6062 (1957). Configuration: Hofheinz, Grisebach, *Ber.* **96,** 2867 (1963); Harris *et al., Tetrahedron Letters* **1965,** 679. There are three erythromycins produced during fermentation, designated A, B, and C; A is the major and most important component. Erythromycins A and B contain the same sugar moieties, desosamine, *q.v.,* and cladinose (3-*O*-methylmycarose). They differ in position 12 of the aglycone, erythronolide, A having an hydroxyl substituent. Component C contains desosamine and the same aglycone present in A but differs by the presence of mycarose, *q.v.,* instead of cladinose. Structure of B: P. F. Wiley *et al., J. Am. Chem. Soc.* **79,** 6070 (1957); of C: *eidem, ibid.* 6074. Synthesis of the aglycone, erythronolide B: E. J. Corey *et al., ibid.* **100,** 4618, 4620 (1978); of erythronolide A: *eidem, ibid.* **101,** 7131 (1979). Asymmetric total synthesis of erythromycin A: R. B. Woodward *et al., ibid.* **103,** 3215 (1981). NMR spectrum of A: D. J. Ager, C. K. Sood, *Magn. Reson. Chem.* **25,** 948 (1987). Biosynthesis: Martin, Goldstein, *Progr. Antimicrob. Anticancer Chemother., Proc. 6th Int. Congr. Chemother.* **II,** 1112 (1970); Martin *et al., Tetrahedron,* **31,** 1985 (1975). Cloning and expression of clustered biosynthetic genes: R. Stanzak *et al., Biotechnology* **4,** 229 (1986). Reviews: T. J. Perun in *Drug Action and Drug Resistance in Bacteria* **1,** S. Mitsuhashi, Ed. (University Park Press, Baltimore, 1977) pp 123-152; Oleinick in *Antibiotics,*

vol. **3,** J. W. Corcoran, F. E. Hahn, Eds. (Springer-Verlag, New York, 1975) pp 396-419; *Infection* **10,** Suppl. 2, S61-S118 (1982). Comprehensive description: W. L. Koch, *Anal. Profiles Drug Subs.* **8,** 159-177 (1979).

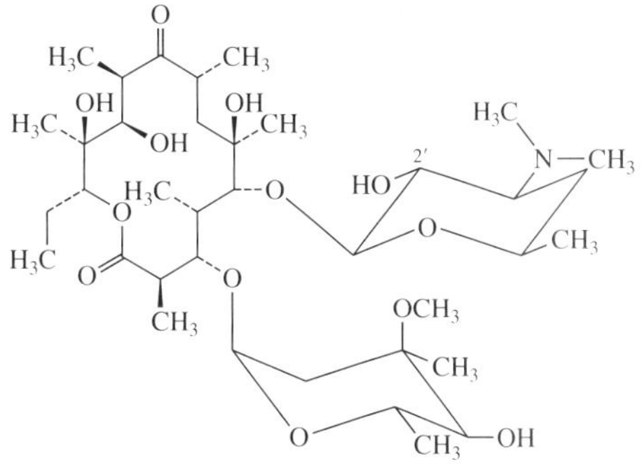

Erythromycin A

Hydrated crystals from water, mp 135-140°, resolidifies with second mp 190–193°. Melting point taken after drying at 56° and 8 mm. $[\alpha]_D^{25} - 78°$ (c = 1.99 in ethanol). uv max (pH 6.3): 280 nm (ε 50). pKa₁ 8.8. Basic reaction. Readily forms salts with acids. Soly in water: ~2 mg/ml. Freely sol in alcohols, acetone, chloroform, acetonitrile, ethyl acetate. Moderately sol in ether, ethylene dichloride, amyl acetate.

Ethylsuccinate, $C_{43}H_{75}O_{16}$, Anamycin, Arpimycin, Durapaediat, E.E.S., E-Mycin E, Eryliquid, Eryped, Erythro ES, Erythro-Holz, Erythroped, Esinol, Monomycin, Paediathrocin, Pediamycin, Refkas, Sigapedil, Wyamycin E. Prepn: Brit. pat. **830,846;** R. K. Clark, U.S. pat. **2,967,129** (1960, 1961 both to Abbott). Hydrated crystals from acetone + water, mp 109–110°. $[\alpha]_D - 42.5°$.

THERAP CAT: Antibacterial.

THERAP CAT (VET): Antibacterial.

Figure E1.2 The *Merck Index* entry for a widely used antibiotic. (Data from S. Budavari, M. J. O'Neill, A. Smith, P. E. Heckelman, and J. F. Kinneary, eds., *The Merck Index,* 12th ed., p. 625, Merck Research Laboratories, Whitehouse Station, NJ, 1996.)

charged form to the neutral free base form. For the purification step, crystallization is a good choice, since hydrated crystals have been obtained from water.

This example illustrates how laboratory data can guide the formulation of a sequence of unit operations to purify a bioproduct. Chapter 11, Bioprocess Design, elaborates further on the selection of unit operations for a bioseparations process.

1.8.2 BASIC PRINCIPLES OF ENGINEERING ANALYSIS

It is said that any good engineering calculation ends with a number bearing a dollar sign. The purpose of engineering analysis, therefore, is to determine "how much" and "how fast." These questions are answered in the form of equations that can be solved for product capacity, processing rate, and possibly product purity. To obtain values for such variables requires a set of *governing equations* derived from the three principal ingredients of engineering analysis: *equilibria, material balance,* and *flux (or transport phenomena).* The translation of processing rates into cost is detailed in Chapter 11.

Material balance

Operating curves (for each unit operation) are built on the basis of material balances and used in consort with equilibrium curves. The material balance equation states

$$\text{Accumulation} = \text{inflow} - \text{outflow} + \text{amount produced} - \text{amount consumed} \quad (1.8.1)$$

Examples of Equilibria

Traditionally, a chemical reaction that is at equilibrium,

$$A + B \rightleftharpoons C \quad (1.8.2)$$

can be characterized by an equilibrium constant

$$K_{eq} = \frac{[C]}{[A][B]} \quad (1.8.3)$$

where all concentrations are in mole fractions, molarity, or molality. Likewise, in an extraction process that has gone to equilibrium,

$$K = \frac{y}{x} \quad (1.8.4)$$

where y is the concentration of a separand in the extract phase, and x is the concentration of the same separand in the raffinate (usually heavy) phase, and K is called the "partition coefficient." In the case of adsorption, the equilibrium constant relates the concentration in the adsorbent phase, $[CS]$, to the concentration in the liquid phase, $[C]$, at equilibrium, where C is a chemical species and S is an adsorption site. When the adsorption equilibrium is linear, for example, the following relation is valid at low concentrations:

$$[CS] = K_{eq}[C] \quad (1.8.5)$$

Examples of Flux Relationships (Transport Phenomena)

Flux relationships abound in chemistry and physics, and they all state that

$$\text{Flux} = \text{coefficient} \times \text{driving force} \qquad (1.8.6)$$

where flux is in units flowing per unit area per unit time, driving force is a gradient down which units flow, and the "coefficient" is a permeability or the inverse of a resistance, derivable from properties of the medium. Thus, we have Ohm's law

$$J_e = CE \qquad (1.8.7)$$

where J_e is current density, C is electrical conductivity (property of the medium), and E is electrical potential gradient. Fick's first law applies for diffusive flux due to a concentration gradient dc/dx in one dimension

$$J_D = -\mathscr{D}\frac{dc}{dx} \qquad (1.8.8)$$

where $\mathscr{D}$, the diffusion coefficient, is also a property of the medium and in some cases calculable from the Stokes–Einstein equation for spheres

$$\mathscr{D} = \frac{kT}{6\pi\mu a} \qquad (1.8.9)$$

where k is the Boltzmann constant, T is absolute temperature, μ is viscosity, and a is particle radius. Equation (1.8.9) is particularly useful when property data are not available.

Of great interest in downstream processing is the empirical law (Darcy's law) of flow through porous medium

$$J_w = L_p\Delta p \qquad (1.8.10)$$

where J_w is the fluid flux, L_p is Darcy's law permeability, and Δp is pressure drop across the porous medium, such as a filter or a column of adsorbent resin—systems very frequently used in bioprocessing.

1.8.3 PROCESS AND PRODUCT QUALITY

The measures of product quality due to processing are purity, fold purification, specific activity, and yield. Purity is defined as follows:

$$\text{Purity} = \frac{\text{amount of product}}{\text{amount of product} + \text{amount of total impurities}} \qquad (1.8.11)$$

Fold purification is the ratio of the purity at any stage in the process to the purity at the start of the purification process. Another measure of purity is

$$\text{Specific activity} = \frac{\text{units of biological activity}}{\text{mass}} \qquad (1.8.12)$$

where units of biological activity are assayed by means of a biological test, such as moles of substrate converted per second per liter or fraction of bacterial cells killed. Examples of analytical measurements are given in Chapter 2. For proteins, the mass in Equation (1.8.12) is usually total protein; on this basis, the specific activity reaches a constant value when the protein is pure.

Yield is given by

$$\text{Yield} = \frac{\text{amount of product produced}}{\text{amount of product in feed}} \tag{1.8.13}$$

Purity is a strictly quantitative measure and not always an expression of the quality of the product. A therapeutic protein can be 99.99% pure but still unacceptable if any pyrogen is present. A pyrogen is any substance that produces a fever. Most pyrogens are exogenous, which means they do not directly cause fever. Exogenous pyrogens cause release of endogenous pyrogenic substances that gain access to the hypothalamic thermoregulatory center, where they initiate fever. On the other hand, if the product is not a therapeutic protein but an industrial enzyme, then practically any impurities that do not inhibit the activity of the product or endanger the user are allowed. Figure 1.23 indicates this principle by comparing the level and range of purity acceptable for three different protein products. In short, two kinds of measurements of purity are required: activity, composition, and structure on the product itself; and host cell materials, degraded product, and excipients (additives) on the impurities.

1.8.4 CRITERIA FOR PROCESS DEVELOPMENT

Approaches to process development (also known as process synthesis) must be made with future scaleup in mind. A biotechnology product production process is a set of connected

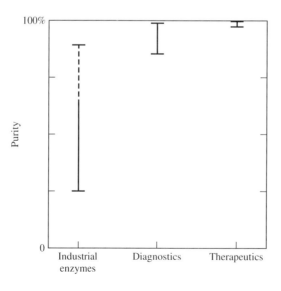

Figure 1.23 Different levels of purity are required for different products.

unit operations that produces a final product able to meet defined levels of quality, yield, and cost. The following criteria should therefore be used in evaluating and developing a bioseparation process:

- Product purity
- Cost of production as related to yield
- Scalability
- Reproducibility and ease of implementation
- Robustness with respect to process stream variables

The integration of unit operations for the efficient synthesis of bioseparation processes is discussed fully in Chapter 11, Bioprocess Design.

1.9 The Route to Market

The route for a biotechnology product being on the market is complex and can be lengthy. Figure 1.24 charts the pathway to market for a recombinant biotechnology product. Note that several years are required for the completion of the route from genetic discovery to marketing.

Besides identifying the market for the product and determining the optimal process for the expected production rate, the personnel involved in this effort must make sure that all the regulatory requirements are met. This is clearly an interdisciplinary task, involving not only engineers and bioscientists but also marketing and regulatory personnel. Often, applications personnel, such as medical doctors and veterinarians, must also be involved. Now we turn to some of the factors that are important to consider on the route to market.

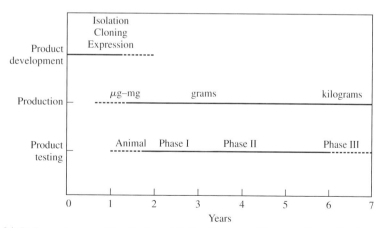

Figure 1.24 Pathway to market beginning with the discovery of a recombinant biopharmaceutical product.

1.9.1 The Chemical and Applications Range of the Bioproduct

As mentioned earlier, modern biotechnology produces the full range of substances for a wide range of applications; some well-known examples help emphasize this point:

Aspartame, a dipeptide nonnutritive sweetener

Paclitaxel, a triterpene from plants used in cancer treatment

Erythropoietin, a peptide hormone that stimulates red cell production in the bone marrow

Oligonucleotides, which can be made to function according to shape and according to catalytic activity

Bacillus thuringiensis (Bt), whole bacterial cells that when dried can be sprayed on crops to prevent insect damage

Note the categories: food, medicine, and agriproducts. While most of the excitement in biotechnology and bioprocessing today is in the area of biopharmaceuticals produced by genetically engineered organisms, a much wider variety of products exists, usually with lower value per mass unit, including biopesticides, novel food products, polymer components, hydrogel polymers, industrial enzymes, commodity enzymes (laundry products), and fine chemicals. To know what product purity, cost, and FDA requirements must be met, the exact application of the proposed bioproduct must be identified early in process development, or even beforehand.

1.9.2 Documentation of Pharmaceutical Bioproducts

Medical products are catalogued in numerous commercial handbooks, including the *Merck Index* and *Physician's Desk Reference* ("PDR"). The official documentation of all compounds sold for medication must appear in the *U. S. Pharmacopeia*. For each substance the following data are given: structural formula, empirical formula, Geneva name (to the extent possible), potency, packaging and storage, reference standards, labeling, chemical identification methods, and assays for identity and purity. Validations of these properties are the responsibility of the manufacturer; in the United States, the FDA grants permits, first to manufacture for testing, then to test, and then to manufacture for sale and to sell.

1.9.3 GLP and cGMP

GLP refers to "good laboratory practice," and cGMP refers to "current good manufacturing practice." These are guidelines of procedure in research and manufacturing mandated by the Food and Drug Act of 1976, which states, in part (Paragraph 210.1): "This chapter contains the minimum current good manufacturing practice for methods to be used in, and the facilities or controls to be used for, the manufacture, processing, packing, or holding of a drug to assure that such drug meets the requirements of the act as to safety, and has the identity and strength and meets the quality and purity characteristics that it purports or is represented to possess."

Once a product has met the requirements of the act, it may be considered an "investigative new drug" (IND) for which a license is issued. The most important matter for the engineer is the fact that the process defines the product. Thus a published description of a pharmacological product in the *U. S. Pharmacopeia* (USP) includes a structural formula and Geneva name (if applicable), method of preparation, methods of analysis, and definitions of units of activity (if applicable).

1.9.4 FORMULATION

Formulation is that postpolishing process whereby a drug or other chemical is converted to a dosage form. It is combined with *excipients* that contribute to its stability under storage conditions, deliverability in medical applications, or final concentration. Thus excipients can be stabilizers (reducing compounds, polymers, etc.), tablet solids or parenteral diluents [a gum, water for injection, poly(ethylene glycol), penetrant oil, etc.], or a liquid diluent (water for injection). Formulation is also subject to cGMP and usually defines a commercial product.

1.10 Summary

Bioseparations is an essential component of the production of bioproducts, the market for which is large and continuing to grow rapidly. Bioproducts encompass human therapeutics and diagnostics, agriculture specialty chemicals, and nonmedical diagnostics.

- The products in this market are identifiable as small biomolecules, macromolecules, and particulate products, including cells.
- Much of the recent effort in bioseparations has focused on proteins produced by recombinant DNA technology. Purification of recombinant proteins can be simplified by the addition through genetic engineering of a peptide "tag" that enables affinity purification, such as the addition of a histidine oligomer that selectively binds to immobilized transition metals. Some recombinant proteins are expressed as insoluble inclusion bodies and must be resolubilized and then refolded. Refolding can sometimes be avoided by expressing the target protein as part of a fusion protein that includes a soluble carrier protein.
- The structure of proteins is established at up to four levels: primary (amino acid sequence), secondary (hydrogen-bonded structures), tertiary (the folding pattern of hydrogen-bonded and disulfide-bonded structures), and quaternary (the formation of multimeric complexes by individual protein molecules). The stability and functions of proteins are established on the basis of these structure determinants.
- Product stability is a criterion for selecting unit operations. Some processes, especially those involving excess heat, bubble formation, pH extremes, or untimely drying, can destroy the function of bioactive substances such as enzymes, hormones, antibiotics, and some vitamins.
- Once synthesized, nearly all bioproducts are separated using a sequence of processes usually consisting of separation of insolubles, product isolation, purification, and polishing.
- Material balance, equilibrium, and transport equations form the scientific foundation of engineering calculations in biochemical separations.

- Purity specified as [product]/([product] + [contaminants]), specific activity (biological activity/mass), and yield (mass after purification/initial mass) are calculated as quality indicators in purification.

NOMENCLATURE

a particle radius (μm)

c solute concentration (M, or g liter^{-1})

C electrical conductivity (ohm^{-1} cm^{-1})

$\mathcal{D}$ diffusion coefficient (cm^2 s^{-1})

E electrical potential gradient (V cm^{-1})

J_D diffusive flux of a solute (mol cm^{-2} s^{-1}, or g cm^{-2} s^{-1})

J_e electrical current density or flux (A cm^{-2})

J_w fluid flux (liter cm^{-2} h^{-1})

k Boltzmann constant (1.3807 $\times$ 10^{-23} J K^{-1})

K partition coefficient (dimensionless)

K_{eq} equilibrium constant (units vary)

L_p permeability coefficient (liter cm^{-2} h^{-1} Pa^{-1})

p pressure (Pa)

T absolute temperature (K)

x linear distance (cm)

x raffinate phase concentration of a separand in extraction (M, or g liter^{-1})

y extract phase concentration of a separand in extraction (M, or g liter^{-1})

μ viscosity (g cm^{-1} s^{-1})

PROBLEMS

The solution to the following problems may require the use of information not in the textbook (which is part of the exercise).

1.1   **Product Molecular Weight and Value** Rearrange the following five bioproducts approximately according to molecular size (left column) and according to cost per kilogram in the right column: recombinant human growth hormone for injection (hGH), lactic acid (LA), riboflavin for animal feed (B$_2$), goat antibodies for AIDS testing kits (gAb), and yeast phospholipase C for laundry presoaks (yPLc).

MW	\$/kg
——————	——————
——————	——————
——————	——————
——————	——————
——————	——————

1.2   **Product Concentrations** A 10^{-3} M solution of gamma globulin (antibodies) is called a "concentrated solution." Why is it called a concentrated solution? Support your answer with a calculation.

1.3   **pH-Dependent Change of Conformation of Poly-L-glutamic Acid** A synthetic polypeptide made up of L-glutamic acid residues is in a random coil configuration at pH 7.0 but changes to α helical when the pH is lowered to 2.0. Explain this pH-dependent conformational transition.

1.4   **Properties for Processing** Riboflavin (vitamin B$_2$) is isolated by filtration and polished by spray drying. The efficiency of both processes (higher yield per unit time) increases with increasing temperature. Find the necessary handbook data and determine the maximum temperature at which you could run these two unit operations.

1.5   **Process Sequencing** Arrange the following products in order of increasing molecular weight (i.e., smallest first), and write after each product the most likely separation process to be useful in its isolation, using the list of processes given in Table P1.5.

1.6   **Properties: Diffusivity** Estimate the molecular diameter and diffusion coefficient for the proteins ribonuclease (MW 13,700 Da), hemoglobin (MW

TABLE P1.5

Product	Process
Insulin	Distillation
Citric acid	Filtration
B lymphocytes	Adsorption
Vaccine virus	Extraction
Gamma globulin	Centrifugation
Bacitracin	Precipitation
Ethanol	Ultrafiltration
Ribozyme	
Hemoglobin	
Riboflavin	

68,000), and urease (MW 480,000), assuming the molecules are spherical and the density of each protein molecule is 1.3 g/cm^3.

1.7 **Preliminary Selection of Purification Steps** Based on information in the *Merck Index,* what do you think are the most likely unit operations that should be used for the isolation and purification of the following bioproducts?
(a) Lincomycin
(b) L-Lactic acid
(c) L-Asparaginase

1.8 **Calculations for the Purification of a Recombinant Protein** The purification of a recombinant protein is carried out starting with 100 liters of a clarified cell lysate (i.e., the cells have been lysed, and the cell debris has been removed to give a clarified solution), which has a total protein concentration of 0.36 mg/ml and a recombinant protein concentration of 2.2 U/ml, where U denotes units of biological activity of the recombinant protein. It is known that the completely pure recombinant protein has a specific activity of 40.0 U/mg. Purification is continued until a chromatography step that yields 2.0 liters of a fraction containing the protein, with a total protein concentration of 1.11 mg/ml and a recombinant protein concentration of 43.2 U/ml.

For the recombinant protein, calculate the starting and ending purity, the starting and ending specific activity, and the percentage yield and fold purification through the chromatography step.

1.9 **Process Synthesis** A process for isolating an antibody against insulin has, as a unit operation, the reaction of the antibody with the antigen in a continuous stirred tank reactor. The reaction product is a precipitate that is continuously removed from the reactor with 10% of the solution, which is mouse serum. Since insulin is an expensive reagent, only stoichiometric amounts can be added to the mouse serum, which contains 8 mg/liter of anti-insulin. If this particular monoclonal antibody precipitates with its antigen in a 1:1 ratio, how many milligrams of insulin must be added to the reactor per hour to process 100 ml of mouse serum per hour? (Assume that equilibrium is achieved in this reactor.) Sketch a flowchart of this process.

1.10 **Product Concentrations** Using the *Handbook of Biochemistry and Molecular Biology* (G. Fasman, ed., CRC Press, Cleveland, 1976) or a similar source or a suitable biochemistry textbook (in other words, looking up information), find the necessary information to determine the amount of material required to make 1 ml of each of the following aqueous solutions:
(a) 0.01 M cytochrome c
(b) 1×10^{-7} M β-galactosidase
(c) 0.01 M porcine insulin
(d) 0.01 M human hemoglobin
(e) 0.1 M streptomycin
(f) 1×10^{-6} M oligonucleotide with 10 nucleotides

Also calculate the concentrations in terms of the following additional standard means of expressing bioproduct concentrations: percent (weight per volume) and milligrams per milliliter. Assuming that the solutions are in pure water, also express the concentrations as mole fractions. Discuss the feasibility of making each one of these solutions.

References

1. Consulting Resources Corporation, Newsletter, Spring, 2000 (data reproduced by permission).
2. Bailey, J. E., and Ollis, D. F. (1986). *Biochemical Engineering Fundamentals,* 2nd ed., McGraw-Hill, New York.

3. Lehninger, A. L. (1975). *Biochemistry,* 2nd ed., Worth, New York.
4. Mazur, A., and Harrow, B. (1971). *Textbook of Biochemistry,* 10th ed., Saunders, New York.
5. Stryer, L. (1988). *Biochemistry,* 3rd ed., Freeman, New York.
6. Mathews, C. K., and Van Holde, K. E. (1996). *Biochemistry,* 2nd ed., Benjamin/Cummings, Menlo Park, CA.
7. Thayer, A. M. (1998). Great expectations. *Chem. Eng. News,* (August 10), vol. 76, p. 19.
8. Weiss, M. (1992). Stability data prove critical for assuring biopharmaceutical shelf life. *Gen. Eng. News,* vol. 14, no. 2, p. 10.
9. Carter, D. C., and Ho, J. X. (1994). Structure of serum albumin. In *Advances in Protein Chemistry,* Vol. 45, V. N. Schumaker, ed., Academic Press, New York, p. 153.
10. Kellner, R., Meyer, H. E., and Lotspeich, F. (1994). Amino acid analysis. In *Microcharacterization of Proteins,* R. Kellner, F. Lotspeich, and H. E. Meyer, eds., VCH, Weinheim, p. 93.
11. Scopes, R. K. (1994). *Protein Purification,* 3rd ed., Springer-Verlag, New York.
12. Maa, Y. F., and Hsu, C. C. (1996). Effect of high shear on proteins. *Biotechnol. Bioeng.,* vol. 51, p. 458.
13. Virkar, P. D., Narendranathan, T. J., Hoare, M., and Dunnill, P. (1981). Studies of the effects of shear on globular proteins: Extension to high shear fields and to pumps. *Biotechnol. Bioeng.,* vol. 23, p. 425.
14. Arnold, F. H. (1991). Metal–affinity separations: A new dimension in protein processing. *Bio/Technology,* vol. 9, p. 152.
15. Smith, D. B., and Johnson, K. S. (1988). Single-step purification of polypeptides expressed in *Escherichia coli* as fusions with glutathione *S*-transferase. *Gene,* vol. 67, p. 31.
16. di Guan, C., Li, P., Riggs, P. D., and Inouye, H. (1981). Vectors that facilitate the expression and purification of foreign peptides in *Escherichia coli* by fusion to maltose-binding protein. *Gene,* vol. 67, p. 21.
17. Davis, G. D., and Harrison, R. G. (1999). New fusion protein systems designed to give soluble expression in *Escherichia coli. Biotechnol. Bioeng.,* vol. 65, p. 382.
18. LaVallie, E. R., DiBlasio, E. A., Kovacic, S., Grant, K. L., Schendel, P. F., and McCoy, J. M. (1993). A thioredoxin gene fusion expression system that circumvents inclusion body formation in the *E. coli* cytoplasm. *Bio/Technology,* vol. 11, p. 187.
19. Southworth, M. W., Amaya, K., Evans, T. C., Xu, M. -Q., and Perler, F. B. (1999). Purification of proteins fused to either the amino or carboxy terminus of the *Mycobacterium xenopi* gyrase A intein. *BioTechniques,* vol. 27, p. 110.
20. Rudolph, R., Bohm, G., Lilie, H., and Jaenicke, R. (1997). Folding proteins. In *Protein Function. A Practical Approach,* T. E. Creighton, ed., IRL Press, Oxford, p. 57.
21. Lilie, H., Schwarz, E., and Rudolph, R. (1998). Advances in refolding of proteins produced in *E. coli. Curr. Opin. Biotechnol.,* vol. 9, p. 497.
22. Blanch, H. W., and Clark, D. S. (1996). *Biochemical Engineering,* Dekker, New York.

Analytical Methods

The development of efficient and reliable processes for bioseparations is dependent on the availability of suitable analytical methods. This means it is important that work on analytical methodology for the bioproduct of interest start at the very beginning of process development. Analytical studies are important throughout the development and scaleup of the process, as changes can occur either to the product or to its associated impurities from what may be thought of as minor changes in the process.

This chapter gives access to the vocabulary and techniques used in quality control and analytical development activities, starting with a description of specifications typically set for a pharmaceutical, and the rationale behind them. Then, before discussing the assays themselves, we describe assay attributes, which can be measured and used to help not only the assay developer but also the biochemist and engineer responsible for developing downstream processes determine the usefulness of the assay. Finally, we turn to assays that are commonly applied in biotechnology, as they apply to biological activity, identity, and purity.

These assays are the ultimate yardsticks by which the process is measured. Purification methods are developed for their ability to remove a contaminant from the product of interest, whether it is a related molecule, a contaminant related to a host organism, such as DNA or endotoxin, or a process contaminant, such as a residual solvent or water. Critical to understanding process performance is an understanding of how the assays that measure these contaminants have been developed, what the assay strengths and limitations are, and what they indicate and why.

2.1 Instructional Objectives

After completing this chapter, the reader should be able to do the following:

- Identify "specifications" that are desired, especially by regulatory agencies, in the definition of a product.
- Define assay attributes in terms of precision, accuracy, specificity, linearity, limits of detection, range, and robustness.
- Identify and select laboratory methods for assaying biological activity.

- Determine enzyme activity from absorbance measurements.
- Identify and select laboratory methods for assaying purity.
- Predict the maximum temperature in electrophoresis gels.
- Define measurements made in microbiological assays.

2.2 Specifications

The specifications that are set for any product that is expected to be sold are important for assuring the quality and consistency of the product. When the product is to be sold as a pharmaceutical, specifications (among other things) must be reviewed and approved by the U.S. Food and Drug Administration. The FDA is interested in a well-specified product, whose manufacturer assures through routine measurements such product attributes as

TABLE 2.1
Example Specification for a Therapeutic Protein[a]

Specification Type and Method	Drug Substance (frozen solution)	Drug Product (10 mg vial, lyophilized)
Identity by HPLC	Coelutes with reference standard	Coelutes with reference standard
Identity by peptide map	Conforms to reference standard	
Identity by amino terminal sequencing	70–90% [1–55] form	
Content by ultraviolet	$\geq$10.0 mg/ml	90–110% label claim
Bioassay: specific activity	$\geq$1,000 IU/mg	$\geq$9000 IU/vial
Purity by RP-HPLC	$\geq$97.6%	
Total and individual related substances by RP-HPLC	Total RS $\leq$2.5% (relative to active) truncated [5–55] form $\leq$1.0%	Total RS $\leq$1250 μg/vial truncated [5–55] form $\leq$150 μg/vial
Multimeric forms by IE-HPLC	$\leq$5.0% multimeric forms	
Residual organic solvents by gas chromatography	Isopropanol $\leq$150 ppm, acetonitrile $\leq$100 ppm	
Trace metals	Copper $\leq$15 ppm	
Host cell proteins	$\leq$200 ppm	
Endotoxin	$\leq$2.5 EU/mg protein	$\leq$200 EU/vial
Moisture		$\leq$5 mg/vial
General inspection	Verify description, appearance, and container integrity	
Physical appearance	Clear, colorless to slightly yellow	White to off-white cake or powder
Solubility		2 mg/ml soluble to clear, essentially colorless solution
pH	4.5–7.0	4.5–5.5
Content uniformity		$\leq$6.0% RSD (n = 10, per USP)
Bioburden	<10 organisms/mg	
Particulates		$\geq$10 μm: $\leq$6,000/container $\geq$25 μm: $\leq$600/container (per USP)
Sterility		No growth (per USP/CFR)

[a]HPLC, high performance liquid chromatography; RP, reversed phase; IE, ion exchange; IU, international units; RS, related substances; EU, endotoxin units, USP, U.S. Pharmacopeia; CFR, Code of Federal Regulations. Data from J. Cohen and M. Busch, "Setting biopharmaceutical specifications," *BioPharm.*, vol. 11, p. 38, August, 1998.

biological activity, purity, identity, physical characteristics (including color and other appearance qualities), and product safety. The example specification for a therapeutic protein shown in Table 2.1 gives examples of the following:

- Identity determination (identity by HPLC, identity by peptide map, and identity by amino terminal sequencing)
- Biological activity (bioassay: specific activity)
- Purity [content by ultraviolet, purity by reversed-phase HPLC (RP-HPLC), total and individual related substances by RP-HPLC, multimeric forms by ion exchange HPLC, residual organic solvents by gas chromatography, trace metals, host cell proteins, endotoxin, moisture, bioburden, particulates, and sterility]
- Physical/appearance qualities (general inspection, physical appearance, solubility, pH, and content uniformity)

The specification usually evolves as the process is developed and the product's properties are learned. It is always best to start with broad specifications that are narrowed as the product is developed for market. Products intended for parenteral human use, however, always entail certain specifications, such as sterility. Safety specifications are of the utmost importance and must be well thought out from the onset of process development.

2.3 Assay Attributes

Analytical methods can be characterized by a set of measurable attributes. Techniques used to measure these attributes give insight into the quality of the results obtained by using the assay. The effectiveness of a method for measuring attributes depends on its precision, accuracy, specificity, linearity, range, and robustness. Not all attributes are appropriate to all methods, as the following subsections explain.

2.3.1 Precision

Precision is a measure of the reproducibility of an assay. Precision is expressed as a relative standard deviation (RSD), defined as the standard deviation divided by the average, also known as the coefficient of variation (CV). Typically, this is converted to a percentage. The precision tells the likelihood that a repeat test will give the same result. Assays that give precision within ±2% are normally acceptable. Assays for which RSD is greater than 5% are generally unacceptable for anything but estimates, which sometimes are good enough.

Precision is usually measured by performing replicate analyses on a reference standard or other well-characterized material. Ideally, all outside factors that can add to the variability of the assay are minimized or reduced. The sample and all reagents used in its preparation should have good stability and should give a result that is typical of the results expected in practice. Demonstrating that a purity assay can give a result of 100% pure on 100% pure material is not very interesting. Demonstrating the ability to reproduce a purity number like 97.3% is much more challenging.

Precision is evaluated based on standard deviation, which is given by the following equation:

$$\sigma = \frac{1}{n-1}\sqrt{\sum_{i=1}^{n}(x_i - \bar{x})^2} \qquad (2.3.1)$$

where $\bar{x}$ is the mean of n measurements, and x_i is an individual measurement. At least three replicates must be made before a meaningful standard deviation can be determined. The standard deviation is an estimate of the true variability of the assay, based on a limited sample (the true variability requires infinite replicates to measure); however, the true value is approached asymptotically as replicates are added. The standard deviation is within 90% of the true variability after 10 measurements.

2.3.2 ACCURACY

Accuracy is a measure of the closeness of the assay result to the "true value." Accuracy is often measured on the basis of the recovery of a spike of pure product added into the sample. Spiking the sample with a known amount of standard allows the analyst to determine the amount of an unknown quantity that is likely to be measured by the assay. The spiked standard should be of known content and purity, because the exact quantity of standard introduced into the sample matrix provides the basis for the assay. Accuracies within 1% are exceptional; assays outside 5% are typically unacceptable.

Accuracy experiments are typically some of the first executed in the qualifying of an assay. Quite often, accuracy is more dependent on the sample preparation and storage conditions than on the analytical method itself. For instance, product material may bind nonspecifically to beakers and plastic pipet tips. When nonspecific binding is found, a variety of different materials, or pretreatments for materials, may be used to increase accuracy. Without good recovery of the analyte (i.e., the substance being analyzed), precision and other attributes are difficult to measure or interpret. A 60% spike recovery will be difficult to reproduce with regularity, for example, while 100% recoveries are very reliable. Therefore, determining and increasing spike recovery at an early stage in assay development and making the determination across the range of concentrations anticipated for the method is desirable.

2.3.3 SPECIFICITY

The ability of the method to distinguish between the analyte and similar components is referred to as the method's specificity. No assay is specific for the target molecule against all possible contaminants, so specificity must be defined with respect to the intent of the method. For instance, an assay designed to quantitatively resolve two related chemical impurities should not also quantitate DNA.

Specificity always means that no other molecules in the sample matrix interfere with the quantitation of the target molecules. The application of this definition varies from assay to assay. For instance, when trying to measure very small amounts of a chemically unrelated contaminant such as endotoxin in a protein product, specificity is the lack of inhibition or enhancement of the endotoxin signal. In the measurement of closely related contaminants, such as the quantity of a des-Met protein variant (a protein without an N-terminal methionine, a common impurity in recombinant protein preparations) in a protein product (or other deletion or misincorporation products) by HPLC, any such contaminants must be sufficiently resolved from the product to be accurately quantitated. No single HPLC technique is specific for all possible product-related impurities that may be found in a protein manufacturing process. For this reason, HPLC techniques are almost always used in pairs called orthogonal methods, so that all contaminants can be resolved.

Identity methods usually require a particular result. Consider the example of a protein product. If N-terminal amino acid sequencing is used to characterize the protein, it will be specific for the protein to the extent that any contaminants with different residues among, for example, the first 20 amino acids, will give failing results. However, N-terminal sequencing does not distinguish contaminants with residue substitutions beyond the point where the sequencing ends. Other identity methods are tryptic (or other enzymatic) maps, specific precipitation or colorimetric assays (such as the enzymatic cleavage of a specific substrate to a pigment product), elemental analysis, total amino acid content, and HPLC retention time. Demonstrating specificity in identity assays consists primarily of demonstrating a lack of competing activities or similar responses at similar concentrations from contaminating substances.

For purity assays, the specificity is usually demonstrated through resolution of known contaminants. For instance, in HPLC or capillary electrophoresis (CE), contaminants are resolved according to a molecular property. Usually, these contaminants are collected and identified, or identified by using a "hyphenated" detection method, such as mass spectrometry (MS) attached to an HPLC or CE (LC-MS, CE-MS).

2.3.4 LINEARITY, LIMIT OF DETECTION, AND LIMIT OF QUANTITATION

The linearity of a method is its ability to produce a response proportional to the concentration of the analyte. Linearity is assessed by creating a standard curve for the analyte(s) by a linear least-squares fit of the response against the concentration (or amount). The correlation coefficient (reported as r^2) is the measure of the linearity and is defined as follows [1]:

$$r^2 = \frac{SS_R}{SS_{yy}} = \frac{\sum_{i=1}^{n}(\hat{y}_i - \bar{y})^2}{\sum_{i=1}^{n}(y_i - \bar{y})^2} \tag{2.3.2}$$

where n is the number of measured values, y_i is the measured value i, $\hat{y}_i$ is the model of the measured value i, and $\bar{y}$ is the mean of the measured values. The square of the correlation coefficient varies over the range $0 \le r^2 \le 1$ and is a measure of the amount of variability in the data explained or accounted for by the linear model. For example, if r^2 is 0.98, then 98% of the variability in the data is accounted for by the model.

Linearity is important because responses from assays for any given analyte amount assayed may not always be identical. It is therefore important to have confidence that the response given is proportional to the actual result sought. Assays become nonlinear at both high and low concentrations, where detectors become saturated (high) or nonspecific responses or noise become a significant proportion of the response (low). It is always important to be making measurements that are bracketed by concentrations on the standard curve, to be assured that results reported are not within these regions of saturation or noise. Extrapolated results are not acceptable.

In purity assays, the linearity of the response of the contaminants is also of interest. Quite often, the response factor (slope of the regression, absorbance, etc.) for the contaminants is not the same as for the main analyte, although it is often assumed to be the same. Authentic samples of impurities are required to prove their amounts in purity assays.

When a standard curve is constructed via regression, a slope, intercept, and r^2 result. The slope is the response factor or signal. The intercept should be zero. If the intercept is not zero, there may be some competing activity in the sample (see Section 2.3.3), or noise may be present. The limit of quantitation (LOQ) and limit of detection (LOD) are precision measurements made at the lower extreme of the linearity curve. One method for determining LOQ and LOD is to determine the noise inherent in the system and then set responses that are multiples of the noise for the limits. The noise inherent in the system may sometimes be assessed by analysis of "blank" samples. Sometimes it is specified by the instrument being used. Frequently, LOD is three times the noise, and LOQ is ten times the noise. Another way is to measure precision with small amounts of analyte, and set LOQ as the concentration above which variability is less than 10% RSD and LOD as the concentration above which precision variability is less than 25% RSD. The rationale for setting these limits is based on the intent of the assay and on the way the assay results are used.

2.3.5 RANGE

The range of an assay refers to the upper and lower limits within which the assay can produce accurate and precise results. "Range" refers not only to the concentration or the amount of the analyte, but also to the range of solution properties associated with the sample, such as pH, buffer composition, and temperature. The assay range may also refer to the sample stability. It is virtually impossible to investigate every permutation of the assay range, especially if interacting factors are considered. The analytical chemist is usually judicious in selecting the factors that may have the largest impact on the assay result.

The assay range is reported in the units that describe the sample attributes for which the assay is valid (e.g., pH 6.5–7.5). It is understood that accuracy and precision of the result do not vary across the assay range. Sometimes a secondary range may be listed, for which precision and accuracy are somewhat less, but still high enough for assay results to be useful.

2.3.6 ROBUSTNESS

While "range" refers to the sample attributes that are acceptable for analysis, "robustness" refers to the assay conditions. In an HPLC method, for instance, flow rate, temperature, pH of the buffers, and solvent content of the buffers may be tested to determine method limits. The method is robust within the limits determined in the test. If the limits determined in robustness testing are not easily achieved in the laboratory, further method development may be required. For example, additional analytical development would be needed when an HPLC buffer pH of 5.5 gives a different result from pH 5.6 and pH 5.7. Again, it is unrealistic to suppose that every factor will be tested and the impact on the result determined. However, critical parameters, or those that are inherently hard to control, need to be examined. Interacting factors will play a role, although the precise role is rarely measured. Statistical design of experiment methodologies is often useful to assess range and robustness.

2.4 Analysis of Biological Activity

Most bioproducts are useful on the basis of their biological activity. It is important, therefore, that the biological activity of the desired bioproduct be preserved during the process of bioseparation. It follows that an assay for biological activity, sometimes called potency, is an essential component of any work on the development of such a process. There are three major types of biological activity assay: animal model assays, cell-line-derived bioassays, and *in vitro* biochemical assays.

2.4.1 ANIMAL MODEL ASSAYS

Animal model assays, which have a relatively long history of use, probably give the best indication of biological activity. These assays, however, have several major disadvantages including a long analysis time (several days to weeks), poor reproducibility of results; and the need for a large number of animals and for appropriate animal facilities and handlers [2]. Sometimes animal model assays must be used because no cell-line-derived or *in vitro* biochemical assay has been developed. One recombinant protein that is assayed by an animal model assay is methionyl–human growth hormone (hGH). In this assay, hypophysectomized female rats are monitored for weight gain over an 11-day period after daily injections with two doses of methionyl-hGH, and the relative biological activity of the test material is obtained by comparing its activity with that of a reference standard material [3].

2.4.2 CELL-LINE-DERIVED BIOASSAYS

Cell-line-derived bioassays are easier to perform than animal model assays; they also give faster results (1–3 days) and are much less expensive and wasteful [2]. Cell-line-derived assays give information on the effect of the bioproduct in a living system, but they are also imprecise because of the variances of living cells. They can be automated, however, and therefore can be repeated enough times to give relatively reproducible results. Two general categories of cell-line-derived bioassays are cell-bound receptor systems and cell-culture-based assays.

Cell-bound receptor systems are used to assay bioproducts, such as proteins, that bind to receptors in a quantifiable and reproducible fashion [4]. It may be necessary to express the receptor on the cell surface for the proper folding of the receptor and thus product binding. An example of this type of assay is the use of human placental cells as a source of human insulin–like growth factor I receptors [5]. These cells can be bound to a solid support and a radiolabeled product used to set up a competitive assay. Sometimes, the cell membranes containing the receptor can be purified and used in place of the cells, although this may lead to additional nonspecific reactions with the disrupted, denatured, and hydrophobic cell membranes.

Cell-culture-based assays can be used when there is a cellular response to the bioproduct, which ideally should be linear and directly related to the desired biological activity [4]. This type of assay has been used, for example, to measure the concentration of nerve growth factor (NGF): when NGF is added to a pheochromocytoma cell line (PC12 cells), the cells stop dividing, sprout neurites, and become electrically excitable [6]. Most cell-culture-based biological assays, however, are based on general effects such as inhibition or enhancement of cell growth. Tests of inhibition of cell growth are routinely used to test the biological activity of antibiotics. One commonly used assay of this type is the dilution test, in which the cells containing the antibiotic are incubated for a fixed period of time, typically 16–20 h, and the minimum concentration of antibiotic that causes inhibition of visible cell growth (called the MIC) is determined [7]. Another type of cell culture inhibition assay for antibiotics is the turbidometric assay, which can be completed in less time than the dilution test and is capable of excellent precision [8]. The turbidity of the cells is measured as absorbance on a spectrophotometer, and the antibiotic concentration is determined from a standard curve.

2.4.3 *In Vitro* Biochemical Assays

In comparison to assays using animals or cells, *in vitro* biochemical assays are simple, fast, precise, and accurate, and thus are usually preferred for measuring biological activity in process development work when they are available. The most prevalent type of biochemical assay is the one that measures enzymatic activity. The result of a test of enzyme activity is the number of *units of activity,* where one unit (U) is defined as the enzyme activity that causes one micromole (μmol) of substrate to react in one minute at specified conditions [9].

There are two general classes of techniques for measuring enzyme activity, stopped methods and continuous methods [10]. In the stopped method, the enzyme is incubated with its substrates for a fixed amount of time, and then the reaction is stopped by one of several ways. The product formed is measured, and the units of activity are then calculated, assuming that the reaction rate is linear from time zero. This assumption of linearity can be checked by carrying out a series of incubations at different times. In the continuous method, multiple readings of the amount of product formed are taken over a period of time, usually on the order of minutes, and the activity is calculated from the slope of the plot of reading versus time. One commonly used variation of the continuous method relies on the principle of coupling: a product that cannot be observed directly is removed or reacted on, either chemically or enzymatically, to form a compound that can be observed.

Other biochemical assays are based on mimicking the bioproduct's action *in vivo*. The activity of the recombinant tissue plasminogen activator (rt-PA), for example, can be measured with an *in vitro* clot lysis assay that can be performed within hours. The rt-PA is added to a synthetic fibrin clot in the presence of plasminogen, and the rt-PA converts the plasminogen to the active enzyme plasmin, which then lyses the clot. The assay is followed spectrophotometrically or visually by monitoring the release of trapped air bubbles. This assay is reported to have an accuracy of 99% and a relative standard deviation of 5% [11].

EXAMPLE 2.1

Coupled Enzyme Assay for Alcohol Oxidase   For the determination of the enzymatic activity of alcohol oxidase (AO) from yeast, the following coupled reactions are utilized [12]:

$$\text{methanol} + O_2 \xrightarrow{\text{AO}} \text{formaldehyde} + H_2O_2$$

$$H_2O_2 + DH_2 \xrightarrow{\text{HP}} 2H_2O + D$$

where DH_2 is a dye precursor, D is the dye, and HP is the enzyme horseradish peroxidase. The dye is monitored by means of a spectrophotometer at a wavelength of 405 nm. To prepare the standard curve for the assay, the absorbance readings shown in Table E2.1 were taken at various concentrations of H_2O_2 in the reaction mix with no AO or methanol added, using a light path length of 10 mm:

TABLE E2.1

H_2O_2 concentration (μmol/liter)	$A_{405\ nm}$
7.1	0.25
9.6	0.36
12.3	0.44
15.1	0.57
18.0	0.66

A sample of clarified yeast cell lysate (i.e., cells lysed, and cell debris removed to give a clarified solution) assayed for AO activity by this method gave a rate of change of A_{405} of 0.082/min in the assay after a dilution of the sample of 1:40,000. Calculate the AO concentration in units per milliliter, where 1 unit (U) is equal to the amount of enzyme producing 1 μmol of H_2O_2 per minute under the conditions of the assay.

We first plot A_{405} versus the concentration, c, of H_2O_2 and find that the data are well fitted by a straight line (Figure E2.1), with the result

$$\Delta A_{405} = 0.0377 \Delta c$$

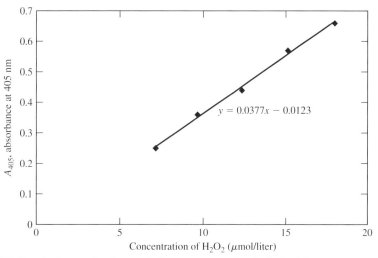

Figure E2.1 Standard curve for the coupled enzyme assay for alcohol oxidase.

From the $\Delta A_{405}/\text{min}$ value for the clarified yeast cell lysate,

$$\frac{\Delta c}{\Delta t} = \frac{\dfrac{\Delta A}{\Delta t}}{0.0377\dfrac{\text{liter}}{\mu\text{mol}}} = \frac{0.082\ \text{min}^{-1}}{0.0377\dfrac{\text{liter}}{\mu\text{mol}}} = 2.18\ \frac{\mu\text{mol}}{\text{liter min}} = 2.18\ \frac{\text{U}}{\text{liter}}$$

Since the sample was diluted 1:40,000 before assaying,

$$\text{Alcohol oxidase concentration in clarified cell lysate} = 40,000\left(2.18 \times 10^{-3}\ \frac{\text{U}}{\text{ml}}\right)$$

$$= 87.2\ \frac{\text{U}}{\text{ml}}$$

2.5 Analysis of Purity

Determinations of purity are inherently defined by the specificity and sensitivity of the methods used to detect the impurities that are present. Impurities can be classified according to whether they are intrinsic to the desired bioproduct (e.g., proteolytically cleaved material from a protein product) or extrinsic to the product (e.g., components of the cell growth medium). It is important that intrinsic purity be evaluated by methods that are orthogonal (i.e., independent of each other), through methods based on different physio-chemical properties [4]. Purity assays also give information about the extent and completeness of the biochemical reaction or reactions leading to the formation of the product, the side products that may have been formed, the efficiency of the purification process, and whatever contamination may have been formed or picked up along the way. The various assays for purity are discussed in this section, with an emphasis on the ones that are the most powerful, such as electrophoretic and HPLC assays.

2.5.1 ELECTROPHORETIC ANALYSIS

Electrophoresis is a separation technique often applied to the analysis of biological or other polymeric samples. This technique is among the most powerful for estimating purity because of its simplicity, speed, and high resolution, and also because there is only a small probability that any of the components being analyzed will be lost during the process of analysis [2].

Electrophoresis has frequent application to analysis of proteins and DNA fragment mixtures and has been increasingly applied to the analysis of nonbiological and nonaqueous samples. The high resolution of electrophoresis has made it an important tool in the advancement of biotechnology. Variations of this methodology are being used for DNA sequencing and fingerprinting, for isolating active biological factors associated with diseases such as cystic fibrosis, sickle cell anemia, myelomas, and leukemia, and for establishing immunological reactions between individual compounds. Electrophoresis is an effective analytical tool because the electric field does not affect a molecule's structure, and it is highly sensitive to small differences in molecular charge, size, and sometimes shape.

The term electrophoresis refers to the movement of a molecule or particle through a stationary fluid under the influence of an electric field (*phorein*: Greek, "to carry"). The study of electrophoresis has included the movement of colloids, fibers, clay particles, latex spheres, and bubbles. While the application of greatest interest for the purpose of this text is the separation of proteins and nucleic acids, the theory is best described in consideration of a distinct particle. The application is extended to these biomacromolecules in that they have colloidal properties, such as a surface charge, an interior with a polarity distinctly different from that of their exterior, and a distinct density (see Chapter 5, Sedimentation). The fundamental principles behind electrophoresis, the commonly used modes of operation, the matrices employed, and the techniques for detection are discussed in detail in the subsections that follow.

Principles

The fundamental principle behind electrophoresis is the existence of charge separation between the surface of a particle and the fluid immediately surrounding it. An applied electric field acts on the resulting charge density, causing the particle to migrate and the fluid around the particle to flow. The electric field exerts a force on the particle's charge or surface potential. This force results in a velocity for the particle that is proportional to the particle's surface potential. Two particles with different velocities will come to rest in different locations after a fixed time in an electric field. The particle velocity is related to the field strength by

$$v = UE \qquad (2.5.1)$$

where v is the particle velocity, E is the field strength or gradient (voltage per length), and U is the apparent electrophoretic mobility. [See Chapter 3, Equation (3.4.4), for the dependence of the electrophoretic mobility on the liquid's dielectric constant and viscosity and on the zeta potential.] There are two contributions to this apparent electrophoretic mobility [13]:

$$U = U_{el} + U_o \qquad (2.5.2)$$

where U_{el} is the electrophoretic mobility of the charged particle and U_o is the contribution from electroosmotic flow.

Electroosmosis is the flow of liquid in a channel to which an electric field has been applied. For instance, when a capillary connects two reservoirs containing an electrolytic fluid (such as salt water) and an electric field is applied between the reservoirs, salt water will flow, typically from the anodic chamber to the cathodic chamber. This is because the SiO^- charge on the glass capillary attracts a positive countercharge that is mobile and flows toward the cathode in an electric field. This occurs until hydrostatic head is built up in the cathodic reservoir, causing a counterbalancing flow through the middle of the capillary. Electroosmotic flow is generally minimized in polymer networks such as gels (where the liquid viscosity is very large or the effective capillary diameter is very small) but is important in open channels such as capillaries.

A schematic of a typical gel electrophoresis apparatus is shown in Figure 2.1. The design of analytical gel electrophoresis equipment is influenced by the generation of heat and electrolysis gases, as discussed next.

Generation of Heat in Electrophoresis Instrumentation

One of the considerations in designing electrophoresis equipment is the dissipation of heat generated by the electric field. The rate of change in temperature in an insulated

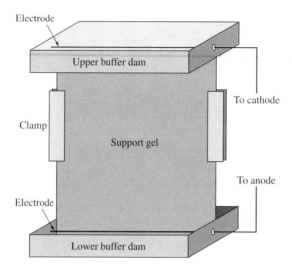

Figure 2.1 Typical gel electrophoresis apparatus. Samples are loaded in wells at the top of the gel, which is supported on both sides by glass plates (not shown), and components contained in the samples are separated as they travel downward toward the anode.

electrophoretic medium is given by

$$\frac{dT}{dt} = \frac{P}{C_p M_e} = \frac{VI}{C_p M_e}$$

(2.5.3)

and the rate of heat generation per unit volume is

$$S = \rho C_p \frac{dT}{dt}$$

(2.5.4)

where P is the power, C_p is the heat capacity of the medium, M_e is the mass of electrophoretic medium, V is the voltage, I is the current, and ρ is the density of the medium. From Ohm's law ($V = IR$, where R is the electrical resistance), it is seen that the temperature change increases as the square of the current. Electrophoresis is thus most conveniently conducted in highly resistive media (low salt concentrations). The addition of polymer matrices, such as gels, also serves to increase the resistivity of the media by increasing the viscosity that is resisting the movement of charge, and creating a more tortuous path through which the charge must migrate.

Heating in electrophoresis causes changes in the viscosity and density of the electrophoretic medium. High temperatures can also damage electrophoretic equipment by warping cooling blocks, melting plastic, or cracking glass plates. When fluids heat unevenly, the hot portion of the fluid tends to rise with respect to cooler portions because of the difference in density. This convection distorts zones and thereby decreases resolution in electrophoretic separations. The ability to remove heat from electrophoretic systems ultimately determines the voltage that can be applied, and the speed and resolution that can be achieved.

Heat is dissipated from electrophoresis equipment by maximizing the surface-to-volume ratio. Most electrophoresis is conducted in thin gels (0.5–1.5 mm) or in capillaries with channels up to 100 μm in diameter. The narrow cross section also reduces the total current, although the current density (A/cm^2) remains independent of scale. The removal of heat from the gel is a typical transport process obeying energy balances and Fourier's law of heat conduction, which states

$$q_x = -k\frac{dT}{dx} \tag{2.5.5}$$

where q_x is the heat flux in the x direction (heat flow per unit area) and k is the thermal conductivity.

EXAMPLE 2.2

Estimation of the Maximum Temperature in an Electrophoresis Gel   A vertical gel electrophoresis system consists of a gel supported on both sides by glass plates, each surrounded by a fluid stream at the same constant temperature on both sides. Develop an expression for the maximum temperature within the gel as a function of the thermal conductivity and thickness of the gel and the glass, the heat transfer coefficient at the outer surface of the glass, and the rate of heat generation per unit volume in the gel.

SOLUTION

A schematic diagram of the cross section of the gel and the glass plates (Figure E2.2) also shows the temperature profile and the thermal conductivities k^{01} for the gel and k^{02} for the glass. We let S be the heat per unit volume per unit time generated uniformly by the electric current in the gel. We begin by making a steady state thermal energy balance over a thin slab of gel thickness Δx, width W, and height H:

$$q_x^{01}\big|_x WH - q_x^{01}\big|_{x+\Delta x} WH + SWH\,\Delta x = 0$$

where q_x^{01} is the heat flow per unit area (heat flux) in the x direction. After dividing by $WH\,\Delta x$ and taking the limit as Δx approaches zero, we obtain

$$\frac{dq_x^{01}}{dx} = S$$

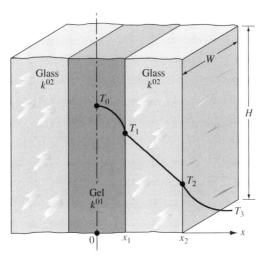

Figure E2.2 Schematic cross-sectional diagram of a vertical electrophoresis system. The glass plates are surrounded on each side by a fluid stream at the same constant temperature, T_3.

Integration gives

$$q_x^{01} = Sx + C$$

where C is a constant of integration. From Figure E2.2, we see that the symmetry of the system gives the boundary condition

$$\text{at } x = 0, \qquad \frac{dT}{dx} = 0$$

From Fourier's law of heat conduction,

$$q_x = -k\frac{dT}{dx}$$

so that we have the boundary condition

$$\text{at } x = 0, \qquad q_x^{01} = 0$$

This gives $C = 0$, so that

$$q_x^{01} = Sx$$

At $x \geq x_1$, the heat flux becomes a constant because there is no heat generation in the glass support plates, which gives the additional boundary condition

$$\text{at } x = x_1, \qquad q_x^{01} = q_x^{02} = q_1 \text{ (a constant)}$$

From the equation for q_x^{01}, this boundary condition gives

$$q_1 = Sx_1$$

Writing Fourier's law of heat conduction for the gel and for the glass, we have

$$-k^{01}\frac{dT^{01}}{dx} = q_x^{01} = Sx$$

$$-k^{02}\frac{dT^{02}}{dx} = q_x^{02} = q_1$$

Integration of these equations gives

$$T_0 - T_1 = \frac{Sx_1^2}{2k^{01}}$$

$$T_1 - T_2 = \frac{q_1(x_2 - x_1)}{k^{02}}$$

Finally, heat is rejected at the outer surface of the glass, and the heat transfer coefficient h is defined as

$$T_2 - T_3 = \frac{q_1}{h}$$

Addition of the previous three equations gives

$$T_0 - T_3 = \frac{Sx_1^2}{2k^{01}} + \frac{q_1(x_2 - x_1)}{k^{02}} + \frac{q_1}{h}$$

Substitution for q_1 and solving for T_0 yields

$$T_0 = T_3 + Sx_1 \left(\frac{x_1}{2k^{01}} + \frac{x_2 - x_1}{k^{02}} + \frac{1}{h} \right)$$

Thus, as we would expect, an increase in the maximum temperature in the gel is caused by increasing the rate of heat generation per unit volume (S, due to current density), increasing the gel thickness ($\Delta x = 2x_1$) or the glass plate thickness ($x_2 - x_1$), and decreasing either of the thermal conductivities (k's) or the heat transfer coefficient h at the surface of the glass plates.

Electrolysis Reactions

The electrodes in electrophoresis equipment are typically constructed from platinum wire, and small ions generally carry the current in the electrophoretic medium. At the interface between the electrodes and the electrophoretic medium, electrolysis reactions occur. For aqueous systems, the following reactions are expected:

$$\text{Cathode} \qquad 2H_2O + 2e^- \longrightarrow 2OH^- + H_2$$
$$\text{Anode} \qquad H_2O \longrightarrow 2H^+ + \tfrac{1}{2}O_2 + 2e^-$$

Electrode chambers are usually open to the atmosphere to permit venting of electrolysis gases. The reactions at the electrodes produce acid (anode) and base (cathode), thus requiring electrophoresis systems to be well buffered. Electrophoresis systems sometimes mix the buffers from the individual electrode reservoirs to equalize the pH.

Modes of Electrophoretic Separation

Electrophoresis can be performed several different ways, of which the most commonly used for bioproducts are denaturing gel electrophoresis (better known as SDS-PAGE: see next subsection), native gel electrophoresis, capillary electrophoresis, and isoelectric focusing. The first three methods are termed simple electrophoresis because they take place in a constant electric field at constant pH and separate according to actual mobility. Isoelectric focusing uses a constant electric field with a pH gradient and allows components to move until they have zero net charge. Simple electrophoresis using either denaturing or native gels, is by far the most common method for the analysis of proteins and DNA. Gel electrophoresis for analysis requires only microgram quantities of protein, usually only a small fraction of what is available.

Denaturing Gel Electrophoresis (SDS-PAGE)   The use of sodium dodecyl sulfate (SDS) with polyacrylamide gel electrophoresis (PAGE) was first described in the late 1960s. The SDS is an ionic surfactant that solubilizes and denatures proteins. Dodecyl sulfate binds a protein through hydrophobic interactions with side chains, and disrupts side chain interactions that cause proteins to fold and interact with other subunits. Nonreduced proteins bind approximately 0.9 to 1.0 g of SDS per gram of protein, or about one dodecyl sulfate molecule for every two amino acid residues.

The SDS-PAGE technique allows separation of molecules according to size or molecular weight. When SDS-treated samples migrate into a gel and are electrophoresed, the principal difference between sample components is molecular size or length, since the ratio of charge to size is virtually the same for all proteins with bound SDS. Smaller molecules travel through the matrix more quickly than those that are larger. The rate at which molecules migrate through a polyacrylamide gel is inversely linear with the logarithm of their molecular size. Thus, denatured samples can be analyzed alongside denatured standards of known molecular weight to estimate the molecular weight of the analyte.

The typical SDS-PAGE results obtained with a gel electrophoresis apparatus presented in Figure 2.2 are for proteins denatured by SDS and stained (left) and also blotted and probed with an antibody, called a "Western blot" (right; see Blotting Techniques later in this chapter). Each band on the gel represents a highly enriched substance that has been separated from the other bands based on slightly different charge or size/shape properties. As shown, several samples are typically analyzed side by side on a gel. Bands that migrate the same distance in different sample mixtures probably have similar mass and charge characteristics. Standards having known mobility or molecular mass are run concurrently with samples whose components are to be identified (column m in Figure 2.2a).

Other dissociating agents may be used to further break down a protein. Urea is often used to disrupt hydrogen bonds. When urea is the only dissociating agent added (no SDS), a protein's intrinsic charge is not affected and separation based on size and charge may be achieved. If a protein contains internal disulfide bonds, a thiol reagent such as β-mercaptoethanol must be used to reduce the sample and break the disulfide bonds. Proteins having reduced disulfide bonds bind approximately 1.4 g of SDS per gram of protein, versus about 1 g per gram when not reduced. Typically, both reduced and nonreduced samples are run in efforts to evaluate internal disulfide bond content.

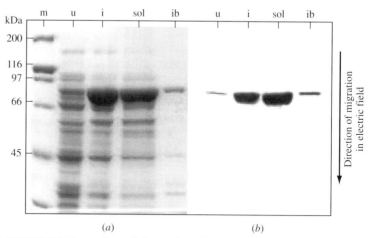

(a)   (b)

Figure 2.2 SDS-PAGE (denaturing gel electrophoresis) and Western blot results for bovine growth hormone (bGH) expressed as a C-terminal fusion to *E. coli* NusA protein. (*a*) Sodium dodecyl sulfate polyacrylamide gel electrophoresis (SDS-PAGE) results with Coomassie blue staining. (*b*) Western blot results obtained by using rabbit anti-bGH polyclonal antibody and visualized by means of chemiluminescence. Fusion proteins were expressed at 37°C in *E. coli* by induction of the *tac* promoter. Equal portions of cell lysate, soluble fraction, and insoluble fraction were loaded. Key: m, markers; u, uninduced whole cell lysate; i, induced whole cell lysate; sol, soluble fraction; ib, inclusion body fraction. (From G. D. Davis and R. G. Harrison, "New fusion protein systems designed to give soluble expression in *Escherichia coli*," *Biotechnol. Bioeng.,* vol. 65, p. 382, 1999. Copyright © 1999. Reprinted by permission of Wiley-Liss, Inc., a subsidiary of John Wiley & Sons, Inc.)

Native Gel Electrophoresis   In this type of electrophoresis, a buffer is used that allows the components separated to be in their native form. When this procedure is used for proteins, a pH in the range of 8 to 9 is desirable so that most proteins will be negatively charged and thus move toward the anode. In this pH range, proteins that are either uncharged or positively charged will either stay in the starting well or move toward the cathode. If the proteins of interest are basic, a lower pH can be used, with the cathode at the bottom of the gel [10].

Native gel electrophoresis is a good way to determine whether a protein of interest forms subunits and, if so, the number of subunits per molecule. The electrophoresis is run under both native and denaturing conditions and the results are compared. An immunoblot may have to also be run to positively identify the protein of interest under both sets of conditions (see the immunoblot in Figure 2.2*b* and Blotting Techniques, later).

Capillary Electrophoresis   Capillary electrophoresis is a highly automated, commercially available technique that bears a strong resemblance to analytical chromatography. First widely used as a support medium for electrophoresis in the early 1980s, capillaries have a small cross-sectional area that minimizes power consumption, while their high surface-to-volume ratio allows efficient dissipation of heat. In practice, electric fields in excess of 300 V cm^{-1} can be utilized without encountering detrimental thermal effects. Capillary electrophoresis separations are rapid (minutes) and produce a chart recording rather than a stained gel for archiving. A typical capillary "electropherogram" is shown in Figure 2.3.

Clearance of sample components from a capillary is a problem that must be considered carefully. Usually, capillary electrophoresis is conducted in a fused-silica capillary that is coated on its inner surface. The capillaries can be open or filled with a gel (see later: Support Media). Capillaries are too expensive to discard after each use, as is done with standard

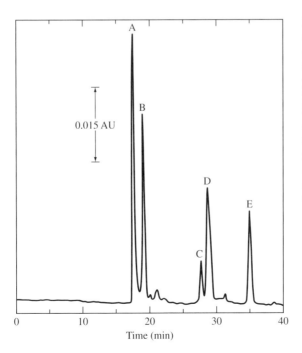

Figure 2.3 Separation of proteins by open tubular capillary electrophoresis, carried out in a 75 cm × 75 μm surface modified capillary at an applied voltage of 75 kV. Peak identities: A, egg white lysozyme; B, horse heart cytochrome c; C, bovine pancreatic ribonuclease a; D, bovine pancreatic α-chymotrypsinogen; E, equine myoglobin. (Data from Y. Walbrooehl and J. W. Jorgenson, "On-column UV absorption detector for open tubular capillary zone electrophoresis," *J. Chromatog.*, vol. 315, p. 135, 1984.)

gels. Therefore, the entire sample must be cleared from the capillary before the capillary is used for another sample, much as in analytical chromatography.

The small quantity of material loaded into the capillary requires that high-sensitivity detection be employed with all capillary electrophoresis instruments. Typical sample volumes range from 5 to 50 nl, less than 1% of the volume typically loaded onto conventional polyacrylamide or agarose gels. For trace analysis, a small number of molecules will be present in the capillary after loading. Detection of these small amounts of material necessitates very sensitive detectors.

Isoelectric Focusing   Isoelectric focusing is a technique used for protein separation that operates by driving proteins to the pH equal to their isoelectric points, where the net charge on the molecule is zero. Resolution depends on the slope of a pH gradient that can be achieved in a gel or capillary.

Ampholytes. An ampholyte is a molecule that can be positively or negatively charged depending on the pH. All amino acids and proteins are ampholytes (or amphoteric). Not only does the sign of the charge of an ampholyte change with pH, but the magnitude of the charge can also vary. The charge on a protein, for example, may vary from +10 or more at low pH, to −10 or more at high pH.

A special class of ampholytes has been synthesized for isoelectric focusing. These ampholytes have an amino end and a carboxyl end that are separated by varying numbers of methylene groups. The farther apart the amino and carboxyl groups, the less they affect the ionization of each other; thus a different isoelectric point is established for each molecular species. These ampholytes, which may be added to an electrophoretic medium, migrate according to their charge under the influence of the applied electric field. When they reach a zone in which the local pH is the same as that ampholyte's isoelectric point, migration stops. The ampholyte molecules buffer themselves and establish the local pH as they migrate through the gel. As the ampholytes reach an isoelectric pH, they establish a spatially distributed and natural pH gradient in the electrophoretic medium.

Isoelectric Focusing Technique. Isoelectric focusing (IEF) is an electrophoretic technique in which amphoteric samples (such as proteins) are separated according to their isoelectric points along a continuous pH gradient. IEF analyses are carried out in various matrices: polyacrylamide, agarose, and capillaries. The agarose or polyacrylamide gels that are used must be prepared with carrier ampholytes bracketing a specific pH range. After some time, the ampholytes separate, and there is a pH gradient that covers the range of all the ampholytes' isoelectric points. A specific class of commercial ampholytes, called Immobilines®, can be added to the matrix, electrophoresed into a gradient, and covalently incorporated into the gel matrix.

Because protein samples are themselves ampholytes, when samples are loaded onto the gel and a current is applied, the compounds migrate through the gel until they come to their isoelectric point, or pI, where they reach a steady state. This technique measures an intrinsic physicochemical parameter of the protein, the pI, and therefore does not depend on the mode of sample application. In practice, IEF has the highest sample load capacity of any electrophoretic technique.

In slab gels, isoelectric focusing typically requires approximately 3 to 30 hours to complete because sample compounds move increasingly slowly as they approach the pH that corresponds to their isoelectric point. This time can be shortened to approximately

30 minutes in capillaries because of the higher electric field employed. Because the gradient ampholytes and the samples stop where they have no mobility, the resistivity of the system increases dramatically toward the end of the experiment, and the current decreases dramatically. For this reason, isoelectric focusing is usually run in constant voltage mode and has the most stringent requirements for cooling of any electrophoretic technique. A plot of current versus time can be used to determine when the system is focused, since the final steady state current is only a few microamperes.

Electrophoretic Instruments

Most electrophoresis equipment shares the basic design shown in Figure 2.1. Electrophoresis equipment typically has two buffer reservoirs, one anodic and one cathodic. The equipment includes an electrophoretic medium (e.g., a gel, paper, or capillary) connecting the two reservoirs, to which a sample is applied. A direct current power supply connects two electrodes suspended in the buffer reservoirs. When an electric potential is applied between the two electrode reservoirs, the sample components migrate through the medium. At completion of an electrophoretic separation, the sample components have been either resolved into several zones along the length of the medium or eluted from the medium into the anodic buffer.

Support Media

The conduction of electric current through an electrophoresis system causes the system temperature to increase if the heat is not dissipated at a rate equal to the rate of its production. Increases in system temperature will increase the electrophoretic mobility of molecules because the viscosity of the matrix decreases with increasing temperature. Additionally, the formation of thermal gradients will result either in convection or a distribution of electrophoretic mobilities for a unique compound, and ultimately in zone spreading. To minimize the influence of convection on the electrophoretic separation, anticonvective support media such as paper, polymer gels, or capillaries are commonly employed.

Separation on paper or polymer gels is influenced not only by electrophoretic mobility, but also by sieving of the molecules as they pass through the network of pores and fibers. The finer the weave of this matrix, the more slowly a molecule travels through it. Therefore, molecular size, as well as charge, can influence the rate of migration in these media.

Paper Electrophoresis   Paper was one of the first matrices used for electrophoresis. In paper electrophoresis, the sample is applied directly to a zone on the dry paper, which is then moistened with a buffer solution before application of an electric field. Dyes are combined with samples and standards to help visualize the progress of the electrophoresis. The movement of samples on paper is best when the current flow is parallel to the fiber axis in the paper.

Some advantages of paper are that it is readily available and easy to handle, requires no preparation, and allows the rapid development of new methodologies. Besides being easy to obtain, paper does not contain many of the bound charges that can interfere with the separation. A disadvantage of paper electrophoresis is that the porosity of commercial paper is not controlled, and therefore the technique is not very sensitive, nor is it easily reproducible.

Polyacrylamide Gels   Polyacrylamide gel electrophoresis is one of the most commonly used electrophoretic methods. Analytical uses of this technique center on protein and nucleic acid characterization (e.g., purity, size, or molecular weight, and composition). Acrylamide is a neurotoxin, however, and the reagents must be combined extremely carefully. An additional drawback is that the gels are not as pliable as most agarose gels.

Polyacrylamide gels are synthesized through the polymerization of acrylamide monomer, $CH_2=CHCONH_2$, and a bifunctional, cross-linking comonomer. Typically, the cross-linking comonomer is N,N'-methylenebisacrylamide (bisacrylamide), $(CH_2=CHCONH)_2CH_2$, although other compounds such as ethylenediacrylate (EDA) or N,N'-diallyltartar diamide (DATD) are sometimes used. The sieving properties of the gel are defined by the network of pores established during the polymerization. As the acrylamide concentration of the gel increases, the effective pore size decreases. By convention, a gel is characterized by two parameters, $\%T$ and $\%C$, where $\%T$ is the mass percentage of total monomer (acrylamide plus comonomer in grams per 100 ml), and $\%C$ is the proportion by mass of monomer (per 100 g of acrylamide) that is the cross-linking agent. Gels typically have $\%T$ values between 3 and 30%.

The most commonly used combination of chemicals to produce a polyacrylamide gel is acrylamide, bisacrylamide, buffer, ammonium persulfate, and tetramethylenediamine (TEMED). TEMED and ammonium persulfate are catalysts for the polymerization reaction. The TEMED causes the persulfate to produce free radicals, initiating polymerization. After the gel has been poured into a prepared form (usually between two vertical, parallel glass plates that are sealed on the bottom and sides, see Figure 2.1), a "comb" can be inserted into the top portion of the gel before polymerization is complete. This comb sets small indentations (wells) into the gel that can be used to apply samples. If the comb is used, samples are typically mixed with a dense liquid such as glycerol to prevent the sample from dispersing into the reservoir buffer. Samples then migrate straight down the gel and appear as columns, or lanes, of separated components at the end of the experiment (see Figure 2.2).

Agarose Gels   Agarose is a polymer extracted from red seaweed. When agar is extracted from the seaweed, it is in two components, agaropectin and agarose. The agarose portion is nearly uncharged, making it desirable for use as an electrophoresis matrix. To prepare an agarose gel, a combination of agarose and buffer is heated until the solid is dissolved. The solution is cooled and then poured into a gel casting apparatus and allowed to gel. The advantages of agarose electrophoresis are that it requires no additives or cross-linkers for polymerization, it is not hazardous, low concentration gels are relatively sturdy, and it is inexpensive. Agarose gels have large pore sizes compared to polyacrylamide gels. Agarose is commonly used for the separation of large molecules such as DNA fragments.

Capillaries   The fused silica capillaries used in capillary electrophoresis are flexible due to an outer polyimide coating and are available in inner diameters ranging from 10 to 300 μm. Fused silica is transparent to UV light, which enables the capillary to serve as its own detection flow cell. Electrostatic interactions with the capillary surface can develop, however, when charged species are being separated. One popular strategy to overcome this problem is to chemically modify the inner capillary surface to produce a nonionic, hydrophilic coating, resulting in the shielding of the silanol functionalities. Surface coatings can reduce or possibly eliminate electroosmosis. In the absence of electroosmosis, movement is solely

TABLE 2.2
Comparison of Electrophoresis Matrices

Matrix	Advantages	Disadvantages
Paper	Contains a low level of bound charge Readily available Easy to handle Requires no preparation	Porosity not controlled Low sensitivity Low reproducibility
Polyacrylamide	Commonly used Useful for proteins and nucleic acids	Acrylamide is a neurotoxin Gels less pliable
Agarose	No additives required Inexpensive Not hazardous Sturdy Used commonly for DNA	Gels have short shelf life
Capillaries	Reusable High resolving power Permit rapid analyses, record data Consumes a small amount of sample Easily automated	Require sensitive detector Relatively expensive equipment Longer development cycle

dependent on the electrophoretic mobilities of the components being analyzed [see Equations (2.5.1) and (2.5.2)], thus eliminating an additional source of variation [13].

Besides the use of open capillaries in capillary electrophoresis, capillaries have been filled with gels to increase resolving power. Placing gels inside capillaries serves mainly to create an anticonvective medium for the minimization of band dispersion [13]. Cross-linked polyacrylamide gels have been widely used as sieving networks for capillary electrophoresis, and linear polyacrylamide, dextrans, polyethylene glycol, polyethylene oxide, and methylcellulose have been used also.

Table 2.2 summarizes the advantages and disadvantages of different electrophoresis matrices.

Detection Techniques

There are many techniques for detecting separated sample components, including chemical stains, photographic media, and immunochemical reactions, and many instrumental techniques. Each detection technique also gives different information about the identity, quantity, and physical properties of the molecules in the mixture. Detection is often a key part of the electrophoretic analysis and usually yields basic information about the mixture being studied.

Most sample components analyzed with electrophoretic techniques are invisible to the naked eye. Thus methods have been developed to visualize and quantify separated compounds. These techniques most commonly involve chemically fixing and then staining the compounds in the gel. Other detection techniques can sometimes provide increased selectivity or sensitivity.

Chemical Staining   The size and type of the compound as well as the electrophoretic matrix dictate and often limit the variety of stains that can be used to help visualize electrophoretic banding patterns. Molecules can be lost during the staining process, so most staining procedures incorporate a "fixing" step, such as a soak in dilute acetic acid for 1 h, either before or in conjunction with staining. Once fixed, molecules can be stained without loss of the separated components.

Two of the more sensitive and more frequently used protein stains are Coomassie brilliant blue (R250 and G250) and silver stain. The Coomassie stains are appropriate for both agarose and polyacrylamide gels. The silver stain is approximately 100 times more sensitive than Coomassie, but its use is more complex and often requires more troubleshooting to obtain the desired results.

To quantitate stained sample components, the gel is scanned with a densitometer. These scans can be compared with equivalent scans of quantitative standards to generate a calibration curve.

Fluorescence   Fluorescence detection, which provides much better detection limits than simple chemical stains, typically involves the covalent binding of a fluorescent residue to the analyte, but any type of specific interaction can be assayed. If the staining process can be performed prior to the electrophoretic analysis, the electrophoretic process can often be followed visually, and the background signal caused by dye trapping in the electrophoretic matrix will be minimized.

Fluorescamine is a popular reagent for labeling of proteins. At room temperature and alkaline pH, fluorescamine can react with primary amines (found on lysine, arginine, and the N-terminus) on the protein to generate a fluorescent derivative. Fluorescamine may alter the charge-to-mass ratio of the analytes, and it may be used to label samples before electrophoresis or to stain the gel afterward, depending on the importance of maintaining the ratio of charge to mass. The reagent ethidium bromide is commonly used to visualize DNA. The ethidium is incorporated into the structure of DNA. Once incorporated, the fluorescence quantum efficiency of ethidium is dramatically enhanced, and the bands can be easily visualized under a UV lamp.

Radioactivity   Another method of visualizing and identifying separation products on a gel is through radioactivity. If a sample is radioactive, the bands that separate during electrophoresis are subsequently radioactive. When the separation is complete, the electrophoretic matrix (e.g., gel or blotting membrane) may be placed against x-ray film until the radiation makes a mirror image of the banding pattern on the film. When this film is developed, the resulting autoradiograph displays the bands.

Immunoelectrophoretic Techniques   The technique of gel electrophoresis has been successfully combined with immunological techniques to further evaluate molecules. A frequently used method of immunoelectrophoresis is a technique known as "crossed immunoelectrophoresis." A sample is first run longitudinally through an agarose gel for a predetermined time. Second, a longitudinal strip of the gel area in which the sample was electrophoresed is typically cut out and placed into a similarly sized area of an antibody-containing gel. As an electrical current is applied to the second gel system, each band of the sample in question electrophoreses sideways through the second gel and forms an antigen–antibody precipitin pattern.

Both agarose and polyacrylamide gel systems have been used for direct immunofixation. In these gels, samples are electrophoresed and then immunofixed either by using strips of cellulose acetate soaked in an antibody or by placing the antibody directly over the sample area of the gel. In a related technique, the proteins on the gel are transferred to a membrane and then are probed with an antibody in the method known as Western blotting, described shortly.

All these techniques are most often, but not exclusively, used in the clinical setting to diagnose abnormalities or to evaluate inheritance patterns of polymorphic proteins.

On-Column/End-Column Detection   Capillary electrophoresis (CE) is characterized by minute amounts of sample contained in small volumes of liquid. To monitor the zones formed by the analysis requires detection methods that are extremely sensitive and do not give rise to extracolumn band broadening or dilution. Most detection methods employed with CE are on-column adaptations of detectors commonly employed with analytical liquid chromatography. By monitoring the zones directly on-column, distortion of the sample zone is minimized and detection sensitivity is maximized.

Absorbance is the most popular detection method employed with CE. Nearly every commercial CE instrument comes equipped with an absorbance detector. For detection by absorbance, the concentration sensitivity for CE tends to be less than for HPLC because capillary width, the cell path length in CE, is usually smaller than in a conventional HPLC flow cell [14].

Enhanced detection sensitivity is achieved with on-column fluorescence and electrochemical detectors. With fluorescence detection, the small dimensions of the capillary helps confine the sample to a small volume. Focusing the excitation beam into the small volume increases the irradiance of the fluorophore and the fluorescence signal. Electrochemical detectors are usually formed by inserting a microelectrode into the capillary, or by butting the electrode up against the capillary outlet. The small dimensions of the capillary ensure that the maximum distance the analyte must travel to reach the surface of the electrode is small, and therefore the electrochemical conversion efficiency is high.

The coupling of CE to mass spectrometry is developing rapidly because of the wealth of chemical information that can be obtained. Most often this coupling is through an on-line electrospray ionization interface. The primary difficulty of using mass spectrometry as the primary detector for capillary electrophoresis is the relative incompatibility of the common electrophoresis buffer salts and solvents with the electrospray interface.

Specialized Techniques

Two-Dimensional Electrophoresis   Two-dimensional (2D) electrophoresis is unique, offering an analytical method that is both reproducible and sensitive. It is referred to as 2D because it employs two different methods of electrophoresis (typically IEF and SDS-PAGE), in two different directions, to separate the sample components. The combination of the two methods gives better resolution of the compounds than could be achieved with either method alone. For example, each method alone may separate up to 100 components of a sample, whereas together they may separate up to 10,000 components. A 2D analytical technique using IEF and SDS-PAGE to separate total protein results in a gel having bands or spots in an apparently random pattern. Each spot represents a unique component present in the sample. The sample spots are lined up on the gel in order of molecular weight in one direction (typically vertical), and isoelectric point in the other (horizontal). Two-dimensional gel electrophoresis is a powerful technique for analyzing very complex samples, such as tissue homogenates, but is not usually applied to purified samples. Only one sample can be analyzed at a time on each gel in 2D gel electrophoresis.

Pulsed Field Gel Electrophoresis   Pulsed field gel electrophoresis was developed to separate large pieces of DNA in agarose gels. DNA had previously been separated by conventional gel electrophoretic techniques. The resolving power of this technique is inversely proportional to the log of the DNA molecular size. Thus as the molecular weight of the DNA increases, the resolution decreases. In pulsed field electrophoresis, the direction of the field is intermittently changed—forward or backward, from side to side, or in some other sequence of directions. This causes DNA chain molecules to reorient in the electric field and entangles the DNA in the polymer matrix. Molecules that are small relative to the space between polymer strands in the matrix exhibit no significant difference in their electrophoretic mobility because they can completely reorient in the alternating field without entangling in the matrix. However, the redirection of the electric field causes molecules larger than the entanglement length to travel in a zigzag pattern, putting kinks in the molecule. The longer the molecule, the more kinks it develops, and the more slowly it travels through the gel. Pulsed field techniques significantly enhance resolution of megabase-sized strands of DNA.

Blotting Techniques   Problems encountered when one is using techniques such as direct immunofixation or application of a ligand to analyze resolved components of a sample mixture on a gel can be circumvented by using blotting techniques followed by staining or autoradiography. It is the inability of some compounds, such as antibodies or ligands, to enter a specific gel matrix that necessitated the development of various blotting techniques. The nucleic acid and protein blotting techniques are useful because they combine electrophoretic analyses and sensitive immunological or hybridization detection tools. These techniques involve the transfer of nucleic acids or proteins, immediately after separation by electrophoresis, from the gel matrix to another matrix. Typically, the other matrix is nitrocellulose paper, nylon, or other high affinity membrane. The mode of transfer is electrotransfer for proteins and capillary transfer for nucleic acids. On nitrocellulose paper or nylon, nucleic acids and proteins are more accessible than in the original gel matrix. This second matrix is then treated with a ligand to identify a specific component of a sample. Table 2.3 summarizes the common electrophoresis blotting techniques. A "Western blot" is shown in Figure 2.2*b*.

TABLE 2.3
Blotting Techniques Commonly Used with Electrophoresis

Technique	Analyte	Membrane	Probe	Detection
Southern blot	DNA	Nitrocellulose	Radiolabeled or nonradiolabeled complementary RNA or DNA	Autoradiography fluorescence, chemiluminescence, or colorimetry
Northern blot	RNA	Nitrocellulose	Radiolabeled or nonradiolabeled complementary RNA or DNA	Autoradiography fluorescence, chemiluminescence, or colorimetry
Western blot	Proteins	Nitrocellulose or nylon	Whole antisera or antibodies	Autoradiography fluorescence, chemiluminescence, or colorimetry

2.5.2 HIGH PERFORMANCE LIQUID CHROMATOGRAPHY (HPLC)

The principles of analytical HPLC, a frequently used method for determination of purity, are identical to those used in large-scale liquid chromatography, so they will not be reviewed here (see Chapter 7). The basic differences between analytical HPLC and large-scale or preparative LC are the size of the stationary phase particles and the sample size. The size of the particles used in HPLC is typically in the range of 1 to 10 μm, much smaller than used in preparative and large-scale conventional LC. While large-scale LC employs sample volumes up to and beyond one column volume, HPLC sample injections range from 0.4 to 4.5% of a column volume. Clearly, in analytical HPLC, a small fraction of the capacity of the bed is used. With this limitation on the amount of material loaded onto the column, "baseline" resolution of contaminants is more likely. Other differences include higher linear velocities and higher pressures in general.

HPLC is a powerful method for the quantitation of impurities. It resolves impurities from the product in a chemistry-specific manner by taking advantage of the binding mechanism inherent between the molecule and the stationary packing. Impurities are usually related to the product, sometimes varying by one amino acid residue, by a stereochemical shift, or by a modification to a residue, such as a methylated histidine. Impurities as low as 0.1% of the total product amount can be detected reliably and can be quantitated at 0.5%. As mentioned earlier, the absolute amount of a contaminant can be known only if its response in the detection method (such as UV) is known, and this can usually be known only by isolating enough of the contaminant to characterize it.

An HPLC chromatogram that indicates the power of this technique is shown in Figure 2.4, which illustrates the separation of hemoglobin (Hb) variants by means of a cation exchange packing. Hemoglobins S and A_0 differ by only a single amino acid substitution in two of the four subunits. The only difference between Hb A_0 and Hb A_{1c} is that the N-terminal valine of two of the chains of Hb A_{1c} is bound covalently to glucose. The resolution of Hb A_{1c} and Hb F requires ammonium sulfate in the mobile phase, which indicates that this separation may be based on a mixed ion exchange and hydrophobic interaction mechanism [15]. This example illustrates the point that even though HPLC is a powerful method, more than one HPLC technique may be required to completely resolve the components in a given sample.

A simple schematic of HPLC equipment is shown in Figure 2.5. A temperature-controlled autosampler holds samples. A robotic arm locates sample vials and extracts a sample of a specified volume with a needle. Then microswitched valves transfer the amount drawn from the vial to the injector. The sample is delivered to the column, along with the solvent or buffer used to develop the chromatography, by a high pressure pump upstream of the injector. The buffer or solvent is usually degassed by vacuum or sparging with helium. The sample and buffer enter the column, where adsorption/desorption phenomena occur (see Chapter 7). The separands elute sequentially from the column and pass through one or more detectors. Detectors range from UV to mass spectrometry devices, depending on the type of information required. All the HPLC components are computer controlled. Computers are required for the large amount of data generated by the detectors, especially when a mass spectrometer or a diode array detector is used, since these detectors generate thousands of data points per second. Computers allow archiving of data for subsequent analysis. Analysis options include peak integration by a variety of methods, peak tangent skimming

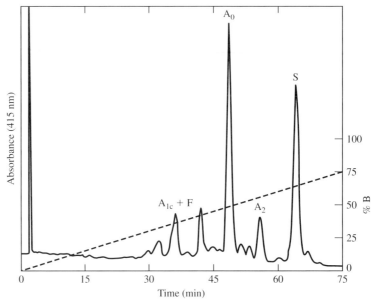

Figure 2.4 Separation of hemoglobin variants by ion exchange HPLC. The column was a SynChropak model CM300 (6.5 μm spherical particles with carboxymethyl groups), with 250 × 4.6 mm i.d.; buffer, 0.03 M bis-Tris/0.0015 M KCN (pH 6.4); 100 min linear gradient (shown as a straight line) from 0 to 0.15 M sodium acetate (*B*); flow rate, 1 ml/min. (Data from F. E. Regnier and K. M. Gooding, "Proteins," in *Chromatography, Part B: Applications,* 5th ed., p. B151, E. Heftmann, ed., Elsevier, Amsterdam, 1992.)

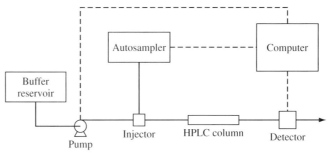

Figure 2.5 Schematic diagram of HPLC equipment. Solid lines represent fluid flows, and dashed lines represent paths of signals to or from the computer.

and dropping to baseline being the two most popular. HETP (height equivalent of a theoretical plate), resolution, skewness, and capacity factor can all be calculated automatically.

2.5.3 MASS SPECTROMETRY

Mass spectrometry is another powerful technique that is now very useful for the analysis of bioproducts and impurities. The result of a mass spectrometry analysis is the molecular weight of the molecule. Two relatively recent developments in mass spectrometry, electrospray ionization (ESI) and matrix-assisted laser desorption/ionization (MALDI), extend the application of mass spectrometry to the analysis of proteins and biopolymers. These methods have the accuracy, sensitivity, and resolving power to detect posttranslational modifications of proteins and peptides [16, 17].

Because ESI mass spectrometry is highly accurate ($\pm0.01\%$) and because it can be easily interfaced with liquid chromatography, it is the method of choice for peptides and proteins having a molecular weight below 70,000 Da [18]. Since this accuracy level translates to an error of ±0.5 Da for a 5000 Da peptide and ±5 Da for a 50,000 Da protein, it is clear that this is a highly useful method for determining whether any modifications have been made to a bioproduct. For a recombinant protein of known theoretical amino acid sequence, for example, this method can easily determine whether any amino acids have been cleaved from either terminus and, if so, which amino acids have likely been cleaved.

Of the various ionization techniques for mass spectrometry, electrospray is the most diverse. ESI can produce singly or multiply charged gaseous ions directly from an aqueous or aqueous/organic solvent system, and it can generate ions from almost every type of charged compound whether it is low or high mass, positively or negatively charged. The sample to be analyzed should be as homogeneous and salt free as possible to maximize sensitivity [18].

2.5.4 COUPLING OF HPLC WITH MASS SPECTROMETRY

The ability of the electrospray ionization/mass spectrometry (ESI-MS) technique to analyze compounds from aqueous and aqueous/organic solutions has established ESI-MS as a convenient mass detector for HPLC. It is possible to directly couple reversed-phase HPLC with ESI-MS, since the solvents and volatile buffer salts commonly used in this type of chromatography can be employed over the whole concentration range [17]. Common effluent from ion exchange HPLC would usually need to be desalted before analysis by ESI-MS. An example of the use of reversed-phase HPLC coupled to ESI-MS is the analysis of a recombinant antimicrobial peptide that was released along with other peptides after the digestion of a fusion protein with cyanogen bromide, which cleaves at methionine residues [19]. This method identified a variant of the antimicrobial peptide that is consistent with a cyanide group added to a histidine residue (molecular weight 25 Da higher than the theoretical value).

2.5.5 ULTRAVIOLET ABSORBANCE

Measurement of the absorbance of ultraviolet (UV) light is a simple, fast, and widely used technique in biotechnology. Molecules absorb light according to the chemical bonds present in the molecule. Typical wavelengths used for detecting molecules with specific bond content are shown in Table 2.4. The amino acids tyrosine and tryptophan have absorbance maxima near 280 nm, so measuring the absorbance at this wavelength has proved convenient for monitoring proteins. Since the content of tyrosine and tryptophan in proteins varies greatly, the amount of absorbance by proteins at 280 nm has a corresponding significant variation, which is accounted for by a change in the molar absorptivity (defined shortly). For small polypeptides having few or no aromatic amino acids, it is common to monitor the absorbance of the peptide bond, having a maximum near 192 nm. Because of

TABLE 2.4
UV Wavelengths Corresponding to Chemical Structures

Chemical structure	Wavelength (nm)
CONH (peptide bond)	192
CH_3CH═$CHCHO$	217
C_6H_6 (benzene)	260
$CH_3(CH$═$CH)_2CHO$	270
$C_{10}H_8$ (naphthalene)	280
$CH_3(CH$═$CH)_3CHO$	312
$CH_3(CH$═$CH)_4CHO$	343
$CH_3(CH$═$CH)_5CHO$	370
$C_{14}H_{10}$ (anthracene)	375

the absorbance by oxygen near this wavelength, measurements for polypeptides must be made away from the peak, and the wavelength 205 nm is typically used [10].

A solution absorbs light according to the Beer–Lambert law:

$$\text{Absorbance} = \log_{10}\left(\frac{\text{incident light}}{\text{transmitted light}}\right) = \varepsilon c L \tag{2.5.6}$$

where ε is the molar absorptivity for the molecule (also called the molar extinction coefficient), c is its concentration, and L is the distance (path length) through which the light passes. The molar absorptivity is dependent on both the molecule and the wavelength of the light. Frequently, the molar absorptivity is measured at an absorbance maximum for the molecule. This assures the maximum sensitivity for the assay, both for LOD and LOQ (see Section 2.3.4), and for quantitation within the standard curve. Frequently, a UV absorbance result is compared against a standard curve, rather than depending on a precisely measured molar absorptivity, so that the calibration of the spectrophotometer and the upper limit of linearity can be ascertained each time the method is used.

EXAMPLE 2.3

Determination of Molar Absorptivity   For the standard curve data for the assay of alcohol oxidase given in Example 2.1, determine the average value of the molar absorptivity.

SOLUTION

From Equation (2.5.6), the molar adsorptivity is given as

$$\varepsilon = \frac{\text{Absorbance}}{c L}$$

Using the path length (L) of 10 mm, we can calculate ε using the above equation at each H_2O_2 concentration given for the standard curve determination (Table E2.3).

The average ε for the five concentrations is 0.00366 liter μmol^{-1} mm^{-1}, which is very close to the literature value of 0.00368 liter μmol^{-1} mm^{-1} [12].

TABLE E2.3

H_2O_2 concentration (μmol/liter)	$A_{405\ nm}$	ε (liter μmol^{-1} mm^{-1})
7.1	0.25	0.00352
9.6	0.36	0.00375
12.3	0.44	0.00358
15.1	0.57	0.00377
18.0	0.66	0.00367

2.5.6 CHNO/AMINO ACID ANALYSIS (AAA)

Sometimes, the best way to know the content of a purified molecule is to directly measure its constituent atoms or signature molecules. This is the case with CHNO (carbon, hydrogen, nitrogen, oxygen) analysis and amino acid analysis (AAA). These methods are typically applied to a dried preparation of a purified molecule, with AAA being applied exclusively to proteins and peptides.

CHNO are measured from a dry powder by a single method or by a combination. It is very typical to do flame ionization/atomic absorption. In this way, the ratio of each of these elements can be determined and compared with the expected ratio in the molecule (including the expected salt form, and water of hydration). The limiting element can then be used, along with the dry weight, to determine the exact amount of the target molecule. This correlation technique can be applied to other atomic species as well, for instance iron quantification in an iron-containing protein, such as hemoglobin.

EXAMPLE 2.4

Calculations Based on CHNO Analysis   You have 10 g of a protein powder. The protein has a molecular weight of 5533 and the chemical formula $C_{242}H_{397}N_{68}O_{78}$. You perform CHNO and find the following mass percentages:

C	50.11%
H	7.45%
N	16.40%
O	26.04%

What are the mass fractions of protein and water in the powder?

SOLUTION

The formula represented by the mass fractions is $C_{3.56}H_{6.31}N_1O_{1.39}$. The formula for the protein, normalized to nitrogen content is $C_{3.56}H_{5.84}N_1O_{1.15}$. There is an excess of 0.24 mol of oxygen, and 0.47 mol of hydrogen, confirming the formula of the excess mass as H_2O. There are x grams of protein with $397(100)/5533 = 7.2$ wt% hydrogen, and $(10 - x)$ grams of water with $2(100)/18 = 11.1$ wt% hydrogen. There is 0.745 g of hydrogen in the powder, so we can write a mass balance on hydrogen as

$$0.072x + 0.111(10 - x) = 0.745$$

Solving for x gives 9.36 g of protein. Therefore,

Mass fraction of protein = 0.936

Mass fraction of water = 0.064

The problem could also be solved based on doing a mass balance for oxygen instead of hydrogen.

Amino acid analysis is similar. The standard hydrolysis procedure uses 6 N HCl at 110°C for 24 h. These conditions, however, destroy 50 to 100% of cysteine/cystine and tryptophan and convert asparagine and glutamine to aspartic acid and glutamic acid, respectively. An alternative hydrolysis method has been developed that preserves tryptophan, and special pretreatments before the acid hydrolysis allow for the quantitative determination of asparagine and glutamine [20].

2.5.7 PROTEIN ASSAYS

Protein assays are also used to determine content in a powder or solution. Protein assays are less specific than CHNO or amino acid analysis because they are based on assumptions about the atomic content of proteins, or the specific binding of some residues and their general prevalence in proteins. For instance, the Kjeldahl assay measures total nitrogen. The assumption is made that there are approximately 1.1 mol of nitrogen per amino acid residue.

The Bradford [21] or Coomassie blue assay depends on the binding of the Coomassie dye to protein. The reaction is based on the shift of the absorbance maximum of the dye from 465 nm to 595 nm, which occurs upon binding of dye to protein. The absorbance at 595 nm is usually compared against a standard protein, such as bovine serum albumin (BSA) or ovalbumin. This method has the advantage of being more rapid (requires 10 minutes) than either the Lowry or the bicinchoninic acid assay.

The Lowry [22] method relies on the reaction between cupric ions in alkaline tartrate solution and peptide bonds. The solution of copper in tartrate buffer is called "Biuret reagent." Addition of a phenolic reagent (Folin–Ciocalteu phenol reagent) causes a purple color, which is measured spectrophotometrically at a convenient wavelength between 550 and 750 nm. The absorbance is proportional to protein concentration and is compared to a standard, typically BSA.

The bicinchoninic acid assay [23] works based on the reaction between copper ions and the peptide bond as well. In this case, copper is reduced from Cu(II) to Cu(I). Bicinchoninic acid is a chromophoric scavenger of Cu(I), forming a purple complex that is measurable at 562 nm versus a protein standard.

2.5.8 ENZYME-LINKED IMMUNOSORBENT ASSAY

Enzyme-linked immunosorbent assays (ELISAs) are used when specific molecules need to be detected at low concentrations (Figure 2.6). ELISAs use an antibody raised against a target molecule, called an antigen. Plates are prepared by immobilizing the antibody to the

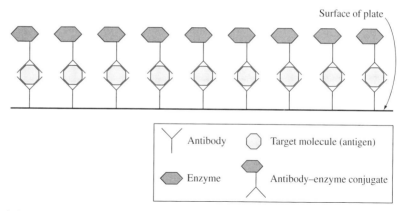

Figure 2.6 Schematic of ELISA assay.

plate surface and washing off unadsorbed antibody. Then the test solution containing the target molecule is incubated with the immobilized antibody and washed to remove non-specifically adsorbed target molecule. The antibody–enzyme conjugate is incubated with the antibody–antigen (target molecule) complex and washed, whereupon enzyme substrate is added. (Enzymes that are conjugated to antibodies are usually those that will give a detectable color change when the enzyme reacts with the substrate.) The titer of target molecule is determined by the intensity of the color change in the well of the plate, usually measured by light absorbance in the visible range.

An important application of the ELISA in bioseparations process development is for the quantitation of the level of host cell proteins, which requires that large numbers of proteins be detected at part-per-million sensitivity [4]. In this case, the antibody attached to the surface must be polyclonal, that is, raised in an animal after injection of the target antigens, not made in cell culture by a hybridoma.* The immunogens are usually derived from a fermentation of the host cell with a transfected plasmid that is identical to the production plasmid except that it lacks the target gene. This cell is fermented as usual, with induction and feeding steps as prescribed, and then purified with the same process. This enriches the host cell proteins that are likely to be produced and then copurified along with the product. Wider cuts are usually made in the chromatography steps, to give a wider range of protein contaminants for the immunization step. The proteins collected are injected into a rabbit or goat, serum is collected after several days, and the immunoglobulins purified. The antibodies are purified to eliminate any reactivity with the product protein. This is important, as the actual samples tested will have large amounts of product protein. The technique does not distinguish between individual contaminants, although the antibodies can be used in Western blots (Section 2.5.1) to determine the exact protein contaminants for which the assay is specific.

*Polyclonal means that there is a mixture of antibodies that have been raised against several antigens or against only one antigen. A monoclonal antibody is specific for a certain region, called an epitope, of a single antigen. An antigen can have more than one epitope.

2.5.9 GAS CHROMATOGRAPHY

Gas chromatography is used to separate and quantify volatile organic compounds. The principles are identical to those in liquid chromatography, with an inert carrier transporting solutes as they adsorb into and desorb from a stationary adsorbent phase. In the case of gas chromatography, the column is generally a coated capillary ranging in length from 30 to 100 m. The carrier is usually helium or another inert gas. The detectors used are usually flame ionization detection (FID) or thermal conductivity detection (TCD). Mass spectroscopy is a common add-on detector, especially when the contaminating species are unknown before the analysis is undertaken. Various injectors are available for the processing of solid, liquid, and gas samples. Liquid samples are injected directly and allowed to evaporate in the oven before being carried downstream into the capillary column. Gas samples can be introduced through a gastight syringe, a pressurized vessel, or through a headspace sampler. Solid samples can be processed with a headspace sampler (where a volatile substance is expected to evolve from the solid sample), or with a thermal desorption unit, when higher temperatures are required, or decomposition of the sample is sought.

Many standard methods are available for gas chromatography, for most common solvents, pesticides, carcinogens, preservatives, monomers, and surfactants. Gas chromatography is commonly used to detect residual solvents in pharmaceuticals and biologicals. There are amino acid derivatization methods that allow chiral separations by gas chromatography. The technique may also be used to assess degree of polymerization and polymer purity when a formulation requires an excipient for stability or bioavailability.

2.5.10 DNA HYBRIDIZATION

DNA hybridization is used to detect small amounts of DNA in a purified product. Of primary concern is DNA from the host recombinant organism. DNA probes that are sequence specific for the host cell, and primarily for the plasmid that carries the product gene, are synthesized with a radioactive or nonradioactive label. The DNA in the sample is adsorbed to a nitrocellulose filter. The sample is then hybridized in the presence of the probes through a series of annealing steps. The amount of DNA in the sample can then be measured by detecting the label on the probe (autoradiography, fluorescence, chemiluminescence, or colorimetry). Detection limits are commonly in the picogram range.

DNA hybridization techniques can also be performed on samples that have been amplified through the polymerase chain reaction (PCR). This technique, which is used more commonly to amplify a gene that is responsible for an interesting activity in a cell homogenate, is applied in this analytical setting only when required detection limits are lower than stated earlier. A DNA fragment serves as a probe, and DNA polymerase with guanine (G), cytosine (C), adenine (A), and thymine (T) serves to make extra copies of the DNA of interest. This technique can be used to amplify the number of copies of DNA by hundreds- to millions-fold.

2.5.11 ICP/MS (AES)

Inductively coupled plasma (ICP) coupled with mass spectrometry (MS) or atomic emission spectroscopy (AES) is used primarily to measure metallic elements in a sample. Inductively coupled plasma is a method for burning a sample and volatilizing the ash in a highly dispersed fashion. The resulting material can be easily analyzed by mass spectroscopy or atomic emission spectroscopy to determine its identity and concentration (in reference to standards). Heavy metal content is important for pharmaceuticals, especially those that will be administered chronically or for long periods of time. Metals can contaminate a product from contact with various metal surfaces (especially low grade steels), but also from contact with resins and filter materials. Metals are also found in excipients and buffering agents, and they may be concentrated over time by various large-scale separation methodologies.

2.5.12 DRY WEIGHT

Dry weight is the least specific of the assays used in biotechnology. However, it has some purposes, especially with crude materials. Before any initial purification, whether it be from a fermentor or a synthetic peptide reactor, a dry weight is often the most informative assay to monitor the behavior of the reaction step and to predict the performance of the first purification step. Dry weight is a simple assay conceptually, but in fact it requires relatively large sample amounts and careful control of sample containers and drying conditions for accuracy and precision.

2.6 Microbiology Assays

Microbiology assays are conducted to determine the safety of the product for human consumption. The history of pharmaceuticals shows that the two most dangerous things in a product are likely to be either the product itself (consider thalidomide, Phen-Fen, chemotherapeutics) or bacterial contamination (consider *Salmonella* sp., *E. coli,* HIV in transfusion blood). Therefore, basic microbiological, and more recently, viral screens have been required for pharmaceutical and food products.

2.6.1 STERILITY

Sterility tests are performed on a select sampling of a batch of sterile product. They consist simply of injecting an amount of the product into a sterile growth-supporting broth, either casein soy digest or thioglycolate medium (usually both). If, after 14 days, the broth is cloudy, the product is not sterile and must be destroyed. Along with this test, the broths are tested with control bacteria to ensure that they support growth (i.e., are "growth promoting"), and samples of the product are tested to make sure they do not inhibit the growth of bacteria themselves (called "bacteriostasis/fungistasis testing"). These three tests, taken together, can be used to demonstrate that certain units from the batch are sterile. Unfortunately, those units have been destroyed in the testing. There are no tests that demonstrate, with complete assurance, that the entire batch is sterile and safe for human use.

Therefore surrogate tests are performed on the process of manufacturing the pharmaceutical. Sterile broth is used in the process, instead of the product, to demonstrate that as the entire batch is processed, contamination is precluded. Bacterial spikes in the form of spore strips are used to challenge sterilization steps, such as autoclaving and irradiating, to demonstrate many logarithms of colony reduction through these processes. These tests form a package of data that give a high level of sterility assurance.

2.6.2 BIOBURDEN

Bioburden testing is the process of determining how much bacterial contamination is present in a sample. Prior to a sterilizing step, the probability that bacteria or mold infest a product in some way is high, especially in a large-scale facility. To determine the bioburden, a sample is filtered on a 0.2 μm porosity filter. The filter is then incubated on an agar plate for 24 to 48 h, and the colonies are counted. The result is a number of "colony-forming units" that were present in the sample prior to filtration. The bioburden value is often used to set the logarithm reduction required from the sterilization step.

2.6.3 ENDOTOXIN

Endotoxin is a lipopolysaccharide that forms some of the cell wall of Gram-negative bacteria. It is pathogenic in humans, causing illness and death in sufficiently high doses. A product that shows no bioburden and is sterile can still contain endotoxin from prior contamination. Endotoxin tests are designed to detect this possibility.

Early endotoxin tests employed rabbits. Rabbits were injected with product, and their temperatures monitored for several days. Rabbits that maintained a normal temperature indicated an endotoxin-free preparation. This test was costly and time-consuming, however, and a new test was developed that used the blood from horseshoe crabs. Called the LAL test, for *Limulus* amoebocyte lysate, this blood forms a clot when exposed to endotoxin. Horseshoe crab blood goes much further than a colony of rabbits, and the test takes only 30 min. More recently, an artificial substrate has been developed that is cleaved in the presence of endotoxin and produces a color. This chromogenic assay is the most convenient, and it gives a direct measure of the concentration of endotoxin. Since endotoxin is not a single chemical with a defined molecular weight, it is measured in units, which are described by any one of the three tests just described.

2.6.4 VIRUS AND PHAGE

A more recent concern in pharmaceutical preparations is the virus, once limited primarily to the blood fractionation industry. However a wider variety of human sources, not the least of which is cell culture, has become available for production of bioactive products in the last three decades. Viruses that infect these sources are well suited for human infection, an occurrence that is relatively rare between species. Furthermore, less is known about the prevention and cure of viral infections, making them more dread than bacterial infection.

Viruses are hard to detect. Viruses in general must be found through incubation with human cells that are susceptible to viruses in general. If the virus is "lytic"—that is, if it lyses its host to release daughter virus particles—the presence of viral contamination is easily detected. A confluent cell layer is treated with the sample and checked after a day for "plaques." The plaques, or areas devoid of cells, indicate the presence of virus. Specific viruses can be identified by several immunological tests and nucleic acid probes [24]. Immunological tests commonly used are ELISAs (Figure 2.6) that detect virus particles and fluorescent antibody methods that use antibodies made against viral antigens. When nucleic acid probes are used, it is sometimes necessary to amplify the target viral DNA by means of the polymerase chain reaction.

To ensure a virus-free product requires much the same assurance as sterility. Steps in the purification process are challenged to determine their "viral clearance." In general, three qualified viral clearance steps are incorporated into any process that has as the product source, human blood or bodily fluid or cells, recombinant or natural.

Phages, a virus of bacteria, pose little threat to humans but can destroy a recombinant production area. When phages have infected manufacturing, they must be detected and eliminated. Phage are detected in the same manner as viruses (i.e., by using bacterial host cells), with lytic phage being more readily detected than nonlytic.

2.7 Summary

Analytical methods are vital to the specification of a product through assays of biological activity, purity, and sterility. The efficient development of bioseparation processes depends on the availability of suitable analytical methods.

- "Specifications" required by regulatory agencies include biological activity, purity, composition, physical characteristics, and safety features.

- Assay attributes consist of precision, the reproducibility of a numerical value of a measurement; accuracy, the closeness of a numerical measurement to its real or universal value; specificity, the extent to which an assay represents the amount of target product and not something else; linearity, the dependence of the assay measurement on the first power of the quantity being measured; limits of detection, the minimum amount of an analyte that can be measured; range, the limits over which the attributes have acceptable values; and robustness, the insensitivity of the attributes to varying conditions, especially those that may be difficult to control.

- Laboratory methods for assaying biological activity include animal model assays, cell-line-derived bioassays, and *in vitro* biochemical assays.

- There are two general classes of techniques for measuring enzyme activity, stopped methods and continuous methods. Product formation in these assays is typically followed spectrophotometrically.

- Purity assays should be sensitive to impurities present in small quantities. The most powerful assays for purity are electrophoretic analysis, high performance liquid chromatography (HPLC), and mass spectrometry (MS). MS with electrospray ionization (ESI) has high accuracy ($\pm 0.01\%$) for molecular weight determination, which enables the detection of posttranslational modifications of proteins and peptides.

- The velocity of a charged particle undergoing electrophoresis is

$$v = UE$$

where v is the particle velocity, E is the field strength (voltage per length), and U is the apparent electrophoretic mobility. There are two contributions to this apparent electrophoretic mobility,

$$U = U_{el} + U_o$$

where U_{el} is the electrophoretic mobility of the charged particle and U_o and is the contribution from electroosmotic flow.

- The most commonly used electrophoretic techniques for bioproducts are denaturing gel electrophoresis (better known as SDS-PAGE), native gel electrophoresis, capillary electrophoresis, and isoelectric focusing. Blotting methods add to the analytical specificity of electrophoretic methods.

- Microbiology assays are used to determine sterility. Sterility tests must be performed by adding sample to a nutrient medium with positive controls consisting of the same nutrient medium spiked with bacteria and negative controls showing that the medium does not inhibit growth.

- Endotoxin testing requires a biological assay, typically the *Limulus* amoebocyte lysate (LAL) assay. Viruses are usually assayed in human cells *in vitro*.

NOMENCLATURE

A	absorbance (dimensionless)
c	concentration of analyte (M, or g liter^{-1})
C_p	heat capacity (J g^{-1} K^{-1})
E	applied field strength (V cm^{-1})
h	convective heat transfer coefficient (kcal m^{-2} h^{-1} °C^{-1})
H	height of gel (cm)
I	current (A)
k	thermal conductivity (kcal m^{-1} h^{-1} K^{-1})
L	path length in optical cell (cm)
M_e	mass of electrophoretic medium (g)
n	total number of measurements in a sample
P	power (J s^{-1})
q	heat flux (kcal m^{-2} h^{-1})
r^2	correlation coefficient
R	electrical resistance (ohms)

S	rate of heat generated per volume by electric current in an electrophoresis gel (kcal h^{-1} cm^{-3})
SS_R	sum of squares of the deviation of the model value from the mean of y (units vary)
SS_{yy}	sum of the squares of the measured value from the mean of y (units vary)
T	temperature (K)
U	apparent electrophoretic mobility (cm^2 s^{-1} V^{-1})
U_{el}	electrophoretic mobility of a charged particle (cm^2 s^{-1} V^{-1})
U_o	contribution to electrophoretic mobility from electroosmotic flow (cm^2 s^{-1} V^{-1})
v	particle velocity (cm s^{-1})
V	voltage (V)
W	width of gel (cm)
x	distance (cm)
x_i	individual measurement, where $i = 1, 2, \ldots, n$ (units vary)

$\bar{x}$ mean of measurements of x (units vary)

y_i individual measurement, where $i = 1, 2, \ldots, n$ (units vary)

$\bar{y}$ mean of measurements (units vary)

$\hat{y}_i$ model value of measured value i (units vary)

Greek Letters

ε molar absorptivity (liter μmol^{-1} mm^{-1})

ρ density (g cm^{-3})

σ standard deviation (units vary)

PROBLEMS

2.1   **Generation of a Specification and Validation Plan** Specifications have been developed for a recombinant protein made from *E. coli* (see Table P2.1). The product has three disulfide bridges, is formed in inclusion bodies, and is purified through three chromatography steps, one of which is affinity chromatography. For each specification, check the assay attributes that apply from the indicated attributes.

TABLE P2.1
Recombinant Protein from *E. coli* with Three Disulfide Bonds, from Inclusion Body, with Affinity Purification

Specification	Precision	Accuracy	Specificity	Linearity	Range	Robustness
Purity by reversed phase, >98.0%, should detect misfolded protein, oxidized product, aggregates, some misincorporations and deletions						
Purity by size exclusion chromatography, >99.0%, for aggregates						
Purity by ion exchange chromatography, >98.0%, orthogonal purity method, should detect deamidations and other degadation products, fragments						
Identity by tryptic map						
Leached antibody from affinity step by ELISA, <10 ppm (depends on toxicity profile)						
E. coli proteins by ELISA, <100 ppm						
Host DNA, <100 ppb by autoradiograph from hybridized strands						
Bioassay, activity between 75 and 125% of standard reference material						
Product content, typically by HPLC, to be within 2.5% of the specified formula						
Endotoxin <1 endotoxin unit/mg (depends on dose, validation specified by the USP)						
Bioburden <100 colony-forming units/g (validation specified by the USP)						
pH, within 0.2 unit of formula (calibrated assay)						
Appearance, a transparent to opalescent solution essentially free of particulate						

2.2   Bioassays   Discuss the relative importance of the bioassay during the different phases of research and development of a new bioactive molecule, including discovery research, laboratory process development, and large-scale process development.

2.3   Enzymatic Activity Assay for L-Alanine Dehydrogenase   L-Alanine dehydrogenase can be assayed by using the following reaction:

L-Alanine + NAD$^+$ + H$_2$O

$$\xrightarrow[\text{L-alanine dehydrogenase}]{} \text{pyruvate} + \text{NADH} + \text{NH}_4^+$$

For this assay, a sample containing this enzyme was diluted 1:1000 and added to a solution containing L-alanine, NAD, and carbonate buffer at pH 10.0 and 25°C. Absorbance readings versus time were taken using a path length of 10 mm and wavelength of 340 nm as shown in Table P2.3.

TABLE P2.3

Time (min)	A_{340}
2.0	0.27
2.5	0.33
3.0	0.40
3.5	0.47
4.0	0.53
4.5	0.60
5.0	0.67

From these data and the molar absorptivity for NADH ($\varepsilon = 6.2 \times 10^2$ liter mol^{-1} mm^{-1} at 25°C and 340 nm), calculate the enzyme concentration in units per milliliter. (*Note:* NADH absorbs at 340 nm, but NAD$^+$ does not.) If the purified enzyme has a specific activity of 30 U/mg and the enzyme solution that was assayed has a protein concentration of 1.1 mg/ml, what is the purity of the enzyme in the sample and what is its specific activity? (Data from H. U. Bergmeyer and M. Grassl, "Handling of reagents," in *Methods of Enzymatic Analysis,* vol. II, H. U. Bergmeyer, ed., p. 102, Verlag Chemie, Weinheim, 1983.)

2.4   Troubleshooting purity methods   A recombinant protein produced in a production scale process typically has a level of host cell protein (HCP) by ELISA

of 30 ppm. The most recent batch showed a shift in HCP to 80 ppm. What steps can you take to confirm this result?

2.5   Operating Problems During the Electrophoresis of Proteins   For the following problems that can occur in the analysis of proteins by gel electrophoresis, what do you think is the cause of the problem, and how do you think the problem can be solved?

(a) "Smile effect"—band pattern curves upward at both sides of the gel
(b) Lateral band spreading
(c) Skewed or distorted bands
(d) Vertical streaking of protein bands

2.6   Protein Charge During Gel Electrophoresis   A solution of 0.1% sodium dodecyl sulfate (SDS) binds so strongly to polypeptide chains that one detergent molecule is sufficient to saturate two amino acid residues (R. Scopes, *Protein Purification,* 3rd ed., p. 297, Springer-Verlag, New York, 1994). For a protein with a molecular weight of 60,000 in 0.1% SDS, estimate the effect on the electrical charge per protein molecule.

2.7   Problem in the Determination of Molecular Weight by Gel Electrophoresis   A purified recombinant protein is analyzed for molecular weight by SDS-PAGE at pH 8.5. From the protein sequence deduced from the gene that was expressed in bacteria, the protein is expected to have a molecular weight of 44,000. However, the molecular weight of the protein is found by SDS-PAGE to be 52,000. Explain the reason or reasons for this difference in molecular weight. What calculation could you make to help explain this discrepancy?

2.8   Temperature in a Vertical Electrophoresis Gel   You are designing a mini–vertical gel electrophoresis system to separate proteins. The gel is 7 cm high, 8 cm wide, and 0.075 cm thick and runs at 200 V and 60 mA. The glass support plates on either side of the gel are each 0.200 cm thick. To finalize the design, you need to decide whether the system should be placed in an air bath or a water bath. Estimate the maximum temperature in the gel for both cases using thermal conductivities and heat transfer coefficients found in the literature. Which bath do you recommend be used?

References

1. Hines, W. W., and Montgomery, D. C. (1990). *Probability and Statistics in Engineering and Management Science,* Wiley, New York.
2. Garnick, R. L., Solli, N. J., and Papa, P. A. (1988). The role of quality control in biotechnology: An analytical perspective. *Anal. Chem.,* vol. 60, p. 2546.
3. Wilhelmi, A. E. (1973). Bioassay. In *Peptide Hormones,* Part II, S. A. Berson and R. Yalow, eds., Elsevier, New York, p. 296.
4. Anicetti, V., and Hancock, W. S. (1994). Analytical considerations in the development of protein purification processes. In *Protein Purification Process Engineering,* R. G. Harrison, ed., Dekker, New York, p. 11.
5. Marshall, R. N., Underwood, L. E., Voina, S. J., Fouschee, D. B., and Van Wyk, J. J. (1974). Characterization of the insulin and somatomedin-C receptors in human placental cell membranes. *J. Clin Endocrinol. Metab.,* vol. 39, p. 283.
6. Karey, K. P., Riss, T. L., Burleigh, B. D., Parker, D., and Sirbasku, D. A. (1988). Human recombinant insulin–like growth factor I. II. Binding characterization and radioreceptor assay development using BALB/c 3T3 mouse embryo fibroblasts. *In Vitro Cell. Develop. Biol.,* vol. 24, p. 1107.
7. Jones, R. N., Barry, A. L., Gavan, T. L., and Washington, J. A. (1985). Susceptibility tests: Microdilution and macrodilution broth procedures. In *Manual of Clinical Microbiology,* 4th ed., E. H. Lennette, ed., American Society for Microbiology, Washington, DC, p. 972.
8. Platt, T. B. (1986). Microbiological assay of antibiotics in body fluids and tissues. In *Modern Analysis of Antibiotics,* A. Aszalos, ed., Dekker, New York, p. 341.
9. Bergmeyer, H. U. (1974). Determination of enzyme activities. In *Methods of Enzymatic Analysis,* Vol. 1, H. U. Bergmeyer, ed., Verlag Chemie, Weinheim, p. 121.
10. Scopes, R. K. (1994). *Protein Purification,* 3rd ed., Springer-Verlag, New York.
11. Carlson, R., Garnick, R. L., Jones, A. J. S., and Meunier, A. M. (1988). The determination of recombinant human tissue–type plasminogen activator activity by turbidimetry using a microcentrifugal analyzer. *Anal. Biochem.,* vol. 168, p. 428.
12. Bergmeyer, H. U., and Grassl, M. (1983). Handling of reagents. In *Methods of Enzymatic Analysis,* vol. II, H. U. Bergmeyer, ed., Verlag Chemie, Weinheim, p. 102.
13. Rabel, S. R., and Stobaugh, J. F. (1997). Capillary electrophoresis. In *Chromatographic Analysis of Pharmaceuticals,* 2nd ed., J. A. Adamovics, ed., Dekker, New York, p. 209.
14. Li, S. F. Y. (1992). *Capillary Electrophoresis,* Elsevier, Amsterdam.
15. Regnier, F. E., and Gooding, K. M. (1992). Proteins. In *Chromatography,* Part B: *Applications,* 5th ed., E. Heftmann, ed., Elsevier, Amsterdam, p. B151.
16. Bahr, U., Karas, M., and Hillenkamp, F. (1994). Analysis of biopolymers by matrix-assisted laser desorption/ionizatin (MALDI) mass spectrometry. In *Microcharacterization of Proteins,* R. Kellncr, F. Lottspeich, and H. E. Meyer, eds., VCH, Weinheim, p. 149.
17. Metzger, J. W., and Eckerskorn, C. (1994). Electrospray mass spectrometry. In *Microcharacterization of Proteins,* R. Kellner, F. Lottspeich, and H. E. Meyer, eds., VCH, Weinheim, p. 167.
18. Siuzdak, G. (1996). *Mass Spectrometry for Biotechnology,* Academic Press, San Diego, CA.
19. Haught, C., Davis, G. D., Subramanian, R., Jackson, K. W., and Harrison, R. G. (1998). Recombinant production and purification of novel antisense antimicrobial peptide in *Escherichia coli. Biotechnol. Bioeng.,* vol. 57, p. 55.
20. Kellner, R., Meyer, H. E., and Lottspeich, F. (1994). Amino acid analysis. In *Microcharacterization of Proteins,* R. Kellner, F. Lottspeich, and H. E. Meyer, eds., VCH, Weinheim, p. 93.

21. Bradford, M. M. (1976). A rapid and sensitive method for the quantitation of microgram quantities of protein utilizing the principle of protein-dye binding. *Anal. Biochem.,* vol. 72, p. 248.

22. Lowry, O. H., Rosebrough, N. J., Farr, A. L., and Randall, R. J. (1951). Protein measurement with the Folin phenol reagent. *J. Biol. Chem.,* vol. 193, p. 265.

23. Smith, P. K., Krohn, R. I., Hermanson, G. T., Mallia, A. K., Gartner, F. H., Provenzano, M. D., Fujimoto, E. K., Goeke, N. M., Olson, B. J., and Klenk, D. C. (1985). Measurement of protein using bicinchoninic acid. *Anal. Biochem.,* vol. 150, p. 76.

24. Brock, T. D., Madigan, M. T., Martinko, J. M., and Parker, J. (1994). *Biology of Microorganisms,* 7th ed., Prentice Hall, Englewood Cliffs, NJ.

Cell Lysis and Flocculation

If a product is synthesized intracellularly and not secreted by the producing cell, or if the product is to be extracted from plant or animal or fungal tissue, it is necessary to remove the product from the cell or tissue by force. The choice of procedure is highly dependent on the nature of the product and the nature of the cell or tissue. It was seen in Chapter 1 that bioproducts represent a wide variety of chemical species. In this chapter we also see that the sources of bioproducts—cells and tissues—are widely varied. For this reason there exists a wide variety of methods for breaking, or *lysing,* cells and tissues, broadly classified as "chemical" and "physical" methods.

Once cells have been suspended and/or broken open, the resulting suspension of solids must be separated from the liquid in which it is suspended. This separation process, filtration and/or sedimentation (the subjects of the next two chapters), is enhanced by having larger particles. Larger particles can be achieved by *flocculation,* a process whereby particles are aggregated into clusters, or *flocs.*

In recent years it has become desirable to isolate specific cell types from mixtures of suspended cells and to deliver the resulting cell subpopulation(s) to a process for which they, and only they, are required. Most examples come from *in vivo* sources such as blood and dispersed tissue cells. This aspect of cell processing, namely, cell purification, places special demands on separation processes that are capable of handling particulate matter under conditions that allow cells to remain alive.

This chapter presents two major elements of cell processing: the science and engineering of cell rupture by physical and chemical methods and the flocculation of cells and subcellular particles in aqueous suspension. First, however, it is helpful to develop a broad appreciation for the variety and compositions of cells that are likely to be encountered in downstream bioprocessing.

3.1 Instructional Objectives

After completing this chapter, the reader should be able to do the following:

• Recognize two classes of cells and the structures of cells that cause cell lysis to be necessary for the recovery of bioproducts.

- Choose between chemical and mechanical cell lysis methods and, within these categories, choose appropriate chemical or mechanical methods for general classes of applications.
- Describe the role of electrokinetic phenomena in the flocculation of cells and insoluble particles.
- Identify mechanisms of flocculation.
- Apply, in general terms, the Schulze–Hardy rule to flocculation problems.

3.2 Some Elements of Cell Structure

Initial guidance for the selection of a cell disruption method depends completely on the cell type. Once cell structure is understood, the procedure required to open the cell and otherwise process it follows a fairly logical route.

3.2.1 PROKARYOTIC CELLS

Prokaryotic cells (cells that do not contain a membrane-enclosed nucleus) are classified as either Eubacteria (commonly called bacteria) or Archaea. Each of these groups has its own industrial potential. For example, Archaea are now looked upon as a source of industrial enzymes that are stable over a wide range of temperatures, pH values, and ionic strengths. Eubacteria, especially *Escherichia coli,* are the workhorses of genetic engineering and prokaryotic biotechnology. The envelope of the "generic" bacterial cell is extremely resistant to fracture.

Typically the bacterial cell envelope consists of an inner plasma membrane that separates all contents of the cell from the outside world, a peptidoglycan (peptides linked to oligosaccharides or polysaccharides) cell wall, and outer membrane. Bacterial cells with a very thick cell wall stain with crystal violet (Gram stain) and are called "Gram positive," while those with thin cell walls stain very weakly—"Gram negative" (see Figure 3.1). The space between the plasma membrane and the peptidoglycan in Gram-negative bacteria is

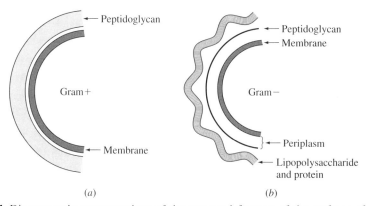

(a)   *(b)*

Figure 3.1 Diagrammatic representations of the structural features of the surfaces of (*a*) Gram-positive and (*b*) Gram-negative bacteria. The membrane is also called the plasma membrane or the cytoplasmic membrane.

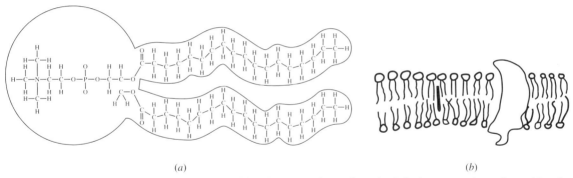

(a)   (b)

Figure 3.2 (*a*) Phospholipid molecule and its outline. (*b*) Cell plasma membrane formed by phospholipids with their polar head groups in contact with aqueous phases. The molecules are about 2.5 nm long. Vertical bar represents a cholesterol molecule, and the large, irregular blob represents a protein molecule that is in contact with both the cytoplasm and the extracellular milieu.

called the "periplasm"; proteins can be secreted across the plasma membrane into the periplasm in these bacteria. In addition, some Eubacteria have an outer capsule composed of mucopolysaccharide (see Figure 1.22). *Pneumococci* are spherical bacteria, some species of which cause diseases; their capsular polysaccharide ("PCP") is used as a vaccine against pathogenic species, and *Pseudomonas* species are known for their ability to form biofilms on this basis.

Most biological membranes are phospholipid bilayers. Figure 3.2*a* is a sketch of the shape of a phospholipid molecule, and Figure 3.2*b* indicates how millions of such molecules form a bimolecular layer between the inside and outside of the cell. Membrane stiffness is conferred by cholesterol and other steroids that partition into the nonpolar layer, while flexibility is conferred by amphiphilic proteins (i.e., proteins that have both hydrophilic and hydrophobic regions on their surface) that partition into the membrane and, in most cases, are exposed to the aqueous phase on both sides, as indicated by the blob in Figure 3.2*b*.

The bacterial cell wall protects the plasma membrane and the cytoplasm from osmotic stress. The isoosmotic external concentration for most cells is 0.3 osmolar (osM). (Osmolarity refers to the molar concentration of all species in solution including ionized species.) Higher concentrations outside the cell draw out water and cause the cytoplasm to shrink—"plasmolysis." The cell wall is rigid, so the cytoplasm collapses within the plasma membrane while the wall does not. Concentrations of salts and neutral solutes lower than 0.3 osM outside the cell force water into the cell—"turgor." Too much turgor can break the cell wall (lysis), a situation that can aid in cell disruption for retained bioproducts.

3.2.2 EUKARYOTIC CELLS

Eukaryotic cells (cells with nuclei and internal organelles) are considerably more complicated than prokaryotic cells, and bioproducts may have to be released from intracellular particles that are themselves coated with membranes and/or consist of large macromolecular aggregates. The eukaryotes include fungi (yeasts are most interesting in bioprocessing, as are antibiotic-producing molds such as *Penicillium*), and, of course, the higher plants and animals, the cells of which are today grown routinely in culture as if they were microorganisms. Cell biologists recognize all cell membranes (including those of bacteria) as a

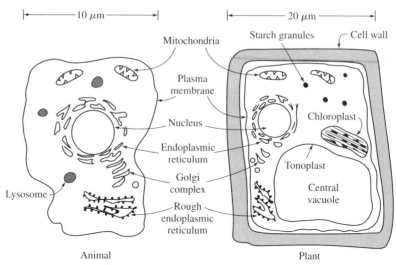

Figure 3.3 Eukaryotic cells. Simplified diagrammatic representation of an animal cell and a plant cell. The lysates of such cells contain the internal structures (organelles) shown.

separate, immiscible, liquid phase relative to the rest of the cell. The eukaryotic cell must thus be viewed as a multiphasic system. The complexity of generic animal and plant cells is indicated in Figure 3.3. Note that animal cells do not have a cell wall, while the cell wall in plants is very thick. The cell membrane of animal cells is easily broken, whereas the cell wall of plants is strong and relatively difficult to break.

It should be mentioned that the living cell is an effective chemical processor and an inspiration to the development of separation methods. Many things that a cell does can be imitated on the laboratory scale. The way solutes and particles get through cell membranes, for example, has stimulated an entire field of research in micellar extraction, liquid film membranes, and organic solvent enzymology.

3.3 Cell Lysis

There are two principal means of lysing cells to obtain their contents: chemical cell lysis and physical destruction via mechanical force. Changes in osmotic pressure, as just mentioned, involve modification of a chemical potential that actually results in a mechanical force. Surfactants and enzymes added to a cell suspension act by dissolving a portion of the cell membrane and/or cell wall. Because chemical lysis conditions are detrimental to some bioproducts, it is sometimes necessary to use pure physical force methods, with high shear at high pressure being most popular. The various cell lysis methods are categorized in Table 3.1.

3.3.1 OSMOTIC AND CHEMICAL CELL LYSIS

Every cell membrane maintains a substantial osmotic gradient; however, a drastic reduction in extracellular concentration of solutes, say from 0.15 M to 0.001 M, will tend to burst

TABLE 3.1
Cell Disintegration Techniques[a]

Method	Technique	Principle	Stress	Cost	Examples
Chemical	Osmotic shock	Osmotic rupture of membrane	Gentle	Cheap	Rupture of red blood cells
	Enzyme digestion	Cell wall digested causing rupture	Gentle	Expensive	*M. lysodeikticus* treated with egg lysozyme
	Solubilization	Detergents solubilize membrane	Gentle	Moderate	Bile salts acting on *E. coli*
	Lipid dissolution	Organic solvent dissolves in membrane, destabilizes	Moderate	Cheap	Toluene disruption of yeast
	Alkali treatment	Saponification of lipids dissolves membranes	Harsh	Cheap	Nucleic acid extractions
Mechanical	Homogenization (blade or pestle type)	Cells chopped in blender or sheared	Moderate	Moderate	Animal tissues
	Grinding	Cells ruptured by grinding with abrasives	Moderate	Cheap	Yeast enzymes
	Ultrasonication	Cells broken by sonic cavitation	Harsh	Expensive	Cell suspensions on small scale
	Pressure cell	Cells broken by shear when forced through small hole	Harsh	Moderate	Large-scale treatment of cell suspensions
	Ball mill	Cells crushed between glass or steel balls or beads	Harsh	Cheap	Large-scale treatment of cell suspensions and plant tissues

[a]See references 1 and 2.

cells that do not have walls, such as animal cells and protoplasts [i.e., cells that normally have walls but have had them removed as a result of enzymatic digestion (lysozyme) or anti-biotics (penicillin)]. If the transmembrane osmotic pressure is due to solute concentration inside the cell and out, then the van't Hoff law can be used to estimate this pressure, which applies to ideal, dilute solutions:

$$\pi = RT(c_i - c_o) \tag{3.3.1}$$

where π = osmotic transmembrane pressure
  R = gas constant
  T = absolute temperature (K)
  $c_i - c_o$ = difference between total solute molarity inside and outside the cell, respectively

Most animal cells are quickly lysed by a rapid transition to distilled water.

Bacterial and plant cells are protected against osmotic lysis by cell walls, as noted earlier. The weakening or partial destruction of these walls can be achieved with chemical agents. These may consist of detergents, chelators, enzymes, solvents, and so on. These are summarized in the upper portion of Table 3.1.

Enzymes and Antibiotics

In the case of bacteria, the enzyme lysozyme (15 kDa), readily available in crude form from hen egg whites and relatively expensive to purchase, can be incubated at a fraction of a milligram per milliliter, in the broth, resulting in sufficient digestion of the cell wall to allow osmotic rupture of the resulting bacterial protoplast. One must be very cautious in the choice of hydrolytic enzymes for cell lysis. Proteases should not be used if the desired bioproduct is a protein. Even a label on a bottle that says polysaccharidase or nuclease does not mean that there is no protease activity inside. Hydrolase preparations, especially from animal sources, tend to be mixtures. Antibiotics that are specific for prokaryotic cell wall synthesis, such as penicillin, can be used to produce protoplasts. Simply incubating a bacterial culture for a few hours in a solution of this type of antibiotic above its minimal inhibitory concentration (MIC) will result in a few cell divisions without cell wall synthesis, and the culture will consist primarily of protoplasts.

Detergents

Nonionic detergents are able to break plasma membranes and are commonly used to lyse cultured animal cells [3]. Nonionic detergents are used because they are far less denaturing for proteins and other biological compounds than ionic detergents. Some nonionic detergents that are commonly used are Triton X-100 (polyoxyethylene [9-10] *p-t*-octyl phenol), octyl β-glucoside, and Tween 20 (PEG-20 sorbitan monolaurate). Many proteins are still active in 1 to 3% Triton X-100. Different proteins react differently to different detergents, so finding the best detergent is usually a trial-and-error procedure. In one published procedure to lyse cultured animal cells, 0.1 to 0.3% Triton X-100 is added to a cell suspension of 10^7 to 10^8 cells/ml [3]. The addition of glycerol to the cell extract can help to stabilize proteins (0.2 vol of 50% glycerol).

Solvents

Solvents have been used to lyse cells, especially eukaryotes. A variety of methods with toluene have been used to lyse yeast cells, but some have not been completely successful with all commercial yeasts [2]. Acetone is often used early in the preparations of biochemicals from animal tissue homogenates. It dissolves cell membranes as well as excess fat, and at appropriate concentrations may aid in precipitating the product if that is desirable (see Chapter 8, Precipitation). A bioproduct that is extracted from cells with acetone and then precipitates is called an "acetone powder" after drying. When the precipitation is carried out below 0°C, proteins retain their native conformation, and powder precipitates can be stored before being used. Proteins can be resolubilized by mixing the acetone powder in an appropriate buffer that is warm [2].

3.3.2 Mechanical Methods of Lysis

Forces for lysing cells mechanically are developed by compression and shear. These are summarized in the lower portion of Table 3.1. Sonication, ball milling, pestle homogenization, and reciprocating milling are used. In the case of bench-scale homogenizers, a tightly fitting pestle (10–100 μm gap) is reciprocated and/or rotated in a glass or steel cylinder.

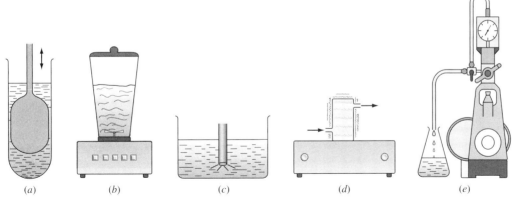

Figure 3.4 Equipment used for breaking cells. (*a*) Hand-operated or motor-driven pestle homogenizer. (*b*) Waring (blade) blender. (*c*) Ultrasonic probe. (*d*) Vibrating glass bead mill. (*e*) Manton–Gaulin pressure-shear cell disintegrator.

The Waring blender of household kitchen fame is also a popular bench-scale homogenizer and has been scaled up for industrial applications (VirTis). At higher scale, shearing devices that pass particle suspensions through small orifices at high pressures (up to 200 MPa) are used as are bead mills and ball mills. Manual (e.g., the French press) and motor-driven unit operations are used. In some motor-driven units the operation is very similar to that of an automobile engine, with the pressure provided intermittently by a reciprocating piston and the sample introduced synchronously under the control of a cam-driven valve. Figure 3.4 is a gallery of these unit operations. Correlations for the scaling of their operation are given in the next sections, which deal with two broad classes of cell homogenizers: high pressure shear homogenizers and bead mills.

High Pressure Cell Homogenizers

The Manton–Gaulin valve-type homogenizer (Figure 3.5) is among the most popular in biotechnology operations. Various designs of valves used in this homogenizer are shown in Figure 3.6. Sample feed enters the valve chamber in pulsatile flow; at each pulse, the valve closes and compresses the cell suspension against an impact ring of hard material serving as the inner wall for the chamber. The valve opens and the lysate escapes, and the cycle is repeated. After N passes through the valve, the extent of disruption is determined on the basis of the amount of soluble protein released. The extent of disruption, symbolized by R, usually depends on a power function of the applied pressure and on the number of passes through the chamber [4]:

$$\ln(1 - R) = -kNP^a \tag{3.3.2}$$

where P is the pressure. The value of a for bacteria and yeast ranges from 0.9 to 2.9 depending on species and culture conditions [5], while k depends on temperature [4].

The commercial high pressure homogenizer made by Microfluidics operates on a different principle. The sample stream is focused at high velocity into two channels that are each 100 μm in diameter. These two streams collide head-on, then turn 90° to exit as one

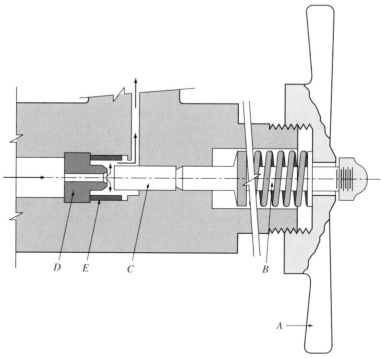

Figure 3.5 Details of a high pressure (Manton–Gaulin) homogenizer valve: A, handwheel for adjusting pressure; B, spring-driven valve rod; C, valve (see Figure 3.6); D valve seat (also see Fig. 3.6); and E, impact ring of hard material. The ring E is sometimes eroded by the impact of cells and debris, which can be abrasive. (Schematic based on a design by Manton–Gaulin APV.)

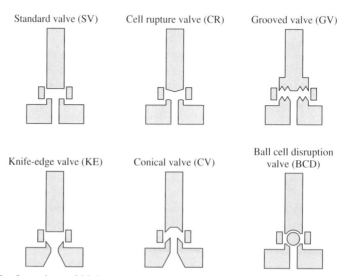

Figure 3.6 Configurations of high pressure homogenizer valves used in Manton–Gaulin homogenizers (Figure 3.5).

stream. The shear and collisions result in cell breakage and the release of soluble protein. For *E. coli* bacteria, the extent of disruption for this homogenizer could be correlated to the number of passes and pressure as follows [6]:

$$\ln(1 - R) = -kN^b P^a \tag{3.3.3}$$

The value of b varies over the range 0.3 to 0.9, depending on the cell type (i.e., recombinant or native), concentration, and specific growth rate. The parameters k and a depend on the type of cells, ranging over $0.3–63 \times 10^{-3}$ MPa^{-a} and 0.6 to 1.8, respectively.

Bead Mills

Classical bead mills, such as those used for producing fine powders of minerals, clays, and pigments have been adapted to the disruption of cells. The unit consists of a horizontal or vertical cylinder filled to some level with (usually) glass beads that are tumbled by rotating agitator disks of various shapes. A generalized bead mill configuration is shown in Figure 3.7.

The governing correlation for a bead mill operated in batch mode is [7]:

$$\ln(1 - R) = -kt \tag{3.3.4}$$

where t is the processing time for a batch. It is obvious that this unit can be operated in continuous mode, in which case the correlation is [7]:

$$\frac{1}{1 - R} = 1 + \left(\frac{kt}{j}\right)^j \tag{3.3.5}$$

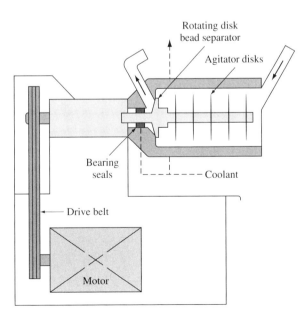

Rotating disk
bead separator

Agitator disks

Bearing
seals

Coolant

Drive belt

Motor

Figure 3.7 Schematic diagram of a typical bead mill, showing rotating agitator disks, sample inlet and outlet above the chamber, and rotating disk for the retention of beads.

where t is the mean residence time (volume of liquid in the mill divided by the flow rate), and j is equivalent to a number of continuous stirred tank reactors (CSTRs) in series and need not be an integer. The parameter j must be determined experimentally from residence time distribution studies. The most important factors that affect the cell disruption achieved in bead mills are chamber design, agitator speed, agitator design, bead size, bead loading, cell concentration, feed rate, and temperature; therefore, the application of Equations (3.3.4) and (3.3.5) will depend on these variables [5].

3.4 Flocculation

After cells have been lysed or bioparticles have been dispersed, it is often useful to hasten the subsequent filtration or sedimentation step by flocculation, that is, by reversibly increasing the size of the particles to be separated. Flocculation occurs as the result of adding a suitable chemical called a "flocculant" or by the selection of naturally flocculating cells for fermentation, as in the case of lager yeast. Flocculants can act by forming interparticle molecular "bridges" between particles, in which case the flocculants are usually polymers or oligomers; they can also act by reducing the repulsive forces between cells, usually by reducing the strength of the electrostatic field.

Flocculation is the attachment of suspended particles to one another when van der Waals interactions (attractive forces between nonpolar particles) are not counteracted by electrostatic repulsion (colloid instability). Flocculation is therefore related to the electrokinetic properties of particles and molecules. *Practically all bioparticles suspended in aqueous solutions are negatively charged.* A suspension of particles that do not aggregate is called a "stable" colloidal suspension. The transition to an "unstable" suspension requires the reduction of the surface charge of the particles. The electrokinetic properties of particles and molecules and how they can be manipulated to bring about flocculation are discussed in this section.

3.4.1 THE ELECTRIC DOUBLE LAYER

The electric double layer (cloud of ions of both signs) that surrounds all charged molecules and particles in solution influences several properties of significance in bioseparations:

  Precipitation (salting out)

  Flocculation (aggregation of particles)

  Electrophoresis (electrophoretic mobility)

  Colloid stability (particles remain in suspension)

  Phase partitioning (charge dependence)

All these processes may be considered simultaneously, in view of Figure 3.8, in which negative (typical of biological particles and molecules) particles approach one another in an electrolyte solution. Attracting forces such as London and van der Waals forces are opposed by the interaction of like charges distributed over each particle. If the charge on the particle can be reduced, then closer approach is possible, allowing the formation of London

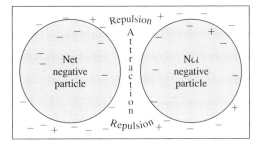

Figure 3.8 Forces between two like-charged particles, suspended in electrolyte solution, as they approach each other.

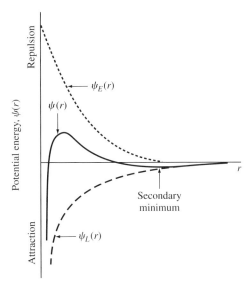

Figure 3.9 Potential energy between two like-charged particles separated by distance r, where $\psi_E(r)$ is the repulsive double-layer (electrostatic) potential, and the attractive potential curve $\psi_L(r)$ is due to London and van der Waals forces. The solid curve $\psi(r)$ is their sum. Where the two sum to values less than zero at the "secondary minimum," weak, reversible coagulation occurs. $\psi_E(r)$ is responsive to a dc electric field leading to electrophoresis or electroosmosis.

and van der Waals and even hydrogen bonds. The formation of such bonds can be understood in terms of the total potential energy diagram (Figure 3.9), in which the electrostatic repulsion curve $\psi_E(r)$ and the attractive potential curve $\psi_L(r)$ sum to values less than zero. Particles attracted to each other at the "secondary minimum" (in Figure 3.9) are still approximately $4/\kappa$ apart, where κ is the *Debye–Hückel constant*.

Figure 3.10 shows the details of the distribution of dissolved ions around a negatively charged, suspended spherical particle. The particle itself is considered to have a net excess of fixed negative charges. An *electric double layer* is created around the particle surface consisting of two regions: an inner region that includes adsorbed ions (the fixed part of the double layer) and a diffuse region in which ion diffusion is weakly affected by the electrostatic potential $\psi_E(r)$, which eventually falls to zero, defining the outer boundary of the double layer. The plane that goes through the fixed layer at about a hydrated ion radius from the solid surface is called the Stern plane (see Figure 3.10).

There are various theories describing the electric double layer. The Gouy–Chapman theory, one of the first, states that the electrical potential as a function of the distance r from

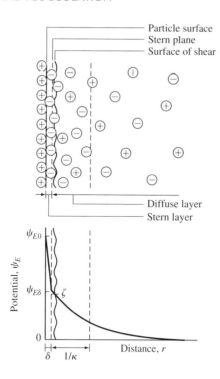

Figure 3.10 Schematic representation of the electric double layer of a model spherical colloidal particle, showing the change in the electric potential ψ_E with the distance r from the surface; ζ is the zeta potential, $1/\kappa$ is the Debye radius, and δ is the distance of the Stern plane from the solid surface (about equal to a hydrated ion radius).

a surface with a uniformly distributed charge is given by [8]

$$\psi_E(r) = \psi_{E0}\exp(-\kappa r) \tag{3.4.1}$$

where ψ_{E0} is the potential at the surface, and κ is the Debye–Hückel constant. According to the theory, this equation is valid at small values of the potential. Equation (3.4.1) shows that the electric potential decreases by $e\ (= 2.72)$ times at a distance of the order of $1/\kappa$. The quantity $1/\kappa$ is called the radius of the ionic atmosphere, or the *Debye radius or length,* and depends on the concentration of dissolved ions according to the equation [8]:

$$r_{DR} = \frac{1}{\kappa} = \left(\frac{\varepsilon R T}{8\pi F^2 \sum c_i z_i^2}\right)^{1/2} \tag{3.4.2}$$

where $\varepsilon =$ absolute dielectric constant of the liquid
$R =$ gas constant
$T =$ absolute temperature
$F =$ Faraday number
$c_i =$ bulk concentration of different ions
$z_i =$ valence of different ions

It should be noted that

$$\sum c_i z_i^2 = 2I \tag{3.4.3}$$

where I is the ionic strength of the solution. The Debye radius, r_{DR}, is typically less than 1 nm for most bioprocessing situations (see Example 3.1).

EXAMPLE 3.1

Dependence of the Debye Radius on the Type of Electrolyte Compare the Debye radius in a solution at 25°C of 0.01 M aluminum acetate to that in a solution of sodium chloride at the same temperature and molarity.

SOLUTION

We use Equation (3.4.2) to calculate the Debye radius for each salt solution. We can calculate ε from

$$\varepsilon = \varepsilon_r \varepsilon_0$$

where ε_r = relative dielectric constant of liquid
ε_0 = permittivity of free space (vacuum)

At 25°C for water, $\varepsilon_r = 78.3$. For 0.01 M $Al(C_2H_3O_2)_3$,

$$\sum c_i z_i^2 = 2I = 0.01(3)^2 + 3(0.01)(1)^2 = 0.12 \text{ M} = 0.12 \times 10^3 \frac{\text{mol}}{\text{m}^3}$$

From Equation (3.4.2),

$$r_{DR} = \left(\frac{\varepsilon RT}{8\pi F^2 \sum c_i z_i^2} \right)^{1/2} = \left[\frac{78.3 \left(8.85 \times 10^{-12} \frac{C^2}{J\,m} \right) \left(8.31 \frac{J}{\text{mol K}} \right) (298 \text{ K})}{8\pi \left(9.65 \times 10^4 \frac{C}{\text{mol}} \right)^2 \left(0.12 \times 10^3 \frac{\text{mol}}{\text{m}^3} \right)} \right]^{1/2}$$

$$= 2.47 \times 10^{-10} \text{ m} = 0.247 \text{ nm}$$

For 0.01 M NaCl,

$$\sum c_i z_i^2 = 0.01(1)^2 + 0.01(1)^2 = 0.02 \text{ M} = 0.02 \times 10^3 \frac{\text{mol}}{\text{m}^3}$$

Substituting this result into Equation (3.4.2) gives

$$r_{DR} = 0.605 \text{ nm}$$

Thus, we have the prediction that the Debye radius will be reduced by more than half when aluminum acetate is used instead of sodium chloride.

Conceptually, the effect of ionic strength on the double layer can be seen in Figure 3.11, in which the vertical dashed line shows the position of the Stern plane, and the potential curves intersect this line at near to the *zeta potential* ζ, which is the potential at the shear plane (see Figure 3.10). The zeta potential, the potential that is measured in an electrophoresis experiment, is the potential that determines the electrophoretic mobility in mobility equations, such as [8, 9]:

$$U = \frac{\varepsilon \zeta}{4\pi \mu} \tag{3.4.4}$$

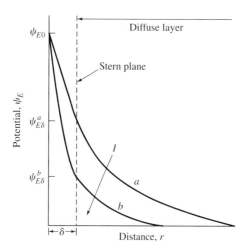

Figure 3.11 The Stern model of the electric double layer showing compression of the double layer due to increasing ionic strength I from curve a to curve b.

where U is the electrophoretic mobility and μ is the viscosity of the liquid. See Equation (2.5.1) for the relationship of electrophoretic mobility to particle velocity and electric field strength.

3.4.2 FORCES BETWEEN PARTICLES AND FLOCCULATION BY ELECTROLYTES

As two particles approach each other, they interact via repulsive (electrokinetic and lubrication) forces and attractive (van der Waals) forces (Figures 3.8 and 3.9). To cause particle aggregation or flocculation to occur is to manipulate these forces. The typical objective is to reduce the electrical repulsion force as much as possible to allow particles to approach close enough to one another to allow the attractive van der Waals force to exceed the repulsive electrical force. This is accomplished in the paradigm of DLVO theory, named after two research groups who studied this problem (Deryagin and Landau of Russia and Verwey and Overbeek of The Netherlands). Figure 3.9 shows a negative (attractive) potential at two locations, at the distance of closest approach and at a "secondary minimum," which is at a distance around $4/\kappa$. Most flocculations are due to attachments between particles in the secondary minimum. If the diffuse double layer extends too far from the particle surface, then van der Waals attraction is impossible. Therefore, from the preceding sections it should be clear that high ionic strength (small $1/\kappa$) will favor particle–particle attachments in the secondary minimum of the potential energy diagram. A useful outcome of the DLVO theory is the following prediction of the critical flocculation concentration of an indifferent electrolyte (an electrolyte that does not chemically adsorb into the Stern layer) [10]:

$$(c_i)_{\text{flocc}} = \frac{9.85 \times 10^4 \varepsilon^3 k^5 T^5 \gamma^4}{N e^6 A^2 z^6} \tag{3.4.5}$$

where $(c_i)_{\text{flocc}}$ = lowest molar concentration of the electrolyte that causes particles to coagulate

ε = absolute dielectric constant of the liquid

k = Boltzmann constant

γ = constant that limits to 1.0 at high potentials and to $ze\psi_{E\delta}/4kT$ at low potentials

$\mathcal{N}$ = Avogadro's number
e = electron charge
A = effective Hamaker constant
z = counterion charge number

The Hamaker constant was originally defined in a theoretical expression for the energy of attraction E_a between spherical particles *in vacuo* as a function of distance between the particles. For equal-sized particles at small interparticle separations, the following relationship was developed [10]:

$$E_a = -\frac{Aa}{12d} \tag{3.4.6}$$

where a is the particle radius and d is the shortest distance between particles.

The presence of a liquid dispersion medium, rather than a vacuum, between the particles lowers the van der Waals interaction energy, and the constant A in Equation (3.4.6) must be replaced by an effective Hamaker constant. The Hamaker constant, however, does not vary greatly from one material to another; for example, this constant has been measured to be 4.3×10^{-20} J for water, 7.8 to 9.8×10^{-20} J for polystyrene, and 8.6×10^{-20} J for natural rubber [11].

The equation for the critical flocculation concentration [Equation (3.4.5)] reveals a number of insights into the requirements for flocculation:

1. The DLVO theory predicts that the flocculation concentrations of indifferent electrolytes depend on z^{-6}. Thus, at the critical coagulation concentration of an indifferent electrolyte, ions having charge numbers 1, 2, and 3 should be in the ratio $1/1^6$, $1/2^6$, and $1/3^6$, which is 1: 0.016: 0.0013, the theoretical counterpart to the empirical Schulze–Hardy rule (see later).

2. For a typical experimental critical coagulation concentration at 25°C of 0.1 M for $z = 1$ and taking 75 mV as a typical value for $\psi_{E\delta}$, the effective Hamaker constant, A, is calculated [10] to be 8×10^{-20} J. This is consistent with direct measurements of A for various materials.

3. Critical flocculation concentrations of electrolytes for particles of a given material should be proportional to ε^3 and independent of particle size.

4. Flocculation concentration is extremely sensitive to temperature (proportional to T^5), indicating the wisdom of carrying out flocculation reactions at low temperature.

5. The critical flocculation concentration does not depend on the concentration of the particles being flocculated [10].

EXAMPLE 3.2

Sensitivity of Critical Flocculation Concentration to Temperature and Counterion Charge Number For the flocculation of bacterial cells, how would the critical flocculation concentration of an indifferent electrolyte be expected to change if the temperature were lowered from 37°C to 4°C? How would this concentration be expected to change if Al^{3+} counterion were used for flocculation instead of Cu^{2+} counterion?

From Equation (3.4.5), it is seen that $(c_i)_{\text{flocc}}$ depends on temperature to the fifth power. Therefore,

$$\frac{(c_i)_{\text{flocc},4°C}}{(c_i)_{\text{flocc},37°C}} = \frac{(277 \text{ K})^5}{(310 \text{ K})^5} = 0.57$$

It can also be seen that $(c_i)_{\text{flocc}}$ depends on the counterion charge to the -6th power, so that

$$\frac{(c_i)_{\text{flocc},Al^{3+}}}{(c_i)_{\text{flocc},Cu^{2+}}} = \frac{(3)^{-6}}{(2)^{-6}} = 0.088$$

Thus, the change in the electrolyte counterion charge would result in a much larger effect on the critical flocculation concentration than the change in temperature.

3.4.3 THE SCHULZE–HARDY RULE

From the first insight enumerated in Section 3.4.2, the dependence of flocculation concentration on z^{-6}, it should be obvious that whenever possible, electrolyte ions having a valence of 3 should be used for flocculation. Since most bioparticles are negative, aluminum salts are one logical choice. This theoretical prediction is empirically manifested in the Schulze–Hardy rule, which confirms that the critical electrolyte concentrations are roughly in the ratio 1000:10:1 for monovalent, divalent, and trivalent ions. An example of an experimental test of this theory is shown in Figure 3.12.

Following the foregoing paradigm, flocculation tests are carried out to experimentally determine critical flocculation concentrations. First a series of about six small test tubes of the suspended solids to be flocculated is prepared. To each is added a measured identical volume of electrolyte solution at a different concentration for each tube. These concentrations, allowing for dilution, are calculated to span the expected critical concentration for flocculation. After the tubes have been allowed to stand for a few minutes, a critical concentration is noted on the basis of the tubes in which flocculation occurred and those in which it did not. Then the flocculation test is completed by setting up a new set of tubes and adding a narrower range of electrolyte concentrations. After a longer standing time, such as 2 h, the suspensions are reagitated to break up large flocs and remix small flocs, thereby creating conditions that approach equilibrium. After another 0.5 to 1.0 h, the samples are observed once again, and the minimum concentration of electrolyte that made a visible change in the appearance of the suspensions is noted and called the critical concentration for flocculation. This concentration is then incorporated into a unit operation.

3.4.4 FLOCCULATION RATE

Besides the critical flocculation concentration of the electrolyte, the rate at which flocculation occurs is of interest, particularly for processes that are going to be scaled up. The theory describing the rate of flocculation was developed by Smoluchowski both for the case of small particles (0.1–10 μm) growing by diffusion of particles and for larger particles that grow by particles colliding and sticking together. Both these cases are described in detail in Chapter 8, Precipitation (Sections 8.3.3 and 8.3.4).

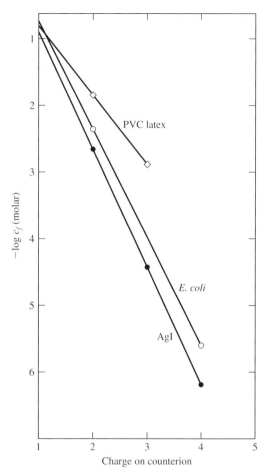

Figure 3.12 Experimental test of DLVO theory and the Schulze–Hardy rule, showing the approximate inverse sixth-power dependence of critical concentration for flocculation (c_f) on ion charge. Three different colloids were flocculated: poly(vinyl chloride) latex particles, *Escherichia coli* bacterial cells, and silver iodide colloidal particles. (Data from W. Stumm and C. R. O'Melia, "Stoichiometry of coagulation," *J. Am. Water Works Assoc.,* vol. 60, p. 514, 1968.)

3.4.5 POLYMERIC FLOCCULANTS

Polyionic polymers can be used as electrolytes in flocculation. They have the added advantage that, when they partition into the double layer, they confer a positive charge on the hydrodynamic unit, and this charge extends beyond the Debye radius, causing the particle to become electrostatically attractive to the double layers of colliding particles. Halverson and Panzer [12] have identified five mechanisms of charge-dependent flocculation:

1. Double-layer compression
2. Specific-ion adsorption
3. Sweep-flocculation enmeshment
4. Polymer charge patch formation
5. Polymer bridging

The first three mechanisms can occur regardless of whether the flocculant is a mineral ion or a polymer. A single cationic polymer molecule can neutralize several charged sites on

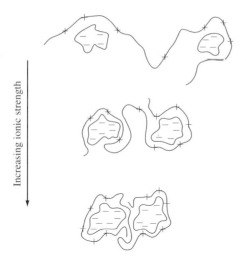

Figure 3.13 Flocculation by charge neutralization and bridging. A cationic polymer is used to collect particles with a net negative charge (characteristic of most bioparticles).

Increasing ionic strength

the particle surface; furthermore, single polymer molecules can be electrostatically attracted to more than one particle simultaneously. This behavior, shown diagrammatically in Figure 3.13, is commonly called "bridging." Figure 3.13 also shows the advantage of choosing a polymer whose charge is opposite to that of the particle.

Polymeric flocculants typically have a high molecular weight and are expensive as process chemicals go (approximately $10/kg). However, they are effective in the parts-per-million concentration range, so the total final cost may not be a serious issue. Polymer flocs usually are stronger than flocs formed by the addition of indifferent inorganic electrolytes. Some flocculating agents may have some toxicity, and this property must always be considered. Both synthetic and natural ionic polymers are available. Synthetic polymers include acrylamide–acrylic acid copolymer (anionic), polyacrylic acid (anionic), alkyl polyamines (cationic), and aminoacrylate–acrylamide copolymer (cationic). Natural ionic polymers include alginates and carboxymethylcellulose (anionic) and tannins, albumins, and diethylaminoethylcellulose (cationic).

Several neutral polymers are also good flocculants. These include poly(ethylene oxide) [aka poly(ethylene glycol)] and polyacrylamide (synthetic), as well as guar gum and its derivatives, locust bean gum, and dextrans (natural). These apparently act by partitioning into the glycocalyx (polysaccharide-containing cell coat) of intact cells owing to strong hydrogen bonding via the hydroxyl groups of the sugars on both the cells and the polymers. Their effectiveness depends on molecular weight, charge, solubility, and so on, and their interactions with cells and particles depend, therefore, on pH, ionic strength, temperature, and dry solids concentration. Unless a great deal is known about the suspended material of interest or extensive experience has been published, the choice of flocculating conditions usually depends on educated trial and error. Typically, the engineer chooses a series of flocculants and tests them at different concentrations and ionic strengths, noting the time to observed flocculation, the clarity of the supernatant after centrifugation, and the quality of the flocculated precipitate.

Real-world tests with undiluted broth must be performed to find both the optimum polymer dose and ionic strength. The optimum final dose will probably be close to that found in this procedure. The flocculant dose for a given dry weight of whole cells (in any volume) should be approximately constant and less than that for a given dry weight suspension of lysate, which may vary according to the degree and mode of lysis.

3.5 Summary

This chapter presents two major elements of cell processing—cell lysis and flocculation. Cell lysis is the process of breaking cell membranes and walls to release the cell contents. Flocculation is the process in which cells and subcellular particles are aggregated into flocs.

- Cells are classified as prokaryotic (no nucleus, such as bacteria) and eukaryotic (fungi, including yeast, and all higher organisms). Cells that do not secrete their products of interest must by subjected to lysis.

- Most prokaryotic cells and all plant and fungal cells possess cell walls that must be ruptured or digested before cells can be lysed and most bioproducts extracted. Animal cells lack cell walls and can be lysed by more benign means such as osmotic shock.

- Mechanical methods of cell lysis include applications of shear through ultrasound, high speed homogenizers, high pressure cells, and grinders such as bead mills. Nonmechanical methods of cell lysis include osmotic shock, detergents, and enzymes.

- Electrokinetic phenomena are important in the flocculation of cells and insoluble particles. Like particles repel each other only when their distance of close approach is of the order of the electric double-layer thickness, or Debye radius, which decreases with increasing ionic strength, which depends, in turn, on z^2, the square of the charge of dissolved ions. In most bioprocessing solutions the Debye radius is less than 1 nm. The critical concentration of ions to induce flocculation depends on z^{-6} according to DLVO theory.

- Above the critical ion concentration, particles approach closely enough to adhere by van der Waals forces. The Schulze–Hardy rule, on the basis of the preceding insight, predicts that required ion concentrations for flocculation are roughly in the ratio 1000:10:1 for monovalent, divalent, and trivalent ions, respectively.

- Ionic polymers induce flocculation by a combination of five mechanisms. Neutral polymers typically form hydrogen bonds with cell surface polysaccharides.

NOMENCLATURE

a constant power of pressure in equations for extent of cell disruption in high-pressure homogenizers [Equations (3.3.2) and (3.3.3)] (dimensionless)

a particle radius (cm)

A effective Hamaker constant (J)

b constant power of pressure for extent of cell disruption in Microfluidics high pressure homogenizer [Equation (3.3.3)] (dimensionless)

c concentration of solute (M, or g liter^{-1})

d shortest distance between particles [Equation (3.4.6)] (cm)

e electron charge (C)

E_a energy of attraction between spherical particles in vacuo (J)

F Faraday number (9.6485×10^4 C g-mol^{-1})

I ionic strength (M)

j constant in equation for extent of cell disruption in a bead mill and equivalent to the number of continuous stirred tank reactors in series [Equation (3.3.5)] (dimensionless)

k constant in equations for extent of cell disruption in homogenizers [Pa^{-a} in Equation (3.3.3); s^{-1} in Equations (3.3.4) and (3.3.5)]

k Boltzmann constant (1.3807×10^{-23} J K^{-1})

N number of passes of fluid through homogenizer (dimensionless)

$\mathcal{N}$ Avogadro's number (6.022×10^{23} molecules g-mol^{-1})

P pressure (Pa)

r distance from surface of spherical particle (nm)

R gas constant (8.3145 J g-mol^{-1} K^{-1})[*]

R extent of cell disruption (dimensionless)

t time (s)

T temperature (K)

U electrophoretic mobility (cm^2 s^{-1} V^{-1})

z valence of ion (dimensionless)

Greek Letters

γ constant in equation for critical flocculation concentration [Equation (3.4.5)] (dimensionless)

δ distance from Stern plane to solid surface (nm)

ε absolute dielectric constant of liquid (C^2 J^{-1} m^{-1})

ε_0 permittivity of free space (vacuum) ($= 8.8542 \times 10^{-12}$ C^2 J^{-1} m^{-1})

ε_r relative dielectric constant of liquid (dimensionless)

ζ zeta potential (J C^{-1}, or V)

κ Debye–Hückel constant (nm^{-1})

μ liquid viscosity (g cm^{-1} s^{-1})

π osmotic pressure (Pa)

ψ total potential (J C^{-1}, or V)

ψ_E electrical potential (J C^{-1}, or V)

ψ_L attractive potential (J C^{-1}, or V)

PROBLEMS

3.1   **Osmotic Cell Disruption**   Use the van't Hoff relationship to estimate the osmotic pressure drop across the membrane of a cell undergoing rupture in a 0.01 M salt solution, assuming the internal salt concentration is 0.2 M and that all salts are fully dissociated. Would you call this a negligible, ordinary, or large pressure drop? Why?

3.2   **Osmotic Disruption of Blood Cells**   It is customary to prepare hemoglobin by the osmotic lysis of red blood cells. What is the van't Hoff pressure drop across the membrane of a red blood cell that is isotonic (0.30 osM) on the inside and submerged in a 0.01% NaCl solution? What is the expected outcome of this situation?

3.3   **Breakage of Yeast Cells in a Valve-Type Homogenizer**   For the breakage of *Candida utilis* yeast cells in a valve-type continuous homogenizer, it is known that the constants in Equation (3.3.2) are $k = 5.91 \times 10^{-4}$ MPa^{-a} and $a = 1.77$ for the operating pressure range of 50 MPa $< P <$ 125 MPa. It is desired that the extent of disruption be ≥ 0.9. Plot how the number of passes varies with operating pressure over the pressure range of 50 to 125 MPa. What pressure range would you probably want to operate in? (Data from C. R. Engler and W. R. Campbell, "Disruption of *Candida utilis* cells in high pressure flow devices, "*Biotechnol, Bioeng.,* vol. 23, p. 765, 1981.)

3.4   **Estimation of the Debye Radius and the Electrostatic Potential**   Estimate the Debye radius for particles in a solution of 50 mM NaCl at 4°C. If the surface electrostatic potential of particles in this solution is 10 mV, estimate the electrostatic potential 1 nm from the surface. How significant is the electrostatic potential at this distance from the surface?

---

[*]Other values of the gas constant R with different units are given in Appendix A.

3.5 **Flocculant Choice**   In a suspension of lysate consisting of a variety of anionic particles, you find that 0.22 M KCl produces the required clarification. However, a downstream adsorption process requires <0.01 M salt, so this will not work. Knowing the Schulze–Hardy rule, you resort to $Al(NO_3)_3$ as flocculant. What concentration of this salt do you expect to need? Will this concentration fulfill the downstream requirement?

3.6 **Ionic Versus Polymeric Flocculant**   You find that you are able to flocculate yeast cell debris in 1 M NaCl in a 100 liter tank at 20°C. The use of this much salt is unsatisfactory for the next downstream step. You then find that a company markets a bridging cationic polymer for $15/kg and that this polymer (molecular weight 9000) is at least the equivalent on a molar basis of a tetravalent indifferent cation. Furthermore, this polymer can be cleared by a downstream cation exchange adsorption step that is part of the process anyway. What will it cost you to use this polymer, per tank, and how difficult would it be to convince your supervisor to pay for it? What would be the cost savings of chilling the tank to 4°C?

References

1. Belter, P. A., Cussler, E. L., and Hu, W.-S. (1988). *Bioseparations,* Wiley, New York.
2. Scopes, R. K. (1994). *Protein Purification,* 3rd ed., Springer-Verlag, New York.
3. Bollag, D. M., Rozycki, M. D., and Edelstein, S. J. (1996). *Protein Methods,* 2nd ed., Wiley-Liss, New York.
4. Hetherington, P. J., Follows, M., Dunnill, P., and Lilly, M. D. (1971). Release of protein from bakers' yeast (*Saccharomyces cerevisiae*) by disruption in an industrial homogenizer. *Trans. Inst. Chem. Eng.,* vol. 49, p. 142.
5. Engler, C. R. (1994). Cell breakage. In *Protein Purification Process Engineering,* R. G. Harrison, ed., Dekker, New York, p. 37.
6. Sauer, T., Robinson, C. W., and Glick, B. R. (1989). Disruption of native and recombinant *Escherichia coli* in a high-pressure homogenizer. *Biotechnol. Bioeng.,* vol. 33, p. 1330.
7. Limon-Lason, J., Hoare, M., Orsborn, C. B., Doyle, D. J., and Dunnill, P. (1979). Reactor properties of a high-speed bead mill for microbial cell rupture. *Biotechnol. Bioeng.,* vol. 21, p. 745.
8. Voyutsky, S. (1978). *Colloid Chemistry,* Mir Publishers, Moscow.
9. Tanford, C. (1961). *Physical Chemistry of Macromolecules,* Wiley, New York.
10. Shaw, D. J. (1980). *Introduction to Colloid and Surface Chemistry,* 3rd ed., Butterworths, London.
11. Hiemenz, P. C., and Rajagopalan, R. (1997). *Principles of Surface and Colloid Chemistry,* 3rd ed., Dekker, New York.
12. Halverson, F., and Panzer, H. P. (1980). Flocculating agents. In *Kirk-Othmer Encyclopedia of Chemical Technology,* vol. 10, 3rd ed. Wiley, New York, p. 489.

Filtration

Filtration is an operation that has found an important place in the processing of biotechnology products. In general, filtration is used to separate particulate or solute components in a fluid suspension or solution according to their size by flowing under a pressure differential through a porous medium. There are two broad categories of filtration, which differ according to the direction of the fluid feed in relation to the filter medium. In *conventional or dead-end filtration,* the fluid flows perpendicular to the medium, which generally results in a cake of solids depositing on the filter medium. In *crossflow filtration* (which is also called *tangential flow filtration*), the fluid flows parallel to the medium to minimize buildup of solids on the medium. Conventional and crossflow filtration are illustrated schematically in Figure 4.1.

Conventional filtration is typically used when a product has been secreted from cells, and the cells must be removed to obtain the product that is dissolved in the liquid. Antibiotics and steroids are often processed by using conventional filtration to remove the cells. Conventional filtration is also commonly used for sterile filtration in biopharmaceutical production. Crossflow filtration has been used in a wide variety of applications, including the separation of cells from a product that has been secreted, the concentration of cells, the removal of cell debris from cells that have been lysed, the concentration of protein solutions, the exchange or removal of a salt or salts in a protein solution, and the removal of viruses from protein solutions.

Filtration often occurs in the early stages of bioproduct purification, in keeping with the process design heuristic "remove the most plentiful impurities first" (see Chapter 11, Bioprocess Design). At the start of purification, the desired bioproduct is usually present in a large volume of aqueous solution, and it is desirable to reduce the volume as soon as possible to reduce the scale and thus the cost of subsequent processing operations. Filtration, along with sedimentation and extraction (see Chapters 5 and 6), is an effective means of accomplishing volume reduction.

In this chapter, conventional and crossflow filtration are first described mathematically as a basis for understanding how these filters perform. Next, filter media and equipment of various types are described, and mechanisms of membrane fouling are presented. The chapter concludes with a discussion of the scaleup and design of the various filtration systems.

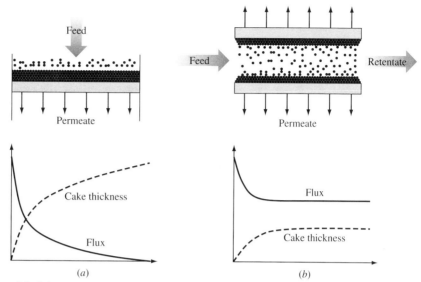

Figure 4.1 Schematic diagrams for (*a*) dead-end or conventional filtration and (*b*) crossflow filtration. For dead-end filtration the thickness of the solids buildup increases and the permeate flux decreases with time, ultimately reaching zero. In crossflow filtration the feed can contain either a soluble or a solid solute, which becomes concentrated at the membrane surface; the permeate flux reaches a constant value at steady state.

4.1 Instructional Objectives

After completing this chapter, the reader should be able to do the following:

- Calculate solvent fluxes and filtration times for conventional filtration using Darcy's law.
- Use washing theory to calculate the removal of solutes from a filter cake.
- Estimate the extent of concentration polarization in crossflow filtration.
- Select filtration media and equipment to meet bioproduct processing requirements.
- Explain the mechanisms of membrane fouling.
- Select filtration unit operations to meet product requirements, consistent with product properties.
- Perform scaling calculations on conventional and crossflow filters.
- Perform process calculations for the four basic modes of operations of crossflow filters.

4.2 Filtration Principles

The theory of filtration has been extensively developed. The goal of this section is to describe the fundamental filtration theory that is important to an understanding of how conventional and crossflow filters operate, which can then be used as a basis for design and scaleup of the unit operations.

4.2.1 CONVENTIONAL FILTRATION

When a slurry containing suspended solids flows against a filter medium by the application of a pressure gradient across the medium, solids begin to build up on the filter medium as shown in Figure 4.1a. The buildup of solids on the filter medium is called a cake. This type of filtration is sometimes referred to as "dead-end" filtration. Darcy's law describes the flow of liquid through a porous bed of solids and can be written as follows:

$$\frac{1}{A}\frac{dV}{dt} = \frac{\Delta p}{\mu_0 R} \tag{4.2.1}$$

where V is the volume of filtrate, t is time, A is the cross-sectional area of exposed filter medium, Δp is the pressure drop through the bed of solids (medium plus cake), μ_0 is the viscosity of the filtrate, and R is the resistance of the porous bed. In this case, R is a combination of the resistance R_m of the filter medium and the resistance R_c of the cake solids:

$$R = R_m + R_c \tag{4.2.2}$$

It is convenient to write the cake resistance R_c in terms of a specific cake resistance α as follows:

$$R_c = \alpha \rho_c \left(\frac{V}{A}\right) \tag{4.2.3}$$

where ρ_c is the mass of dry cake solids per volume of filtrate. Thus, the resistance increases with the volume filtered. Combining Equations (4.2.1), (4.2.2), and (4.2.3), we obtain

$$\frac{1}{A}\frac{dV}{dt} = \frac{\Delta p}{\mu_0[\alpha\rho_c(V/A) + R_m]} \tag{4.2.4}$$

For the case of zero filtrate at time zero, integration of this equation yields

$$\frac{t}{V/A} = \frac{\mu_0\alpha\rho_c}{2\Delta p}\left(\frac{V}{A}\right) + \frac{\mu_0 R_m}{\Delta p} \tag{4.2.5}$$

This is a convenient form of the integrated equation, since a plot of $t/(V/A)$ versus V/A should give a straight line. Filtration data plotted in this manner for *Streptomyces griseus* fermentation broth are shown in Figure 4.2. One noteworthy feature of the plots in Figure 4.2 is that most of the lines connecting the data at given pH values are not perfectly straight. Of the variables constituting the slope in Equation (4.2.5), the only one that could change appreciably with Δp is α, the specific cake resistance. The slopes are increasing with the volume filtered, indicating that α is increasing. However, α can increase if the cake is compressed. Cakes are typically compressible when cells and other biological materials are being filtered.

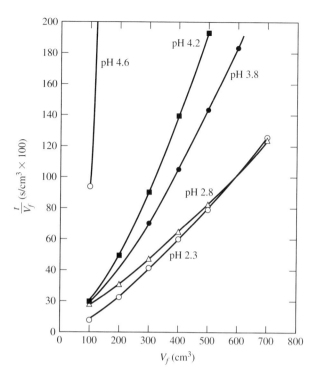

Figure 4.2 Filtration data for *Streptomyces griseus* broth with $\Delta p = 2.0$ bar. The filter medium was of cotton cloth, and diatomaceous earth filter aid was added to the broth. (Data from S. Shirato and S. Esumi, "Filtration of a culture broth of *Streptomyces griseus*" *J. Ferment. Technol. (Japan),* vol. 41, p. 87, 1963.)

EXAMPLE 4.1

Batch Filtration   A Büchner funnel 8 cm in diameter is available for testing the filtration of a cell culture suspension, which has a viscosity of 3.0 cp. The data in Table E4.1 were obtained with a vacuum pressure of 600 mm Hg applied to the Büchner funnel.

The cell solids on the filter at the end of filtration were dried and found to weigh 14.0 g. Determine the specific cake resistance α and the medium resistance R_m. Then estimate how long it would take to obtain 10,000 liters of filtrate from this cell broth on a filter with a surface area of 10 m^2 and vacuum pressure of 500 mm Hg.

TABLE E4.1

Time (min)	Volume of filtrate (ml)
26	100
96	200
197	300
342	400
537	500
692	600
989	690 (end)

SOLUTION

According to Equation (4.2.5), we can plot $t/(V/A)$ versus V/A and obtain α from the slope and R_m from the intercept. We see that the data are reasonably close to a straight line.

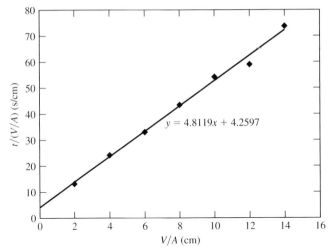

Figure E4.1 Plot of batch filtration data for the determination of α and R_m.

A linear regression of the data in this plot gives the following results (Figure E4.1):

$$\text{Slope} = 4.81 \, \frac{\text{s}}{\text{cm}^2}$$

$$\text{Intercept} = 4.26 \, \frac{\text{s}}{\text{cm}}$$

From these values, we can calculate α and R_m:

$$\alpha = (\text{slope}) \left(\frac{2 \Delta p}{\mu_0 \rho_c} \right)$$

$$= \left(4.81 \, \frac{\text{s}}{\text{cm}^2} \right) \left(\frac{2 \times 600 \, \text{mm Hg} \times \dfrac{1.33 \times 10^3 \, \text{g}}{\text{cm s}^2 \, \text{mm Hg}}}{\dfrac{0.03 \, \text{g}}{\text{cm s}} \times \dfrac{14 \, \text{g}}{690 \, \text{ml}}} \right)$$

$$= 1.26 \times 10^{10} \, \frac{\text{cm}}{\text{g}}$$

$$R_m = (\text{intercept}) \left(\frac{\Delta p}{\mu_0} \right)$$

$$= \left(4.26 \, \frac{\text{s}}{\text{cm}} \right) \times \left(\frac{600 \, \text{mm Hg} \times \dfrac{1.33 \times 10^3 \, \text{g}}{\text{cm s}^2 \, \text{mm Hg}}}{\dfrac{0.03 \, \text{g}}{\text{cm s}}} \right)$$

$$= 1.13 \times 10^8 \, \text{cm}^{-1}$$

This is a typical value of R_m for a large-pore (micrometer-sized) filter.

To determine the time required to obtain 10,000 liters of filtrate using a filter with an area of 10 m^2, we must make the assumption that α does not change at the new pressure drop of 500 mm Hg. We use Equation (4.2.5) and solve for time:

$$t = \frac{\mu_0 V}{\Delta p A}\left(\frac{\alpha \rho_c V}{2A} + R_m\right)$$

We calculate the two components of this equation as follows:

$$\frac{\mu_0 V}{\Delta p A} = \frac{\dfrac{0.03\,\text{g}}{\text{cm s}} \times 10^4\,\text{liters} \times \dfrac{1\,\text{m}^3}{10^3\,\text{liter}} \times \dfrac{100\,\text{cm}}{1\,\text{m}} \times \dfrac{1\,\text{h}}{3600\,\text{s}}}{500\,\text{mm Hg} \times \dfrac{1.33 \times 10^3\,\text{g}}{\text{cm s}^2\,\text{mm Hg}} \times 10\,\text{m}^2} = 1.25 \times 10^{-9}\,\text{cm h}$$

$$\frac{\alpha \rho_c V}{2A} + R_m = \frac{1.26 \times 10^{10}\,\dfrac{\text{cm}}{\text{g}} \times \dfrac{14\,\text{g}}{0.69\,\text{liter}} \times 10^4\,\text{liters}}{2 \times 10\,\text{m}^2\left(\dfrac{100\,\text{cm}}{\text{m}}\right)^2} + 1.36 \times 10^8\,\text{cm}^{-1}$$

$$= 12.8 \times 10^9\,\text{cm}^{-1} + 0.14 \times 10^9\,\text{cm}^{-1} = 12.9 \times 10^9\,\text{cm}^{-1}$$

and finally

$$t = (1.25 \times 10^{-9}\,\text{cm h})(12.9 \times 10^9\,\text{cm}^{-1}) = 16.1\,\text{h}$$

Thus, this filter is probably undersized for the volume to be filtered. In addition, from this calculation we see that at the end of the filtration,

$$\frac{R_m}{R_c + R_m} = \frac{0.14}{2(12.8) + 0.14} = 0.005$$

Therefore, the filter medium is contributing very little of the resistance to filtration, a typical situation in a lengthy dead-end filtration.

The specific resistance of the cake is directly affected by Δp_c, the pressure drop across the cake. Studies have shown that the relationship between specific resistance and pressure drop commonly takes the form [1]:

$$\alpha = \alpha'(\Delta p_c)^s \qquad (4.2.6)$$

where α' and s are empirical constants. The power s has been called the "cake compressibility factor" and has been found to range from zero for incompressible cakes such as sand and diatomite (see Section 4.3.1, Conventional Filtration) to near unity for highly compressible cakes.

For products that are recovered in the filtrate, it is often necessary to wash the filter cake with water or a salt solution to maximize the removal of dissolved product from the cake. Frequently, the wash must be done with more than the volume of liquid in the cake to obtain the recovery desired. This is because some of the product is in stagnant zones of the cake, and transfer into the wash liquid from these zones occurs by diffusion, which takes place at a slower rate than the convective flow of wash through the cake. Data for the washing of filter cakes have been correlated by Choudhury and Dahlstrom [2] using the following equation:

$$R' = \left(1 - \frac{E}{100}\right)^n \qquad (4.2.7)$$

where R' is the weight fraction of solute remaining in the cake after washing (on the basis that $R' = 1.0$ prior to washing), E is the percentage wash efficiency, and n is the volume of wash liquid per volume of liquid in the unwashed cake. The percentage wash efficiency E is therefore 100% minus the percentage solute remaining in the cake after 1.0 volume of wash liquid per volume of liquid in the unwashed cake (or $n = 1.0$). Choudhury and Dahlstrom found the wash efficiency E to vary from a minimum of 35% to a maximum of 86%, with the lower values applying to cakes through which the wash liquid flows relatively rapidly.

Assuming that the liquid viscosity and the pressure drop through the bed of solids are the same during washing as during the filtration of the solids, the washing rate per cross-sectional area can be found from the filtrate flow rate per unit area given in Equation (4.2.4) at the end of the filtration. Thus, for negligible filter medium resistance for filtrate volume V_f at the end of time t_f to form the cake, this yields

$$\text{Washing rate per cross-sectional area} = \left(\frac{1}{A}\frac{dV}{dt}\right)_{V=V_f} = \frac{\Delta p}{\mu_0 \alpha \rho_c (V_f/A)} \qquad (4.2.8)$$

If V_w is the volume of wash liquid applied in time t_w, then

$$V_w = t_w \left(\frac{dV}{dt}\right)_{V=V_f} \qquad (4.2.9)$$

Using the definition of $(dV/dt)_{V=V_f}$ from Equation (4.2.8), we obtain

$$\frac{V_w}{A} = t_w \left(\frac{1}{A}\frac{dV}{dt}\right)_{V=V_f} = \frac{t_w \Delta p}{\mu_0 \alpha \rho_c (V_f/A)} \qquad (4.2.10)$$

At the end of the filtration, the integrated form of the filtration equation [Equation (4.2.5)], with R_m neglected, can be written

$$\frac{V_f}{A} = \left(\frac{2\Delta p t_f}{\mu_0 \alpha \rho_c}\right)^{1/2} \qquad (4.2.11)$$

Substituting this expression for V_f/A in Equation (4.2.10) and simplifying gives

$$\frac{V_w}{A} = \left(\frac{\Delta p}{2\mu_0 \alpha \rho_c t_f}\right)^{1/2} t_w \qquad (4.2.12)$$

From Equations (4.2.11) and (4.2.12), the ratio of t_w to t_f is

$$\frac{t_w}{t_f} = \frac{2V_w}{V_f} \qquad (4.2.13)$$

It is helpful to write t_w/t_f in terms of n, the ratio of the volume V_w of wash liquid to the volume V_r of residual liquid in the cake:

$$\frac{t_w}{t_f} = 2\frac{V_w}{V_r}\frac{V_r}{V_f} = 2nf \qquad (4.2.14)$$

where f is the ratio of V_r to the volume V_f of filtrate at the end of filtration. The ratio f can be determined by a material balance. Thus, for a given cake formation time t_f, a plot of wash time t_w versus the wash ratio n will be a straight line.

4.2.2 CROSSFLOW FILTRATION

As illustrated in Figure 4.1b, the fluid in crossflow filtration flows parallel to the membrane surface, resulting in constant permeate flux at steady state. Crossflow filtration can be divided into two categories depending on whether the component being filtered is soluble or insoluble. When dissolved species such as proteins are being filtered, ultrafiltration membranes are generally used. The ultrafiltration membrane is selected so that the species of interest will not pass through these membranes. The retained species is carried to the surface of the membrane by the convective flow of fluid, and the concentration of the species builds up next to the membrane surface. The concentration of the species can be so high that it precipitates on the membrane surface, further impeding the flow of fluid through the membrane. The resulting layer of solids on the membrane surface has been called a "gel layer." In addition, even without precipitation, the increased osmolarity near the membrane surface creates a solvent gradient that opposes the applied transmembrane pressure Δp. On the other hand, when suspended particles are present, these particles are carried to the membrane surface and form a kind of cake at the surface. In this situation, microfiltration membranes are generally utilized. These membranes let dissolved components pass through but retain particles above a certain size. There are instances, such as in the crossflow filtration of cells that have been ruptured, where the layer at the membrane surface contains both suspended particles and precipitated solutes. More complete information about ultrafiltration and microfiltration membranes is given in Section 4.3.

Dissolved Species

In crossflow filtration, a solution under pressure flows across the surface of a membrane. As a result of this pressure, fluid is forced through the membrane. This flow toward the surface of the membrane causes species dissolved in the solution also to be carried toward the membrane's surface. For a solute that is rejected by the membrane, there will be a concentration gradient of this solute across a stagnant boundary layer next to the surface of the membrane, as indicated in Figure 4.3. The elevation of the solute concentration at the membrane surface (c_w) compared to that in the bulk solution (c_b) is known as *concentration polarization.*

At steady state, the rate of convective mass transfer of solute toward the membrane surface must be equal to the rate of mass transfer of solute by diffusion away from the membrane surface, which is described by [3]:

$$Jc = -\mathscr{D}\frac{dc}{dx} \qquad (4.2.15)$$

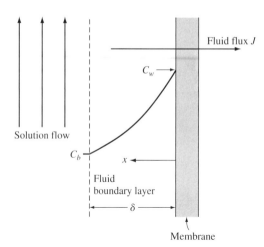

Figure 4.3 Schematic representation of the boundary layer in crossflow filtration with a dissolved solute in the feed. A solution flows parallel to the membrane surface, and fluid flows through the membrane under the influence of pressure. A fluid boundary layer forms next to the membrane surface, creating a gradient in the concentration c of solute.

where J is the transmembrane fluid flux, c is the concentration of the solute, and $\mathcal{D}$ is the diffusion coefficient of the solute. For a boundary layer thickness of δ, the solution of Equation (4.2.15) is

$$J = \frac{\mathcal{D}}{\delta} \ln \frac{c_w}{c_b} \tag{4.2.16}$$

which can be also written as

$$\frac{c_w}{c_b} = \exp\left(\frac{J\delta}{\mathcal{D}}\right) \tag{4.2.17}$$

The term $\mathcal{D}/\delta$ can also be defined as a mass transfer coefficient k. The ratio c_w/c_b is sometimes called the *polarization modulus* and indicates the extent of concentration polarization. From Equation (4.2.17), it can be seen that the polarization modulus is particularly sensitive to changes in J, δ, and $\mathcal{D}$ because of the exponential functionality involved. For high molecular weight solutes (small $\mathcal{D}$) and membranes with high solvent permeability (high J), concentration polarization can become severe, with $c_w/c_b > 10$. At high concentration polarization levels, the solubility of the solute can be exceeded, resulting in the precipitation of the solute and the formation of a solids or gel layer on the membrane surface.

Experimental data have been obtained to support this simple model of concentration polarization. Figure 4.4 shows data for the ultrafiltration of the protein casein and Dextran with a molecular weight of 110,000. For both species, a plot of ultrafiltration flux versus log of bulk concentration gives a straight line for bulk concentration varying by an order of magnitude. Estimates of the concentration of solute at the membrane surface can be obtained from the intercept of the straight line in Figure 4.4 at zero flux. These estimates of c_w appear to be physically reasonable.

Correlations have been developed for the mass transfer coefficient k. For laminar flow, boundary layer theory has been applied to yield analytical solutions (known as the Leveque or Graetz solutions) for k [4]:

$$k = 0.816 \left(\gamma_w \frac{\mathcal{D}^2}{L}\right)^{1/3} \tag{4.2.18}$$

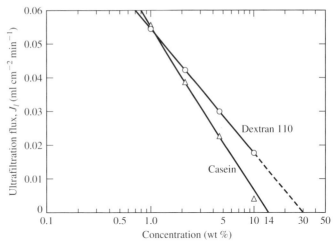

Figure 4.4 Decline in ultrafiltration flux with increasing concentration for solutions of casein and Dextran (MW 110,000). Data were obtained in thin-channel recirculating flow cells. (Data from W. F. Blatt, A. Dravid, A. S. Michaels, and L. Nelsen, "Solute polarization and cake formation in membrane ultrafiltration: Causes, consequences, and control techniques," in *Membrane Science and Technology,* J. E. Flinn, ed., p. 63, Plenum Press, New York, 1970.)

where γ_w is the fluid shear rate at the membrane surface and L is the length of the flow channel over the membrane. The constant 0.816 is applicable for the gel-polarized condition of concentration polarization where the solute concentration at the wall is constant. For a rectangular slit of height $2h$ and bulk stream velocity u_b,

$$\gamma_w = \frac{3u_b}{h} \tag{4.2.19}$$

and the equation for a circular tube of diameter D is

$$\gamma_w = \frac{8u_b}{D} \tag{4.2.20}$$

For turbulent flow, empirical correlations have been developed for the mass transfer coefficient. These correlations are based upon dimensional analysis of the equations of change for forced convection mass transfer in a closed channel, which gives [5]:

$$\mathrm{Sh} = \frac{kD_h}{\mathscr{D}} = f(\mathrm{Re}, \mathrm{Sc}, L/D_h) \tag{4.2.21}$$

where $\mathrm{Re} = \dfrac{D_h u_b \rho}{\mu}$ = Reynolds number

$\mathrm{Sc} = \dfrac{\mu}{\rho \mathscr{D}}$ = Schmidt number

D_h = equivalent diameter of the channel

$\quad = 4\left(\dfrac{\text{cross-sectional area}}{\text{wetted perimeter}}\right)$

u_b = bulk stream velocity

ρ = density of fluid

μ = viscosity of fluid

A typical correlation that has been developed for this Sherwood number is [6]:

$$\text{Sh} = 0.082\,\text{Re}^{0.69}\text{Sc}^{0.33} \qquad (4.2.22)$$

Note that the L/D_h term in the generalized dimensionless analysis solution is left out of the empirical correlation in Equation (4.2.22).

EXAMPLE 4.2

Concentration Polarization in Ultrafiltration   Equipment is available for ultrafiltration of a protein solution at constant volume to remove low molecular weight species (achieved by the addition of water or buffer to the feed in an operation called diafiltration—see later, Example 4.5). The flow channels for this system are tubes 0.1 cm in diameter and 100 cm long. The protein has a diffusion coefficient of 9×10^{-7} cm^2/s. The solution has a viscosity of 1.2 cp and a density of 1.1 g/cm^3. The system is capable of operating at a bulk stream velocity of 300 cm/s. At this velocity, determine the polarization modulus for a transmembrane flux of 45 liters m^{-2} h^{-1}.

SOLUTION

We see from Equation (4.2.17) that we can determine c_w/c_b if we know $J\delta/\mathcal{D} = J/k$, where J is as given in the problem statement. We can determine k from either Equation (4.2.18) or Equations (4.2.21) and (4.2.22), depending on whether the flow is laminar or turbulent. We first need to calculate the Reynolds number to characterize the flow regime:

$$\text{Re} = \frac{D_h u_b \rho}{\mu} = \frac{0.1\,\text{cm} \times \dfrac{300\,\text{cm}}{\text{s}} \times 1.1\,\dfrac{\text{g}}{\text{cm}^3}}{\dfrac{0.012\,\text{g}}{\text{cm s}}} = 2750$$

In the calculation of the Reynolds number, we see that the equivalent diameter of the channel, D_h, is the same as the diameter of the tubes. The flow is turbulent, since the Reynolds number is greater than 2000. For turbulent flow, we need to know Sc, the Schmidt number:

$$\text{Sc} = \frac{\mu}{\rho \mathcal{D}} = \frac{\dfrac{0.012\,\text{g}}{\text{cm s}}}{1.1\,\dfrac{\text{g}}{\text{cm}^3} \times 9 \times 10^{-7}\,\dfrac{\text{cm}^2}{\text{s}}} = 1.21 \times 10^4$$

From Equations (4.2.21) and (4.2.22),

$$k = \frac{\mathcal{D}\,\text{Sh}}{D_h}$$

$$= \frac{\mathcal{D}}{D_h}(0.082)\,\text{Re}^{0.69}\text{Sc}^{0.33}$$

$$= \left(\frac{9 \times 10^{-7}\,\dfrac{\text{cm}^2}{\text{s}}}{0.1\,\text{cm}}\right)(0.082)(2750)^{0.69}(12{,}100)^{0.33} = 3.88 \times 10^{-3}\,\frac{\text{cm}}{\text{s}}$$

From Equation (4.2.17), the polarization modulus can be estimated:

$$\frac{c_w}{c_b} = \exp\left(\frac{J\delta}{\mathscr{D}}\right) = \exp\left(\frac{J}{k}\right)$$

$$= \exp\left(\frac{45\,\dfrac{\text{liters}}{\text{m}^2\,\text{h}} \times \dfrac{10^3\,\text{cm}^3}{\text{liter}} \times \dfrac{1\,\text{m}^2}{10^4\,\text{cm}^2} \times \dfrac{1\,\text{h}}{3600\,\text{s}}}{3.88 \times 10^{-3}\,\dfrac{\text{cm}}{\text{s}}}\right) = 1.38$$

Thus, the concentration polarization is not severe.

Concentration polarization at the surface of the membrane can become great enough that it creates a significant resistance to fluid flow. This effect on the transmembrane fluid flux J can be modeled by using Darcy's law [Equation (4.2.1)] with the flow resistance made up of the sum of the membrane resistance R_m and the resistance R_p of the polarized boundary layer and any gel layer next to the surface of the membrane. For membranes, a correction needs to be made for the osmotic pressure $\Delta\pi$ of the solute that is being filtered, which results in the following equation:

$$J = \frac{\Delta p - \sigma\,\Delta\pi}{\mu_0(R_m + R_p)} \tag{4.2.23}$$

where Δp is the pressure difference between the bulk fluid and the permeate, μ_0 is the viscosity of the permeate, and σ is the reflection coefficient for the solute. A reflection coefficient of 1.0 indicates no passage of solute through the membrane, while a coefficient of zero indicates free passage of the solute with the solvent through the membrane. For an ideal dilute solution,

$$\Delta\pi = RTc_w \tag{4.2.24}$$

where R is the ideal gas constant, T is the absolute temperature, and c_w is the solute concentration at the surface of the membrane.

Data for J versus Δp for the ultrafiltration of serum albumin in a stirred filtration cell are shown in Figure 4.5 as a function of both cell stirrer speed and protein concentration. At high values of Δp for 3.9 and 6.5% protein concentrations, the ultrafiltration flux J is constant with increasing Δp. Constant flux as Δp increases can be interpreted as a condition in which the solute concentration at the membrane surface c_w has reached a solubility limit; this behavior is supported by Equation (4.2.16), where J becomes constant when c_w reaches a constant.

Suspended Particles

In the crossflow filtration of mixtures with suspended particles, the fluid flux toward the membrane carries particles to the membrane surface, where they are rejected and form a cake layer that is analogous to the gel layer in the ultrafiltration of dissolved species. As indicated in Figure 4.6 for crossflow filtration with microporous filtration membranes, the cake layer initially grows with time, thus reducing the permeate flux and constricting the channel. At steady state conditions, the layer reaches a constant thickness, which is relatively

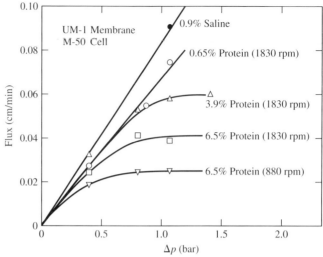

Figure 4.5 Ultrafiltration flux as a function of pressure drop in a stirred cell with varying protein concentration and stirring rate. (Data from E. S. Perry, ed., *Progress in Separation and Purification,* vol. 1, p. 318, Wiley, New York, 1968.)

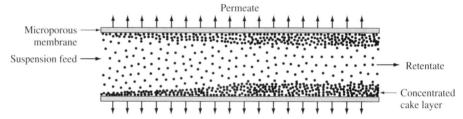

Figure 4.6 Illustration of the buildup of a cake layer at the membrane surface in crossflow filtration with suspended solids in the feed.

thin if the shear exerted by the fluid flowing tangentially to the membrane surface is high enough.

The theory for the crossflow filtration of dissolved species has been found to hold only for very small suspended particles up to approximately 1 μm in size [7]. Beyond this size, experimental membrane flux values are often one to two orders of magnitude higher than those obtained from the Leveque or Graetz theory given by Equation (4.2.18) [8]. Two different theories have been developed to more accurately predict membrane fluxes for particles larger than 1 μm, depending on the size of the particles: a shear-induced diffusion theory for small particles and an inertial lift theory for larger particles.

The shear-induced diffusion theory was developed by Zydney and Colton [9], who proposed that the concentration polarization model could be applied to the crossflow filtration of suspended particles if the Brownian diffusivity was replaced by a shear induced hydrodynamic diffusivity. In shear-induced hydrodynamic diffusion, the particles are randomly displaced from the streamlines in a shear flow and collide with other particles. In their model, Zydney and Colton employed an approximate relationship for the shear-induced diffusion $\mathcal{D}_s$ of spherical particles measured by Eckstein et al. [10] for $0.2 < \phi < 0.45$

$$\mathcal{D}_s = 0.3\gamma_w a^2 \qquad (4.2.25)$$

where ϕ is the particle volume fraction in the bulk suspension, a is the particle radius, and γ_w is the fluid shear rate at the membrane surface. Substituting for the diffusivity in the equation for the laminar flow mass transfer coefficient in the concentration polarization model [Equation (4.2.18)], we obtain

$$k = 0.366\gamma_w \left(\frac{a^4}{L}\right)^{1/3} \qquad (4.2.26)$$

where L is the tube or channel length. Comparing this equation for k with the one for dissolved species in laminar flow, we see that the mass transfer coefficient is much more strongly dependent on the shear rate at the wall. Shear-induced diffusion has been found to dominate for particles up to 30 to 40 μm in size, while inertial lift dominates for larger particles [7].

Inertial lift arises when the Reynolds number based on the particle size is not negligible. Inertial lift produces a velocity that carries particles away from the membrane surface. For fast laminar flow of dilute suspensions with thin fouling layers, the steady state transmembrane flux predicted by the inertial lift theory is [7]:

$$J = \frac{0.036\rho_0 a^3 \gamma_w^2}{\mu_0} \qquad (4.2.27)$$

where ρ_0 and μ_0 are the density and viscosity, respectively, of the permeate. Thus, for transport by inertial lift, the flux is strongly dependent on the particle size and shear rate at the membrane surface and not dependent at all on the length of the filter or on the concentration of particles in the bulk suspension.

4.3 Filter Media and Equipment

In the development of a filtration process, much effort is often devoted to the evaluation of the filter media and equipment that are available. The filter media and equipment selected can have a large impact on the process economics, both in terms of the capital outlay required and the operating expenses. Filter media and equipment considerations are discussed in this section for both conventional (or dead-end) and crossflow filtration.

4.3.1 CONVENTIONAL FILTRATION

Filter Media

There are many different filter media [11, 12], so the choice of the type to use can be difficult. The choice, however, is heavily governed by the type of filtration equipment being used. The filter media most commonly used in the conventional filtration of biotechnology products are woven fabrics, metal fabrics or screens, and rigid porous media. Sterile filtration in biopharmaceutical production is usually carried out in the conventional or dead-end mode, and the medium is either a membrane filter or a depth filter, depending on whether a liquid solution or air is being filtered.

Woven fabrics are available in a wide variety of materials. The chemical resistance of the fabric is a primary consideration for the filtration of biotechnology products. There should be minimal or no leaching of components from the fabric during filtration, and the fabric should be able to withstand any chemicals used during cleaning. For filtration involving a filter cake, woven fabrics are the most common type of filter medium. The minimum size of particles trapped in woven fabrics is $10\,\mu$m.

Metal fabrics or screens are available in several types of weave in nickel, copper, brass, aluminum, steel, stainless steel, Monel, and other alloys. Metal media are especially desirable when good resistance to leaching and corrosion is required. The smallest size of particles trapped in woven wire media is $5\,\mu$m.

Rigid porous media are available in sheets and tubes. A wide variety of materials are available for rigid porous media, including sintered stainless steel and other metals, silica, porcelain, and some plastics.

For the sterile filtration of solutions, an asymmetric membrane filter (see Section 4.3.2) is used in cartridge style. Typically, these membranes are made of cellulose esters or other polymers and have pore sizes of 0.22 or 0.45 μm. In the sterile filtration of air, two types of filters are typically found. High efficiency particulate air (HEPA) filters are large, high throughput ventilation filters used to reduce the particulate load in a room. HEPA filters are typically depth type filters that can collect microbes and other airborne particulates at high flow rates. Depth filters consist of compacted beds of pads of fibrous material such as glass wool. The other type of filter for the sterile filtration of air is a vent filter, installed on all equipment that needs to be filled or drained, such as product tanks. Vent filters are typically hydrophobic asymmetric membrane filters, rated at 0.22 μm.

Filter Aids

It is possible to significantly improve the filtration operation by adding a powdered solid, called a filter aid, to a feed that is to be filtered. The filter aid can be either added directly to the feed or applied to the filtration equipment. The objectives of the first approach, the admixture of the filter aid powder to the suspension, are to maintain the pores in the filter cake open, to make the cake less compressible, and to provide faster filtration. In the second approach the filter medium is precoated with a layer of the filter aid. Rotary vacuum filters can be operated with "precoat" (see the next section).

A successful filter aid must be highly permeable. Other important properties of filter aids that should be considered are surface properties, chemical resistance, and compatibility with the product.

Two of the most widely used types of filter aid are diatomite (kieselguhr or diatomaceous earth) and perlite. Diatomite is the skeletal remain of diatoms, which are a type of algae. The skeletal structure of diatomite particles gives them a high intrinsic permeability. Diatomite consists mostly of silica and thus is insoluble in both strong acids and strong alkalies. Perlite, a glassy volcanic material consisting mainly of aluminum silicate with some combined water, expands on heating during processing to give highly porous particles. Perlite is sometimes used for rough filtrations where high flow rates are desirable and less significance is given to filtrate clarity.

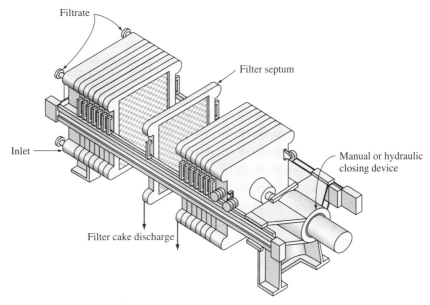

Filtrate

Filter septum

Inlet

Manual or hydraulic
closing device

Filter cake discharge

Figure 4.7 Plate-and-frame filter press.

Equipment

A large variety of conventional filtration equipment is available. This equipment can be classified according to the driving force for filtration, whether by pressure, vacuum, or gravity. Conventional filtration equipment may be designed for batch, semicontinuous, and continuous operation.

One of the most common types of batch filter is the plate-and-frame filter press (Figure 4.7). This filter consists of a varying number of filter chambers with either medium-covered recessed filter plates, or medium-covered plates alternated with frames that provide space for the cake. The chambers are closed and tightened by a hydraulic ram or screw that pushes the plates and frames together. The feed to be filtered enters the filter press under pressure and fills each chamber simultaneously. The liquid passes through the filter medium, and the solids are retained on the medium. The clear filtrate is taken off at a discharge outlet. Mechanized systems are available for opening and closing of the filter press at the end and commencement of each filtration cycle. Because the filter cake is open to the atmosphere at the end of each cycle, this filter should not be used when the filter cake poses a biohazard.

Other types of batch filter are the horizontal or vertical leaf filter, plate filter, tray filter, tube or candle filter, and Nutsche filter. These devices are similar in that they are all pressure filters, with the filter medium being enclosed within a pressure vessel. Because these filters are enclosed, they are advantageous when one is working with materials that are biohazards. A vertical leaf filter is shown as an example in Figure 4.8. The filter leaves are covered with the filter medium on either side, and the filtrate passing into the inside of each filter leaf flows into a filtrate manifold. The Nutsche filter, consisting of one horizontal medium-covered filter plate dividing half of a tank, is the simplest type of pressure filter.

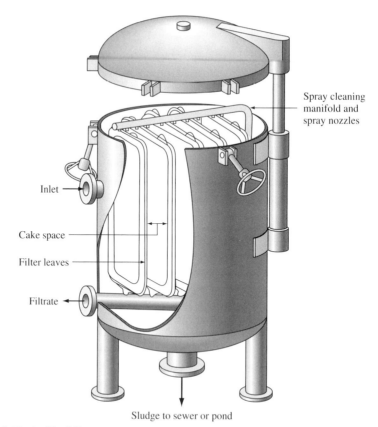

Figure 4.8 Vertical leaf filter.

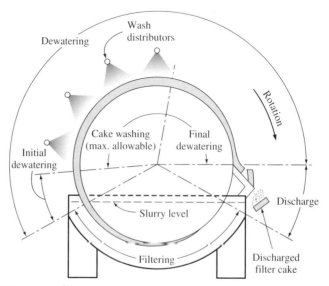

Figure 4.9 Rotary vacuum filter.

The most common type of semicontinuous or continuous filter for biotechnology products is the rotary vacuum filter. The operation of a rotary vacuum filter is illustrated in Figure 4.9. The drum rotates slowly about its horizontal axis and is partially submerged in a slurry reservoir. The surface of the drum is covered with the filter medium, and a vacuum is maintained below the medium on the inside of the drum. Liquid is sucked through the filter medium, and solids in the feed are retained on the medium. Filtration is followed by washing and dewatering. Several different types of cake discharge can be used: simple knife discharge, advancing knife discharge (with precoat filtration), belt or string discharge, and roller discharge. Because the feeds for biotechnology applications tend to be slimy and nongranular, precoat filtration is commonly used during rotary vacuum filtration. In precoat filtration, the filter medium is coated with a thick layer of diatomaceous earth or other type of filter aid. The solids being filtered penetrate a small distance into the precoat. The knife is adjusted to remove all the feed solids on the filter, including any precoat penetrated with feed solids to give a fresh surface of precoat. When the thickness of the precoat becomes small, a fresh batch of precoat is applied to the drum.

4.3.2 CROSSFLOW FILTRATION

Filter Media (Membranes)

Filter media for crossflow filtration are generally referred to as membranes. There are two general categories of membranes: ultrafiltration membranes and microporous membranes. The separation between ultrafiltration (UF) and microfiltration (MF) is based on the pore size of the membrane, with membranes having pores 0.1 μm and larger considered to be microporous. The pore sizes of UF membranes are in the range of 0.001 to 0.1 μm [13], although these membranes are usually classified by their molecular weight cutoff (MWCO), which is the molecular weight of a globular solute at which the solute is rejected by the membrane. Typically, a 90% rejection level is used for establishing the MWCO. Ultrafiltration membranes can be obtained down to a MWCO level of 1000 Daltons (Da) and up to as high as 1,000,000 Da. There is also a category of membranes called reverse osmosis (RO) or hyperfiltration membranes that pass only water and a very low flux of solutes. A very good RO membrane rejects 99.7% or more of sodium chloride [11].

Three basic structures are commonly used for membranes: homogeneous, asymmetric, and composite. These three types are illustrated in Figure 4.10. The homogeneous structure has no significant variation in pore diameter from the filtering surface to the other side. In the asymmetric structure, there is a thin layer next to the filtering surface that has very small pores. Below this thin layer is a much thicker layer that has much larger pores and serves as structural support for the membrane. The composite membrane is similar to the asymmetric membrane in having a thin layer containing very small pores next to the filtering surface;

(a)   (b)   (c)

Figure 4.10 Three commonly used membrane structures: (*a*) homogeneous: uniform pore profile through filter, (*b*) asymmetric: finer filtering surface faces feed suspension, and (*c*) composite: two types of materials used.

however, the thin and thick layers of this membrane are made of two different types of material.

Filtration membranes are made from a wide variety of polymers and inorganic materials. The polymers that are used include cellulose acetate, polyamide, polyether, polycarbonate, polyester, polypropylene, polyethylene, regenerated cellulose, poly(vinyl chloride), poly(vinylidene fluoride) (PVDF), poly(tetrafluoroethylene) (PTFE), acrylonitrile copolymers, and polysulfones. The inorganic materials used include ceramics, zirconium oxide, borosilicate glass, stainless steel, and silver.

Equipment

Crossflow filtration membranes are available in a variety of configurations. The membranes are housed in a physical unit called a *module*. The membrane module must satisfy a number of mechanical, hydrodynamic, and economic requirements, the most important of which are the following [13]:

Mechanical: obtain effective (physical) separation of the feed and permeate streams; provide the necessary physical support for the membrane (including the ability of the module to endure the required pressure drops and any backflushing).

Hydrodynamic: minimize pressure drops through the module (to reduce pumping costs); optimize solute mass transfer (reduce concentration polarization); minimize particulate plugging or fouling; avoid dead spots (for sanitary design).

Economic: maximize membrane packing density (ratio of membrane area to module volume); minimize manufacturing costs; permit easy access for cleaning and/or membrane replacement; provide sufficient chemical resistance and operational lifetime; incorporate modularity of design for easy scaleup, staging, or cascading.

Several of these criteria are in mutual opposition: for example, modules with high membrane packing density tend to be highly susceptible to plugging with particulates. Therefore, the choice of a particular module involves balancing these criteria to arrive at the most economic system for each particular application.

There are five types of module configuration for crossflow filtration (Figure 4.11): hollow fiber, tubular, flat plate, spiral wound, and rotating. The tubular module has the same general configuration as the hollow fiber module, but the tubes have much larger diameters than the fibers. Some key characteristics of these modules are compared in Table 4.1.

Hollow fiber modules consist of an array of narrow-bore, self-supporting fibers that generally have an asymmetric membrane structure. The dense skin layer is usually on the lumen side of the fiber, but it can be placed on the outside. The feed flows through the lumen of the fibers when the dense skin layer is on the lumen side, and the flow is typically laminar. Since the hollow fibers are self-supporting, they can be cleaned by backflushing, that is, by reversing the direction of the permeate flow. One disadvantage of hollow fiber modules is that the entire module usually needs to be replaced upon the rupture of even a single fiber. Also, to avoid plugging of the small-diameter fibers, feed streams generally need to be prefiltered.

Except for some inorganic membranes, tubular membranes are not self-supporting and are usually cast in place within a porous support tube made of fiberglass, ceramic, plastic,

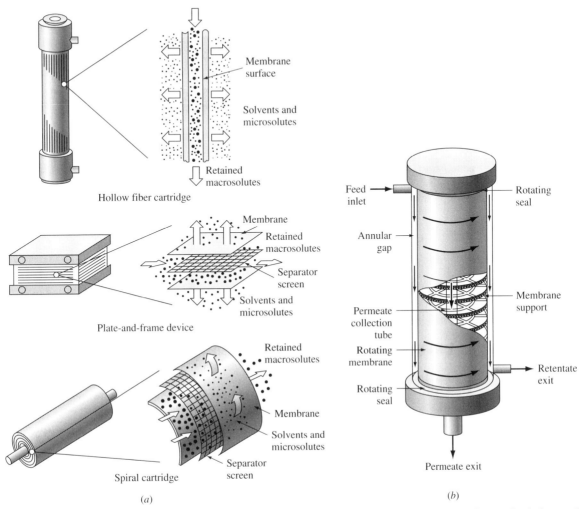

Figure 4.11 Schematic representations of filter modules. (*a*) Hollow fiber, plate, and spiral-wound membrane modules. (*b*) A rotating cylinder module.

TABLE 4.1
Comparison of Key Characteristics of Crossflow Membrane Modules[a]

Module type	Channel spacing (cm)	Packing density (m²/m³)	Energy costs	Particulate plugging	Ease of cleaning
Hollow fiber	0.02–0.25	1200	Low	High	Fair
Tubular	1.0–2.5	60	High	Low	Excellent
Flat plate	0.03–0.25	300	Moderate	Moderate	Good
Spiral wound	0.03–0.1	600	Low	Very high	Poor–fair
Rotating	0.05–0.1	10	Very high	Moderate	Fair

[a]See reference 13.

or stainless steel. The feed flows through the inside of the tubes, while the permeate flows radially outward across the membrane and support tube. Inorganic membranes are usually constructed in a honeycomb monolith unit in which the membranes are arranged in a parallel array. Tubular systems are typically operated in the turbulent flow regime. Because of the large diameter of the tubes in these systems, the pumping costs are relatively high. The chief advantages of tubular modules are that they are highly resistant to plugging by particulates, and they can be cleaned by backflushing. The diameter of the tubes can be selected to avoid prefiltration of the feed.

Most flat plate systems are in a rectangular configuration. The flow channels can be open or can have separator screens to improve the mass transfer characteristics. Units with open channels are usually operated in laminar flow, while those with separator screens are often operated with flow in the turbulent regime. One disadvantage of these modules is that backflushing is often impractical because the membranes are effectively supported only on one side.

Spiral-wound modules are constructed from flat sheet membranes separated by spacer screens. The feed solution is fed into one end of the module and flows through the separator screens along the surface of the membranes. The retentate is then collected on the other end of the module. The permeate spirals radially inward, eventually to be collected through a central tube. The flow in these systems tends to be turbulent. The main disadvantage of spiral-wound modules is that they are susceptible to fouling by particulates because of the narrow and irregular flow through the spacers.

One effective rotating-type module is the rotating cylinder device originally proposed by Hallstrom and Lopez-Leiva [14]. The feed flows into a thin annular region between two concentric cylinders (Fig. 4.11b). Either cylinder or both can be porous and can have a membrane bound to the surface. With the membrane on the inner cylinder, permeate is collected in the central chamber. When the membrane is on the outer cylinder, permeate is collected in a separate annular permeate region adjacent to the outer porous cylinder. The inner cylinder is rotated at high speed (typically > 3000 rpm) to induce the formation of Taylor vortices, which effectively mix the fluid near the surface of the membrane and thereby increase the mass transfer coefficient for the solutes/particles in the feed. The vortices also help reduce the thickness of the cake or gel layer at the membrane surface. One advantage of this system is that the rate of mass transfer is determined almost entirely by the rate of rotation of the inner cylinder, which effectively decouples the mass transfer characteristics from the feed flow rate. As a result, the rotating devices can be operated at very low flow rates and with minimal pumping costs. Since, however, the energy requirements for rotating the device are usually very high, capital costs are high, and scaleup is very difficult, these systems have been limited to small-scale operation.

4.4 Membrane Fouling

Fouling of membranes results from physical and/or chemical interactions between the membrane and various components that are present in the process stream. Fouling leads to a decline in the permeate flux and a change in the membrane selectivity. Because of the high affinity of proteins for solid surfaces, proteins are the leading contributor to fouling in bioseparations by UF and MF. The flux decrease caused by proteins can be attributed to one or more of the following processes [7]:

- Protein adsorption, which involves the interaction of the proteins and the membrane and occurs with no convective flow through the membrane
- Protein deposition, which is the addition of more protein that is associated with the membrane, over and above the protein that would be adsorbed in a nonflowing system
- Protein accumulation on the membrane surface exposed to the flowing process stream as a result of concentration polarization, which can lead to formation of a gel layer

This last effect is governed by the mass transfer coefficient [see Equations (4.2.18) and (4.2.22)], and thus can be minimized by increasing the fluid shear rate at the filtering surface of the membrane. These three processes can lead to pores in membranes that are constricted and even completely blocked, as illustrated in Figure 4.12.

   Protein adsorption studies have been performed for both MF and UF membranes to determine the extent to which the pores are constricted by adsorbed protein. To eliminate concentration polarization effects, these studies were done in the absence of flow. The work with MF membranes and a variety of proteins showed that there was approximately monolayer adsorption throughout the internal pores [7]. Protein adsorption was reduced on the more hydrophilic membranes such as the hydrophilic PVDF type, with only a fraction of monolayer adsorption. A study of the adsorption of bovine serum albumin (molecular weight 69,000 Da) on UF membranes indicated the following extents of protein coverage within the ultrathin skin portion of the membrane: monolayer adsorption for the 300,000 Da MWCO membrane, only about half a monolayer for the 100,000 Da MWCO membrane, and no measurable protein adsorption for the 50,000 Da MWCO membrane [15]. Thus, protein fouling within the membrane can be completely eliminated by making the pore size small enough.

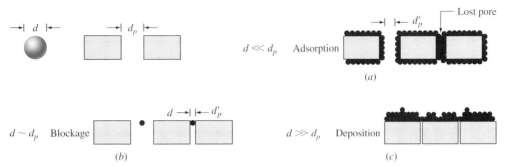

Figure 4.12 Mechanisms of membrane fouling. (*a*) Pore narrowing and constriction. (*b*) Pore plugging. (*c*) Solute deposition and formation of a gel or cake layer. Dimensions: d, protein or particle diameter, d_p, clean pore diameter, d'_p, effective pore diameter in the presence of adsorbed proteins or particles.

4.5 Scaleup and Design of Filtration Systems

Experimental testing is generally needed in the scaleup and design of conventional and crossflow filtration systems. This is especially true for bioprocesses because of the unique features of each bioprocess that is being considered. For example, in the filtration of ruptured cells, there can be great differences in the filtration characteristics of ruptured cells of different types; these differences can even exist for cells of the same class (e.g., yeasts). Some of the more common small-scale tests are described, as well as considerations in the design of filtration systems.

4.5.1 CONVENTIONAL FILTRATION

Small-scale testing procedures are described for the two basic types of conventional filtration, vacuum filtration and pressure filtration. A large majority of all continuous filters use vacuum as the driving force for filtration. Occasionally, however, an enclosed pressure filter is needed in a bioprocess because of the presence of biohazards. The design of sterile filtration systems is also discussed.

Vacuum and Pressure Filtration

To develop data for scaling up the rotary vacuum filter (Figure 4.9), tests with a filter leaf are usually performed. The objective is to simulate the operation of a large-scale rotary vacuum filter (see the stages of operation indicated in Figure 4.9). A laboratory system for doing vacuum filter leaf tests with a precoat is shown schematically in Figure 4.13; filter leaves that do not use a precoat are similar. To do a filter leaf test with the leaf shown in Figure 4.13, the leaf should first be precoated with a filter aid. The precoat is trimmed off until it is flat with the top of the cake ring of the leaf. The leaf is then submerged in the broth being filtered for the amount of time that the rotating drum will be submerged. To simulate the washing segment of the rotating drum, the leaf is submerged in water or other wash liquid for the amount of time of washing for the rotating drum. The leaf is removed from the wash liquid, and air is allowed to be pulled through the cake to dewater it. The cake ring is then screwed down to expose precoat that has not been penetrated by the broth solids, and the solids above the top of the ring are cut off with a knife. The leaf now has a fresh surface of precoat and is ready for another cycle of filtration, washing, and dewatering. This procedure can be repeated for as many cycles as desired. The volume of filtrate and weight of the cake can be measured, and analyses can be performed on the filtrate and cake. For example, if it were desired to recover a protein in the filtrate, the cake solids and the filtrate (plus wash) should be analyzed for the protein.

| EXAMPLE 4.3 | **Rotary Vacuum Filtration**   It is desired to filter a cell broth at a rate of 2000 liters/h on a rotary vacuum filter (see Figure 4.9) at a vacuum pressure of 70 kPa. The cycle time for the drum will be 60 s, and the cake formation time (filtering time) will be 15 s. The broth to be filtered has a viscosity of 2.0 cp and a cake solids (dry basis) per volume of filtrate of 10 g/liter. From laboratory tests, the specific cake resistance has been determined to be |

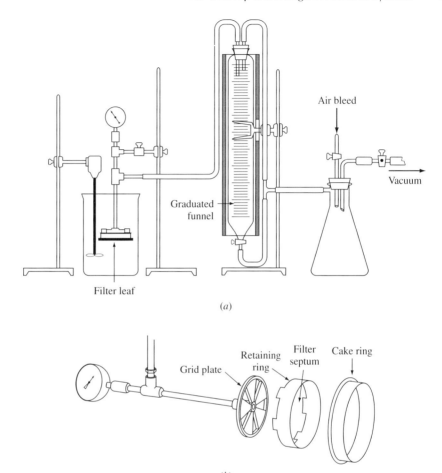

Air bleed

Vacuum

Graduated
funnel

Filter leaf

(a)

Grid plate

Retaining
ring

Filter
septum

Cake ring

(b)

Figure 4.13 (a) Filter leaf apparatus and (b) detail of a precoat filter leaf.

9×10^{10} cm/g. Determine the area of the filter that is required. The resistance of the filter medium can be neglected.

SOLUTION

We can use the integrated form of the filtration equation, Equation (4.2.5), with $R_m = 0$:

$$\frac{t}{V/A} = \frac{\mu_0 \alpha \rho_c}{2 \, \Delta p} \left(\frac{V}{A} \right)$$

We solve for A^2 to obtain

$$A^2 = \frac{\mu_0 \alpha \rho_c V^2}{2 \, \Delta p t}$$

In applying this equation, it is helpful to focus on the area of the drum that is submerged, which is where the cake is being formed and where filtrate is being obtained. Thus, A is the area of that part of the drum that is submerged. We can calculate the volume of filtrate that needs to be collected during the cake formation time of 15 s:

$$V_{15\,s} = \frac{2000\,\text{liters}}{\text{h}} \times 15\,\text{s} \times \frac{1\,\text{h}}{3600\,\text{s}} = 8.33\,\text{liters}$$

We use this volume of filtrate with $t = 15$ s in the equation for A^2 to obtain

$$A^2 = \frac{\dfrac{0.02\,\text{g}}{\text{cm s}} \times 9 \times 10^{10}\,\dfrac{\text{cm}}{\text{g}} \times 10\,\dfrac{\text{g}}{\text{liter}} \times (8.33\,\text{liters})^2 \times \dfrac{\text{cm s}^2\,\text{kPa}}{10^4\,\text{g}} \times \dfrac{10^3\,\text{cm}^3}{\text{liter}} \times \left(\dfrac{\text{m}}{10^2\,\text{cm}}\right)^4}{2 \times 70\,\text{kPa} \times 15\,\text{s}}$$

$$= 0.595\,\text{m}^4$$

$$A = 0.771\,\text{m}^2$$

The area A' of the entire rotary vacuum filter can be calculated from the cake formation time and the total cycle time as

$$A' = \left(\frac{60\,\text{s}}{15\,\text{s}}\right) \times 0.771\,\text{m}^2 = 3.08\,\text{m}^2$$

This is a medium-sized rotary vacuum filter, with possible dimensions of 1.0 m diameter by 1.0 m long.

EXAMPLE 4.4

Washing of a Rotary Vacuum Filter Cake   For the filtration in Example 4.3, it is desired to wash a product antibiotic out of the cake so that only 5% of the antibiotic in the cake is left after washing. We expect the washing efficiency to be 50%. Estimate the washing time per cycle that would be required.

SOLUTION

From Equation (4.2.7) for the washing efficiency of a filter cake

$$R' = \left(1 - \frac{E}{100}\right)^n$$

where R' is the weight fraction of solute remaining in the cake after washing (on the basis of $R' = 1.0$ before washing), E is the percentage wash efficiency, and n is the volume of wash liquid per volume of liquid in the unwashed cake. Substituting $R' = 0.05$ and $E = 50\%$ into this equation gives

$$n = 4.32$$

From Equation (4.2.14), the relationship between the washing time t_w and the cake formation time t_f is given by

$$t_w = 2nf t_f$$

where f is the ratio of volume V_r of residual liquid in the cake to the volume of filtrate V_f after time t_f. Thus, we need to estimate the volume of residual liquid in the filter cake to determine t_w. At the end of the 15 s cake formation time,

$$\text{Weight of dry cake solids filtered} = 10\,\frac{\text{g}}{\text{liter}} \times 8.33\,\text{liters} = 83.3\,\text{g}$$

Assuming the cake is 70 wt% water, which is typical for filter cakes, we find

$$\text{Weight of water in the filter cake} \cong \frac{70}{30} \times 83.3 = 194\,\text{g}$$

Thus,

$$f = \frac{V_r}{V_f} = \frac{0.194}{8.33} = 0.0233$$

$$t_w = 2nft_f = 2(4.32)(0.0233)(15\,\text{s}) = 3.0\,\text{s}$$

From this result, the estimated washing time is 20% of the cake formation time.

For small-scale tests of pressure filtration, a bomb filter may be used (Figure 4.14). The bomb filter contains a filter leaf, typically 2 in. by 2 in., that is covered with an appropriate filter medium. Enough slurry is placed in the bomb to cover the filter leaf. Air may be sparged into the slurry to provide gentle agitation. After the filtration has been started at

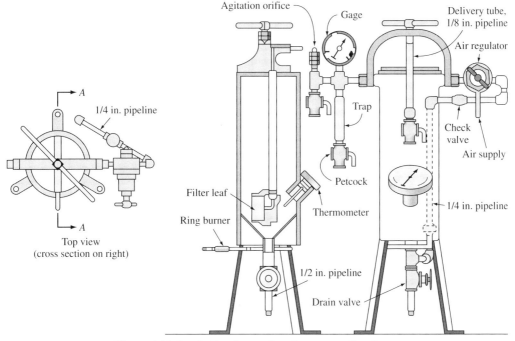

Figure 4.14 Bomb filter for small-scale pressure filtration tests.

a given pressure, the filtrate volume is recorded as a function of time. After a given time for filtration of the slurry, the cake can be washed with a wash liquid and dewatered by flowing air through the cake. Analyses can be performed on the cake and filtrate plus wash.

The data obtained from small-scale filtration tests can be used to calculate the filtration rate and area of the large-scale unit. For units in which the filter medium is continuously washed, it is recommended that filtration rates calculated from small-scale filtration data be multiplied by a factor of 0.9 [11]. For units in which the filter medium is not continuously washed, this factor should be reduced to 0.8. Before scaling up the filtration area, the relationship between the nominal filtration area and the active filtration area should be determined. The nominal filtration area is based on the overall dimensions of the filtering surface. The active filtration area may be less than the nominal area because part of the filtration surface may be blinded by mechanical dividers or by painting.

Considerations for the Design of Sterile Filters

Sterile filters are an important application of conventional filters in the pharmaceutical industry, where many products are sold in sterile form. It is sometimes desirable that other processing steps besides the final one be carried out under sterile conditions, which requires the filtration of both liquids and air.

Liquid Sterile Filtration   In typical sterile liquid filtration applications, there is no appreciable filter load, no cake buildup, and no contribution to filtration resistance from the filtrate that is reliable or reproducible. Further, there is no opportunity to add filter aids to form a cake, since this is usually the last step in a clean process.

Sizing of the filter is almost always based on the maximum allowable load in the solution to be filtered, and the desired sterility assurance limit. The sterility assurance limit is the calculated probability of a single unit of product containing a single microorganism. This is expected to be a maximum of 10^{-3} for aseptic processes, and one usually designs in an extra order of magnitude at least. One determines the maximum load on the filter by knowing the bioburden or viral load specification of the purified pool from the preceding step, or by knowing the acceptable microbial or viral levels in the raw materials that make up that pool. The retentive requirement for the filter is then divided by the retention capability (per area) of the filter to be used. The resulting filter area is the minimum required to meet the sterility goal of fewer than one unit dose contaminated with one organism per thousand doses manufactured. The next step is to ensure that the filter is large enough to allow reasonable processing times. Typically, one does not want to filter for more than one shift (8 h). At 24 h, questions will be raised about "grow-through" (where retained microbes will have colonized the filter and "grown through" to the other side). The filtration time for the filter sized is determined by finding the clean water flow rate for the filter, either through laboratory experiment or through the vendor's published materials, and adjusting for the viscosity of the solution to be filtered relative to water (a correction for the effect of temperature on viscosity may be necessary). Once the flow rate is known, the time to filter a given size batch can be determined. The area of the filter can be adjusted up to reduce the filtration time, but cannot be adjusted down without jeopardizing the sterility assurance.

For the ultimate filtration step, in which the product is rendered sterile, a depth filter is often added upstream of the membrane filter. This protects the sterile filter from fouling during use. It is very risky to replace a sterile filter once in place, with all the connections downstream sterilized. Depth filters with high "dirt loads" can keep particulate matter from reaching the membrane surface, keeping it clean for the removal of microbes. Two sterile filters are sometimes used in series, for added sterility assurance. This gives very high probability that dosage units will not be contaminated, but the arrangement sometimes becomes difficult to manage in other respects. For instance, multiple sterility-assurance tests must be performed, a requirement that adds manipulations and increases chances for breeches in sterility.

Filters used for the purpose of sterilizing pharmaceutical products must be tested to ensure that they have been properly installed and have no defects that could allow sterility to be compromised. The two most common tests for a sterile filter for an aqueous product are the "bubble point" and the "pressure hold" or "diffusion" test. In both tests, a wetted, sterilized filter is pressurized with air or nitrogen. In the bubble point test, this pressure is increased until the lowest pressure is found at which air passes freely through the filter. It is assumed that this is the pressure required to overcome the surface tension and viscosity of water in the largest membrane pore (where the forces are the weakest). As the pore size of the membrane decreases, the bubble point increases. This test will also detect a sliced or unseated O-ring or other seal. The pressure hold test pressurizes a wetted filter to a pressure lower than the bubble point and measures either the total flow of gas through the membrane (by diffusion) or the rate of pressure decay. Both should be low, and both should meet the manufacturer's specifications for an intact, properly installed sterile filter.

Compatibility and extractable tests need to be performed on each filter type, in addition to testing each installed filter for sterility assurance. Compatibility means that the filter does not adsorb an active ingredient or any other type of active excipient, such as an antibiotic or preservative. Such adsorption to the filter medium can change the strength (or biological potency) of the solution after the filtration step. Extractables are chemicals that can leach out of the filter medium, such as monomers from the membrane or support medium, filter coatings or mold release agents, and storage solutions, such as ethanol. These chemicals would be considered impurities in the final product and should be removed. When the filter is used in the final sterilization step, impurities leaching into the product cannot be allowed. Compatibility and extractables should be tested in a sample of the product solution. When this is not possible, because product or excipients would interfere with the analysis of the impurities, a placebo or similar solution should be used.

Air Sterile Filtration   Air filters are also sized based on the anticipated flow rate. Airflows are specified with one or more of the following criteria: pressure differential between areas, air changes per hour, or linear flow rate. Pressure differentials are usually measured between rooms, or between rooms and corridors. These are specified when it is important for air to flow from a "cleaner" area to a "dirtier" area when doors, windows, or passthroughs are opened between them. Pressure differentials are typically 0.2 to 0.6 in. of water ($= 0.5$–1.5 cm of water). Air changes per hour are often specified for rooms based on hazardous materials or operations the room is designed for. Air changeover is the airflow rate divided by the volume of the room. Air changeovers are different from "makeup rate," which is the fraction of fresh air injected in the room's air handler.

Clean rooms often have very high changeover rates but low makeup rates. Rooms handling solvents or hazardous materials have high makeup rates (also called "once-through air") but low air cleanliness requirements. The changeover rate varies with the hazard of the material. Linear flow rate is usually specified in hoods and smaller spaces. Linear flow rates of 5 to 20 ft/s (= 2.4–6.1 m/s) are found in laminar flow hoods, in aseptic areas where pharmaceuticals are sterile processed, and in clean rooms where microelectronics are manufactured. Linear flow rates are established to prevent airborne microbes and particulate matter from settling in or on the materials being processed. The flow rate required for the room or area being ventilated can be determined by one or more of the specifications just cited.

The air quality of a room or area is typically classified according to the number of airborne particles that can be measured in a cubic meter. The most typical classifications are class 100 (<100 particles/m^3), considered to be "clean" or "aseptic," followed by class 10,000, class 100,000, and unclassified. Classes 10,000 and 100,000 are suitable for most biotechnology processing operations. High efficiency particulate air (HEPA) filters are routinely tested for integrity by spraying a diisooctyl phthalate (DOP) aerosol on the feed side of the filters. The typical specification for a DOP test is a 10^3 reduction in aerosol particles.

Vent filters filter air on tanks and drainable equipment. Vent filters typically follow the same specification as the liquid filters used in the process. Vent filters are also sized based on the maximum airflow rate anticipated. The maximum flow rate is often encountered when the tanks are being filled or cleaned. Vent filters must allow flow in both directions, which is a unique requirement among filters discussed in this chapter.

Integrity testing of vent filters is still a developing technology. These filters are typically hydrophobic, so bubble points and pressure holds are not appropriate. The DOP tests will not work because vent filters are typically membrane filters (which are the most compact), while HEPA filters are depth filters. Testing bubble points in solvents is feasible, and often performed, but is considered a destructive form of measurement.

4.5.2 CROSSFLOW FILTRATION

As for conventional filtration, scaleup of crossflow filtration requires data from laboratory or pilot-scale units. There are, however, more different types of designs possible for crossflow filtration compared to conventional filtration because of the various modes of operation that can be used for crossflow filtration and because there is often a choice between batch and continuous operation for crossflow filtration.

Crossflow filtration modules are available from manufacturers for carrying out laboratory or pilot plant tests. The determination of the size of a plant unit can be done by a direct scaleup of the filtration area based on the feed or output flow rate. For this scaleup, however, it is important that the following variables be kept constant [16]:

- Inlet and outlet pressures
- Crossflow (or tangential) velocity
- Flow channel sizes (height and width)
- Feed stream properties—test slurries should be representative of the actual process streams
- Membrane type and configuration—test data from one design cannot directly be used to design another geometry

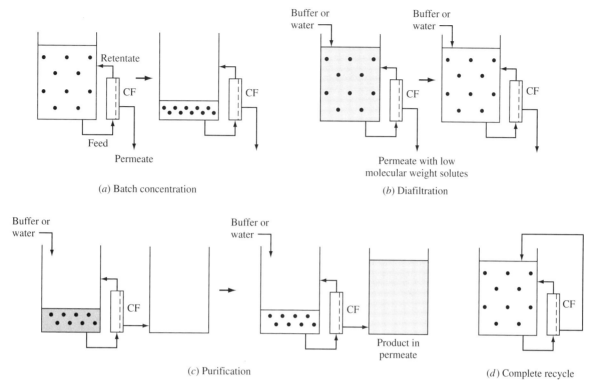

(a) Batch concentration

(b) Diafiltration

(c) Purification

(d) Complete recycle

Figure 4.15 The four basic modes of operation of crossflow filtration (CF = crossflow filter): before the arrow, start of the operation; after the arrow, end of the operation.

It is also important to make an assessment of the rate of fouling of the membranes. Tests should be performed for several cycles of operation to estimate the fouling rate and to determine how fouling can be kept to a minimum.

There are four basic modes of operation of crossflow filtration (Figure 4.15): batch concentration, diafiltration, purification, and complete recycle. In the batch concentration mode, the retained stream containing the product suspended particles or dissolved macromolecules is reduced in volume. In diafiltration, the volume of the retained stream is maintained constant by the continuous addition of water or buffer, which results in the removal of low molecular weight solutes into the permeate. Diafiltration is commonly used when salt removal or exchange is desired. In the purification mode, a low molecular weight product passes into the permeate and is thus separated from higher molecular weight impurities; or, the product can be retained and impurities removed in the filtrate. In the complete recycle mode, both the retained stream and the permeate are returned to the feed tank. Systems may be operated in complete recycle at start-up to reach steady state, saturate the membranes, test for leaks and blockage, and adjust the feed rate.

Diafiltration is similar to dialysis in that salts are exchanged or removed in both modes. Dialysis is a laboratory method that involves placing the feed in a bag made with an ultrafiltration membrane. The bag is placed in a large volume of the new solvent desired, and the new solvent is mixed until the system comes to equilibrium or near equilibrium. Complete equilibrium takes 3 to 5 h, while 90% equilibrium can be reached in 2 to 3 h [17]. For a complete change of solvent, it is necessary to do the dialysis at least twice. It is frequently the job of the engineer to scale up a dialysis done in the laboratory to a diafiltration in the pilot plant or plant. For scaling and planning, it is customary to perform material balances on solids mass and liquid volume around any component of the operation.

EXAMPLE 4.5

Diafiltration Mode in Crossflow Filtration   It is desired to use a crossflow filtration system to desalt 1000 liters of a protein solution containing NaCl. The system is capable of operating at a transmembrane flux of 30 liters $m^{-2} h^{-1}$. To remove 99.99% of the salt, determine the time required and the volume of water required using a crossflow filtration unit with a membrane area of $100 \, m^2$.

SOLUTION

We operate the system in the diafiltration mode, which means that the volume of the solution being desalted is maintained constant. A material balance on the salt in the retained volume gives

$$V \frac{dc_s}{dt} = -Qc_s(1 - r_s)$$

where V = volume of solution being desalted
c_s = concentration of salt in volume V
r_s = fraction of salt rejected by the membrane
Q = filtration rate (volume/time)

For this case we assume that there is no rejection of salt by the membrane, so that $r_s = 0$. Integrating the differential equation with the initial time equal to 0 gives

$$\ln \frac{c_s}{c_{s0}} = -\frac{Qt}{V}$$

where c_{s0} = initial salt concentration. Solving for t, we obtain

$$t = -\frac{V \ln \dfrac{c_s}{c_{s0}}}{Q} = -\frac{(1000 \, \text{liters})(\ln 0.0001)}{\left(30 \, \dfrac{\text{liters}}{m^2 \, h}\right)(100 \, m^2)} = 3.07 \, h$$

$$\text{Volume of water required} = (3.07 \, h)\left(30 \, \frac{\text{liters}}{m^2 \, h}\right)(100 \, m^2) = 9210 \, \text{liters}$$

Thus, a relatively large volume of wastewater will be generated by this process.

In designing a diafiltration process, a decision must be made about the concentration of retained product at which to operate. As this concentration is increased, the filtration flux will decrease according to Equation (4.2.16), and the total volume of filtrate will decrease for the removal of a given percentage of a low molecular weight solute. This leads to an optimum

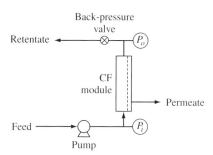

Figure 4.16 Basic components of a crossflow filtration system.

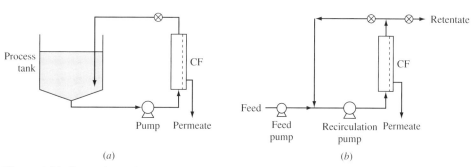

Figure 4.17 Comparison of (*a*) batch and (*b*) single-stage continuous (feed-and-bleed) crossflow filtration systems.

concentration to minimize the time required, which can be determined mathematically if the relation between filtrate flux and concentration in the bulk fluid (c_b) is known [18].

The basic components in the design of a crossflow filtration system are shown in Figure 4.16. A pump flows the feed through the filtration module to give a permeate and a retentate or retained stream. The pump needs to be sized to provide the desired flow velocity and pressure. The transmembrane pressure is controlled by a back-pressure valve on the retentate stream exiting from the filtration module. Thus the transmembrane pressure drop is estimated by

$$\Delta p_{TM} = \frac{1}{2}(p_i - p_o) - p_p \tag{4.5.1}$$

where p_i and p_o are the retentate pressures in and out of the module, respectively, and p_p is the pressure of the outlet permeate. In designing a crossflow filtration system, it is important to minimize the occurrence of gas–liquid interfaces, since bioproduct denaturation, especially of proteins, can occur at these interfaces in the presence of mechanical shear and turbulent flow (see Section 1.4.7).

Crossflow filtration systems may be designed to operate in the batch or the continuous mode (Figure 4.17). In a batch system, feed is pumped through the filtration module and then back to the feed tank. In a variation of this mode (called semibatch) for diafiltration, fluid is continuously added to the feed tank to keep the feed volume constant (Figure 4.15*b*).

In the continuous mode of operation, also sometimes called the "feed-and-bleed" mode, or "retentate bleed" mode, feed is added to a recirculation loop by the feed pump, and concentrate exiting in the retained stream is withdrawn from the system so that the

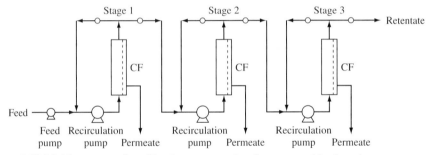

Figure 4.18 Multistage crossflow filtration system using the retentate bleed mode.

TABLE 4.2
Comparison of Batch and Continuous Ultrafiltration Systems[a]

System[b]	Flux (liters m^{-2} h^{-1})	Total area (m^2)
Batch	33.1 (average)	136
Continuous		
One-stage	8.1	555
Two-stage	31.1	243
	8.1	
Three-stage	38.7	194
	23.4	
	8.1	
Five-stage	44.7	165
	35.6	
	26.4	
	17.3	
	8.1	

[a]See reference 18.
[b]System design for 10× concentration factor and feed rate of 5000 liters/h.
Flux from $J = 20 \ln(30/c_b)$.

concentration factor (i.e., concentration in the retentate divided by the concentration in the feed) is at the desired value. When steady state has been achieved, the concentrate will be at its maximum concentration, which means that the filtration flux will be at a minimum throughout the run [18]. It is generally more economical to use a multistage system in a continuous process (Figure 4.18). As more stages are added, the average filtration flux approaches that for a batch system, and thus the total filtration area decreases as more stages are added. This is illustrated in Table 4.2, where batch ultrafiltration operation is compared with continuous operation using one, two, three, and five stages.

Because of the economic advantages of continuous operation and reduced tankage, this scheme is preferable to batch operation for most large-scale ultrafiltration operations [11]. Another advantage of continuous operation is that it permits the minimization of the residence time of the product in the crossflow filtration unit, which is important for products that are sensitive to heat or shear. Dairy and food proteins are generally processed continuously, while most pharmaceutical and biological products are processed using batch operation [18].

4.6 Summary

Filtration separates particulate or solute components in a fluid suspension or solution according to size by flowing under a pressure differential through a porous medium. The two broad categories of filtration are conventional (or dead-end) filtration and crossflow filtration. Fluid flows perpendicular to the filter medium in conventional filtration and parallel to the membrane in crossflow filtration.

- Filtration can be used to accomplish a number of objectives, including the removal of cells or cell debris, the concentration of cells or protein solutions, and the removal or exchange of salts.

- The calculation of solvent fluxes and filtration times for conventional filtration is accomplished using various forms of Darcy's law, which states in general that the flux of filtrate is equal to pressure drop divided by viscosity and resistance, $J = \Delta p / \mu_0 R$, where R is a combination of the resistance R_m of the filter medium and the resistance R_c of the cake solids.

- The time to wash a filter cake to remove all but a desired minimal level of solute can be estimated on the basis of the washing efficiency (E), the filtration time and volume of filtrate before washing, and the volume of residual liquid in the cake.

- At steady state during the crossflow filtration of a stream containing a dissolved solute, the fluid flux of permeate through the membrane is given by

$$J = \left(\frac{\mathscr{D}}{\delta}\right) \ln\left(\frac{c_w}{c_b}\right)$$

where $\mathscr{D}$ is the solute diffusion coefficient, δ is the mass transfer boundary layer thickness next to the membrane, and c_w and c_b are the solute concentrations at the wall and in the bulk fluid, respectively. The term $\mathscr{D}/\delta$ is a mass transfer coefficient that can be predicted for both laminar and turbulent flow.

- For suspended particles greater than 1 μm in crossflow filtration, the boundary layer analysis for dissolved solutes gives low predictions of the fluid flux, and a shear-induced diffusion theory and an inertial lift theory have been used to predict fluxes for this situation, depending on the particle size.

- A variety of filter media, membranes, and equipment is available for conventional and crossflow filtration.

- Fouling of membranes by proteins can be attributed to one or more of the following processes: (1) protein adsorption, which involves the interaction of the proteins and the membrane and occurs with no convective flow through the membrane, (2) protein deposition, which is the addition of more membrane-associated protein over and above the protein that would be adsorbed in a nonflowing system, and (3) protein accumulation on the membrane surface exposed to the flowing process stream as a result of concentration polarization, producing a solids or gel layer on the membrane surface.

- The scaleup of a conventional filtration can be performed with values of the specific cake resistance α and medium resistance R_m determined in laboratory filter leaf tests, assuming that feed stream properties are constant and that α does not change with a change in pressure drop upon scaleup. A crossflow filter can be scaled up directly based on membrane area, provided pressures, crossflow velocity, flow channel sizes, feed stream properties, and membrane type and configuration are held constant.

- There are numerous special considerations for the design and operation of sterile filters for liquids and air.

- There are four basic modes of operation of crossflow filtration: batch concentration, diafiltration, purification, and complete recycle. A multistage continuous process is generally more economical than a batch process in large-scale operations.

NOMENCLATURE

a radius of spherical particles (cm)

A filter cross-sectional area (cm^2)

c concentration of solute (M, or g liter^{-1})

D diameter of tube (cm)

D_h equivalent diameter of channel (cm)

$\mathscr{D}$ diffusion coefficient of solute (cm^2 s^{-1})

$\mathscr{D}_s$ shear-induced diffusion coefficient for a suspended particle (cm^2 s^{-1})

E washing efficiency for a filter cake (%)

f ratio of volume of residual liquid in filter cake to volume of filtrate at end of filtration (dimensionless)

h half-height of rectangular flow channel (cm)

J transmembrane fluid flux (liter m^{-2} h^{-1})

k mass transfer coefficient ($= \mathscr{D}/\delta$) (cm s^{-1})

L length of flow channel over membrane (cm)

n volume of wash liquid per volume of liquid in unwashed filter cake (dimensionless)

p pressure (kPa)

R overall resistance to filtration (cm^{-1})

R gas constant (0.082058 liter atm g-mol^{-1} K^{-1})

R_c resistance of filter cake (cm^{-1})

R_m resistance of filter medium (cm^{-1})

R' weight fraction of solute remaining in the cake after washing (on the basis that $R' = 1.0$ before washing)

Re   Reynolds number ($= D_h u_b \rho / \mu$) (dimensionless)

Sc   Schmidt number ($= \mu / \rho \mathscr{D}$) (dimensionless)

Sh   Sherwood number ($= k D_h / \mathscr{D}$) (dimensionless)

t time (s)

T temperature (K)

u_b bulk fluid velocity (cm s^{-1})

V volume of filtrate (conventional filtration) or volume of retentate (diafiltration) (liters)

Greek Letters

α specific cake resistance (cm g^{-1})

α' resistance parameter for a compressible cake (cm g^{-1} Pa^{-s})

γ_w shear rate at membrane surface (s^{-1})

δ boundary layer thickness (cm)

Δp pressure drop (mm Hg)

$\Delta \pi$ osmotic pressure (mm Hg)

Θ angle of drum rotation of rotary vacuum filter after entering broth (rad)

μ viscosity of fluid (g cm^{-1} s^{-1})

μ_0 viscosity of filtrate or permeate (g cm^{-1} s^{-1})

ρ density of fluid (g cm^{-3})

ρ_c mass of dry cake solids per volume of filtrate (g liter^{-1})

σ membrane reflection coefficient for a solute (dimensionless)

ϕ particle volume fraction in bulk suspension (dimensionless)

ω rate of drum rotation of rotary vacuum filter (rpm)

PROBLEMS

4.1 Conventional Batch Filtration In a laboratory test of the filtration of a cell culture, the following equation was obtained to describe the filtration:

$$\frac{t}{V/A} = K_1 \left(\frac{V}{A}\right) + K_2$$

where $K_1 = 0.18\,\text{min/cm}^2$ and $K_2 = 0.017\,\text{min/cm}$.

The cell slurry has a viscosity of 2 cp, and the filter cake solids (dry basis) per volume of filtrate was 13 g/liter. The pressure drop for the filtration was 610 mm Hg (0°C). Determine the specific cake resistance α and the medium resistance R_m. Evaluate the relative significance of the medium resistance after 5 min of filtration.

4.2 Batch Filtration with Change in Pressure In a filtration of suspended cells under full vacuum using a 20 cm Büchner funnel, you collect data for the volume of filtrate as a function of time (see Table P4.2). The concentration of cells was 100 g/liter (mass of dry cake solids per volume of filtrate), and the viscosity of the filtrate was 1.0 cp.

(a) What is the resistance of the filter medium?
(b) What is the specific resistance of the cake layer?
(c) Being disappointed in the amount of time required, you decide to give the vacuum filter a boost by pressurizing the funnel to 1 atm gage pressure. The result is the same. Concerned that you may have done the wrong thing, you then remove the funnel pressure and reduce the vacuum in the flask to 0.5 atm. The result is the same. Determine the dependence of the specific resistance of the cake layer as a function of the pressure drop. Comment on what is happening in the filtration.

TABLE P4.2

Time (min)	Volume filtered (liters)
0	0
20	4
30	5
40	6.5
50	7
60	7.5
70	7.8

(d) What is the most effective change you can make to decrease the time required to filter a given volume?

4.3 Filtration on a Filter with Uniform Cylindrical Pores A depth filter has a membrane resistance of $R_m = 10^{12}\,\text{cm}^{-1}$ and retains all particles greater than $0.4\,\mu\text{m}$ in diameter. Design a filter having uniform straight cylindrical pores with identical performance (including identical flux) to that of this depth filter using a polycarbonate film $200\,\mu\text{m}$ thick.

(a) How many pores per square centimeter must the filter with straight cylindrical pores have?
(b) Is a filter with straight cylindrical pores feasible in this situation?

Hint: Assume that the flow rate Q through each pore can be described by the Hagen–Poiseuille equation for laminar flow

$$Q = \frac{\pi \left(\dfrac{D}{2}\right)^4 \Delta p}{8\,\mu L}$$

where D and L are the pore diameter and the length, respectively.

4.4 Scaleup of a Rotary Vacuum Filter A rotary vacuum filter is available with an area of $200\,\text{m}^2$ and vacuum pressure of 75 kPa. Filter leaf tests have been performed on a cell broth with a viscosity of 5 cp. The leaf tests gave a specific cake resistance of $1 \times 10^{11}\,\text{cm/g}$ and a medium resistance of $1 \times 10^8\,\text{cm}^{-1}$. The cake solids (dry basis) per volume of filtrate was 15 g/liter. It is desired to operate the large filter with a cycle time of 45 s and a cake formation time of 10 s. What is the filtration rate expected for the rotary vacuum filter? How significant is the medium resistance?

4.5 Accumulated Cake Solids and Cake Thickness for a Rotary Vacuum Filter For filtration with a rotary vacuum filter, the following variables are known: filtrate viscosity (μ_0), pressure drop (Δp), specific cake resistance (α), mass of cake solids on a dry basis per volume of filtrate (ρ_c), and rate of drum rotation in revolutions/time (ω).

(a) It is desired to know the mass X of accumulated cake solids on the dry basis per unit area of the filter. For the case of negligible medium resistance, determine X as a function of the foregoing variables and Θ, the angle that the drum has rotated from a given point after entering the broth.

(b) It is necessary that the thickness of the cake be limited to a value h. Determine h in terms of the same variables in part a, the bulk density of the cake (ρ_b), and the weight fraction of water in the cake (x_w).

4.6 **Continuous Crossflow Filtration System Operating at Steady State**   It is desired to concentrate a cell slurry continuously at a flow rate Q_f and cell concentration x_f to a concentration x_r in a steady state operation with a crossflow microfilter. The process variables are shown in the system diagram (Figure P4.6). The concentration of cells in the permeate is zero. If the dependence of the permeate rate Q_p on cell concentration x entering the pump is known, describe how you would determine the recycle ratio R and concentration x.

4.7 **Crossflow Filtration Operating in Complete Recycle Mode**   A crossflow filtration unit is operated in the complete recycle mode.
(a) Sketch the flow diagram for the system, defining flow stream parameters for all flow streams. Write a material balance around the tank.
(b) The membrane has a molecular weight cutoff (MWCO) of 50,000 Da. (MWCO refers to the molecular weight of globular solute at which the solute is 90% rejected by the membrane.) Your product is a globular molecule with a molecular weight of 50,000 Da. Write a relationship that gives the mass of product per unit time returning to the tank through the permeate return line in terms of the starting concentration of product and the volumetric flow rates.
(c) What are the reasons for operating in the complete recycle mode?

4.8 **Concentration Polarization during Ultrafiltration**   A system is available for the ultrafiltration of protein solutions. On Monday, you filter a globular protein with 12,000 Da molecular weight, and it is fully retained. On Tuesday, you filter a 120,000 Da

molecular weight globular protein at the same molar concentration and using the same membrane. The transmembrane flux was the same on both days. Assuming identical boundary layer thicknesses for both filtrations, in which case will the polarization modulus c_w/c_b be greater? What is the relationship between the polarization moduli for the two cases?

4.9 **Concentration and Diafiltration with an Ultrafiltration System**   A 2000-liter volume of a 30 g/liter protein solution needs to be concentrated to 200 g/liter. Also, the initial salt concentration must be reduced 100-fold. The relationship between the permeation rate and the bulk protein concentration has been determined experimentally as follows:

$$Q = 500 \ln\left(\frac{270}{c_b}\right)$$

where Q = permeation rate (liters/h)
      c_b = bulk protein concentration (g/liter)

Determine the operation batch volume that will give the minimum time for diafiltration. Then determine the times required for (a) concentration to the volume at which diafiltration is to be carried out, (b) diafiltration, and (c) concentration to the final concentration desired.

4.10 **Time Required for Ultrafiltration**   A protein solution at a concentration of 0.5 g/liter and volume of 1000 liters must be concentrated on a crossflow ultrafilter to a concentration of 10 g/liter. The ultrafilter has an area of 100 m^2 and operates at 5°C with an inlet feed pressure of 16.0 bar, outlet feed pressure of 14.0 bar, and back pressure on the permeate of 1.4 bar. The protein has a molecular weight of 20,000 Da and will not pass through the ultrafiltration membranes. The feed has a viscosity of 1.2 cp. The membrane resistance (R_m) of the ultrafilter has been determined to be 2×10^{13} cm^{-1}. The Reynolds number for flow within the ultrafilter is large enough to render concentration polarization negligible. Determine the time required to perform the ultrafiltration.

4.11 **Concentration at the Membrane Surface of a Crossflow Filter**   You operate a crossflow filter at a very low Δp and low flux until the transmembrane solvent flow becomes zero as a result of concentration polarization. Show how you would determine the concentration of solute c_w right at the membrane surface on the feed (retentate) side. Assume that c_w is not high enough to form a precipitate or gel layer.

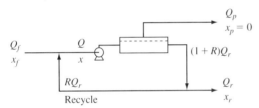

Figure P4.6 Process flow diagram for continuous operation of a crossflow filter. Flow rate is denoted by Q and cell concentration by x.

References

1. Grace, H. P. (1953). Resistance and compressibility of filter cakes. *Chem. Eng. Prog.,* vol. 49, p. 303.
2. Choudhury, A. P. R., and Dahlstrom, D. A. (1957). Prediction of cake-washing results with continuous filtration equipment. *AIChE J.,* vol. 3, p. 433.
3. Michaels, A. S. (1968). New separation technique for the CPI. *Chem. Eng. Prog.,* vol. 64, p. 31.
4. Blatt, W. F., Dravid, A., Michaels, A. S., and Nelsen, L. (1970). Solute polarization and cake formation in membrane filtration: Causes, consequences, and control techniques. In *Membrane Science and Technology,* J. E. Flinn, ed., Plenum Press, New York, p. 47.
5. Bird, R. B., Stewart, W. E., and Lightfoot, E. N. (2002). *Transport Phenomena,* 2nd ed., Wiley, New York.
6. Calderbank, P. H., and Moo-Young, M. B. (1961). The continuous phase heat and mass-transfer properties of dispersions. *Chem. Eng. Sci.,* vol. 16, p. 39.
7. Belfort, G., Davis, R. H., and Zydney, A. J. (1994). The behavior of suspensions and macromolecular solutions in crossflow microfiltration. *J. Membrane Sci.,* vol. 96, p. 1.
8. Porter, M. C. (1972). Concentration polarization with membrane filtration. *Ind. Eng. Chem. Prod. Res. Dev.,* vol. 11, p. 234.
9. Zydney, A. L., and Colton, C. K. (1986). A concentration polarization model for the filtrate flux in cross-flow microfiltration of particulate suspensions. *Chem. Eng. Commun.,* vol. 47, p. 1.
10. Eckstein, E. C., Bailey, P. G., and Shapiro, A. H. (1977). Self-diffusion of particles in shear flow of a suspension. *J. Fluid Mech.,* vol. 79, p. 191.
11. Perry, R. H., Green, D. W., and Maloney, J. O., eds. (1997). Liquid–solid operations and equipment (Section 18) and Membrane separation processes (Section 22). In *Perry's Chemical Engineers' Handbook,* 7th ed., McGraw-Hill, New York.
12. Rushton, A. (1996). Filter media. In *Solid–Liquid Filtration and Separation Technology,* A. Rushton, A. S. Ward, and R. G. Holdich, eds., VCH Publishers, Weinheim, Germany, p. 107.
13. Zeman, L. J., and Zydney, A. L. (1996). *Microfiltration and Ultrafiltration,* Dekker, New York.
14. Hallstrom, B., and Lopez-Leiva, M. (1978). Description of a rotating ultrafiltration module. *Desalination,* vol. 24, p. 273.
15. Robertson, B. C., and Zydney, A. L. (1990). Protein adsorption in asymmetric ultrafiltration membranes with highly constricted pores. *J. Colloid Interface Sci.,* vol. 134, p. 563.
16. Datar, R. V., and Rosen, C.-G. (1993). Cell and cell debris removal: Centrifugation and crossflow filtration. In *Biotechnology,* vol. 3, *Bioprocessing,* H.-J. Rehm and G. Reed, eds., VCH Publishers, Weinheim, p. 486.
17. Scopes, R. K. (1994). *Protein Purification,* 3rd ed., Springer-Verlag, New York.
18. Tutunjian, R. S. (1985). Ultrafiltration processes in biotechnology. In *Comprehensive Biotechnology,* vol. 2, *The Principles of Biotechnology: Engineering Considerations,* M. Moo-Young, ed., Pergamon Press, Oxford, p. 411.

Sedimentation

Sedimentation is the movement of particles or macromolecules in an inertial field. Its applications in separation technology are extremely widespread. Extremes of applications range from the settling due to gravity of tons of solid waste and bacteria in wastewater treatment plants to the centrifugation of a few microliters of blood to determine packed blood cell volume ("hematocrit") in the clinical laboratory. Accelerations range from $1 \times g$ in flocculation tanks to $100,000 \times g$ in ultracentrifuges for measuring the sedimentation rates of macromolecules. In bioprocessing, the most frequent applications of sedimentation include the clarification of broths and lysates, the collection of cells and inclusion bodies, and the separation of fluids having different densities.

Unit operations in sedimentation include settling tanks and tubular centrifuges for batch processing, continuous centrifuges such as disk centrifuges, and less frequently used unit operations such as field-flow fractionators and inclined settlers. Bench-scale centrifuges that accommodate small samples can be found in most research laboratories and are frequently applied to the processing of bench-scale cell cultures and enzyme preparations. Certain high speed ultracentrifuges are used as analytical tools for the estimation of molecular weights and diffusion coefficients.

The chapter begins with a description of the basic principles of sedimentation, followed by methods of characterizing laboratory and larger scale centrifuges. Two important production centrifuges, the tubular bowl centrifuge and the disk-stack centrifuge, are analyzed in detail to give the basis for scaleup. Ultracentrifuges, important for analytical and preparative work, are then analyzed. The effect of flocculation of particles on sedimentation is presented, and sedimentation at low accelerations, where diffusion can be a factor, is discussed. The chapter concludes with a description of centrifugal elutriation.

5.1 Instructional Objectives

After completing this chapter, the reader should be able to do the following:

- Determine the velocity of a sedimenting particle and calculate sedimentation times, equivalent times, and sedimentation coefficients in gravitational and centrifugal fields.

- Perform engineering analyses and scaling calculations on tubular bowl and disk-stack centrifuges.
- Choose the appropriate centrifuge for particular liquid–solid separations.
- Calculate molecular weight from ultracentrifuge data.
- Explain the behavior of sedimenting flocs.
- Discern the relative importance of diffusion in a sedimentation operation.
- Explain how inclined sedimentation, field-flow fractionation, and centrifugal elutriation work.

5.2 Sedimentation Principles

The most frequently encountered inertial fields are gravitational acceleration, $g = 9.8 \text{ m/s}^2$, or centrifugal acceleration, $\omega^2 R$, where R is the distance of the particle from the center of rotation, and ω is angular velocity (rad/s). The same theory can be used to describe sedimentation in both types of inertial fields.

5.2.1 EQUATION OF MOTION

Nearly all analyses of separations by sedimentation begin with the equation of motion of a particle of radius a and density ρ having mass $m = (4/3)\pi a^3 \rho$ in an inertial field moving in the R direction. Assuming the particle is spherical,

$$m\frac{dv}{dt} = F_{iR} + F_{bR} + F_{dR}$$

$$= \left(\frac{4}{3}\right)\pi a^3 \rho \omega^2 R - \left(\frac{4}{3}\right)\pi a^3 \rho_0 \omega^2 R - 6\pi \mu a v \qquad (5.2.1)$$

where subscripts designate forces in the R direction due to i, inertial acceleration, b, buoyancy due to the density of the medium ρ_0 through which the particle sediments, and d, the Stokes drag force, which under conditions of creeping flow is proportional to the velocity v and viscosity μ [1]. Solving Equation (5.2.1) for velocity in a centrifugal field at steady state (all forces balanced, so $dv/dt = 0$) gives

$$v = \frac{2a^2(\rho - \rho_0)\omega^2 R}{9\mu} \qquad (5.2.2)$$

commonly called the "centrifuge equation." If particles are allowed to sediment only in the presence of gravity, then the inertial acceleration is $g = 9.8 \text{ m/s}^2$, and $\omega^2 R$ in Equation (5.2.2) is replaced by g. If, however, the particle is moving outward from the center of rotation in a centrifuge, R is not constant and is related to the velocity by $v = dR/dt$, which gives from Equation (5.2.2) after rearrangement,

$$\frac{dR}{R} = \left[\frac{2a^2(\rho - \rho_0)\omega^2}{9\mu}\right] dt \qquad (5.2.3)$$

with the following initial condition:

$$\text{at } t = 0, \quad R = R_0 \qquad (5.2.4)$$

This equation can be integrated to give

$$\ln\left(\frac{R}{R_0}\right) - \frac{2a^2(\rho - \rho_0)\omega^2 t}{9\mu} \tag{5.2.5}$$

This is a useful equation that relates time to the distance traveled by the particle.

5.2.2 SENSITIVITIES

Creeping flow conditions are usually satisfied in sedimentation. Calculating the Reynolds number for spherical particles

$$\mathrm{Re} = \frac{2a v \rho}{\mu} \tag{5.2.6}$$

typically gives $\mathrm{Re} < 0.001$, since particle velocity is of the order of a few micrometers per second. Table 5.1 shows the sedimentation velocities for important bioparticles and biomolecules calculated from Equation (5.2.2) with $\rho_0 = 1.0$ g/cm^3, $\mu = 0.01$ g cm^{-1} s^{-1} (poise), and representative values of ρ and a.

It is clear from Table 5.1 that gravitational sedimentation is too slow to be practical for bacteria, and conventional centrifugation is too slow for protein macromolecules. In the case of true particles, flocculation (see Section 5.6) is often used to increase the Stokes radius a, while ultracentrifugation (see Section 5.5) is used in macromolecular separations.

When particle density and solvent density are equal, the sedimentation velocity v is zero, and the process is called *isopycnic* or *equilibrium* sedimentation. This fact is exploited in the determination of molecular densities and in the separation of living cells. A density gradient or a density shelf is employed in such cases. Densities of representative cells, organelles, and biomolecules measured by this method are given in Table 5.2 [2–5]. The density of *Amoeba proteus* cells is low because these cells contain fat vacuoles of low density and have no cell wall.

An example of a density shelf used for the separation of cells is the preparation of lymphocytes by sedimentation. The goal of this separation is to remove erythrocytes from a leukocyte population on the basis of a density shelf. By combining Ficoll, a high molecular weight polymer, and Hypaque, a heavily iodinated benzoic acid derivative, in appropriate proportions in aqueous buffers, it is possible to achieve a density around 1.07 g/cm^3 in isotonic solutions. At this density most white blood cell subpopulations will float and nearly all red blood cells will sediment.

TABLE 5.1

Calculated Settling Velocities and Reynolds Number for Example Bioproducts (Assumes $\rho_0 = 1.0$ g/cm^3 and $\mu = 1.0$ cp)

Bioparticle or biomolecule	Sedimentation radius, a (μm)	Density, ρ (g/cm^3)	Dimensionless acceleration (G), $\omega^2 R/g$	Sedimentation velocity, v (cm/h)	Reynolds number, Re
Yeast cell	2.5	1.1	1	0.5	7×10^{-6}
Bacteria cell	0.5	1.1	1	0.02	6×10^{-8}
Protein	0.005	1.3	10^4	0.06	2×10^{-9}

TABLE 5.2
Measured Values of the Density of Representative Cells,
Organelles, and Biomolecules

Cell, organelle, or biomolecule	Density, ρ (g/cm^3)	Ref.
Escherichia coli	1.09[a]	2
Bacillus subtilis	1.12	3
Arthrobacter sp.	1.17	4
Saccharomyces pombe	1.09	2
Saccharomyces cerevisiae	1.11[a]	2
Amoeba proteus	1.02	2
Murine B cells	1.06[a]	2
Chinese hamster ovary (CHO) cells	1.06	2
Peroxisomes	1.26[a]	5
Mitochondria	1.20[a]	5
Plasma membranes	1.15[a]	5
Proteins	1.30[a]	5
Ribosomes	1.57[a]	5
DNA	1.68[a]	5
RNA	2.00[a]	5

[a]Average value.

TABLE 5.3
Effect of Particle Volume Fraction ϕ on the
Particle Sedimentation Velocity for
Spherical Particles

ϕ	v_c/v
0.01	0.95
0.05	0.79
0.10	0.61
0.20	0.35

When the concentration of sedimenting particles increases, the sedimentation velocity has been found to decrease, a phenomenon known as "hindered settling." This effect has been quantified by the following expression for particles of any shape [6]:

$$v_c = v(1 - \phi)^n \tag{5.2.7}$$

where v_c is the sedimentation velocity of particles in a concentrated suspension, v is the velocity of individual particles [Equation (5.2.2)], ϕ is the volume fraction of the particles, and n is a function only of the shape of the particle and of the Reynolds number. For spherical particles with Re < 0.2 (usually satisfied during sedimentation), the exponent n has been found to be 4.65. Equation (5.2.7) may also be applied to particles of any size in a polydisperse system, using the volume fraction for all the particles in the calculation [7].

The magnitude of the hindered settling effect for spherical particles as a function of the particle volume fraction ϕ can be seen in Table 5.3. Note that hindered settling can be significant for particle concentrations of a few percent or greater.

5.3 Methods and Coefficients

While the fundamental principles of sedimentation are important to a basic understanding of this subject, other less theoretical methods have been developed for the actual practice of sedimentation. Equilibrium sedimentation, the sedimentation coefficient, equivalent sedimentation time, and sigma analysis are some of the more important of these methods.

5.3.1 EQUILIBRIUM SEDIMENTATION

Some products can be isolated on the basis of their density. In ultracentrifugation this is true of nucleic acids, and isopycnic (same-density) sedimentation was the method originally used to demonstrate the semiconservative replication of DNA [8]. When $\rho = \rho_0$ in Equation (5.2.2), then $v = 0$ and all inertial motion stops; therefore, if the density of the solution is known at the location where motion stops, the density of the solute or suspended particle is known. In most applications inertial motion is arrested in a density gradient, so that the density of the medium increases below the arrest point, and the buoyant force [Equation (5.2.1)] is greater than the inertial force, which causes the particle to return to its isopycnic level. Centrifugation can therefore be used analytically to determine particle or macromolecule density, as discussed earlier.

In practice, there are at least three routes to the establishment of conditions for isopycnic sedimentation—the creation of a region in the sedimentation vessel where $\rho_0 \geq \rho$. One method is to layer solutions of decreasing density, starting at the bottom of the vessel and proceeding until the vessel is filled. The resulting gradient is like a staircase until diffusion smooths it out. Another is to centrifuge at extremely high speed, resulting in isothermal stratification of a density-forming solute, such as CsCl. Such a gradient is not necessarily linear. The most widely used method of forming a density gradient is the gradient mixing method, in which two cylindrical containers, one containing a concentrated solution and the other containing a dilute solution and a stirring apparatus, are linked as in Figure 5.1 to produce an outflow with a linear salt gradient. For these gradient mixers, the time-dependent solute concentration is as follows [9]:

$$c(t) = c_{1,0} + \frac{Q(c_2 - c_{1,0})}{2V_0} t \tag{5.3.1}$$

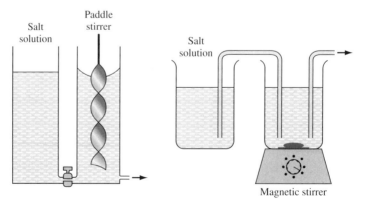

Figure 5.1 Two types of linear gradient mixer.

where Q is outflow rate due to pumping or gravity feed, V_0 is initial volume in each cylinder, $c_{1,0}$ is the initial concentration of solute in the mixed chamber, and c_2 is the concentration of solute in the nonmixed chamber (constant). By programming Q, one can set up a variety of salt gradients.

5.3.2 SEDIMENTATION COEFFICIENT

When a body force is applied, velocity through a viscous medium is usually proportional to the accelerating field (examples are electric, magnetic, and inertial). In the case of sedimentation, the resulting constant, a property of both the particle and the medium, is the *sedimentation coefficient,* which is defined as

$$s \equiv \frac{v}{\omega^2 R} \qquad (5.3.2)$$

Comparing this equation with Equation (5.2.2), we see that

$$s = \frac{2a^2(\rho - \rho_0)}{9\mu} \qquad (5.3.3)$$

which defines s in terms of only properties of the particle and the medium. This coefficient is usually expressed at $20°C$ and under conditions (viscosity and density) of pure water as

$$s_{20,w} \text{ (s)}$$

The sedimentation coefficient is often expressed in svedberg units, where 10^{-13} s $= 1$ svedberg unit (S), named after the inventor of the ultracentrifuge, The Svedberg.

EXAMPLE 5.1

Application of the Sedimentation Coefficient In 1974 D. E. Koppel measured the sedimentation coefficient ($s_{20,w}$) for the smaller ribosomes from *Escherichia coli* at 70 S (Koppel, D. E., *Biochemistry, 13,* 2712, 1974). Estimate how long it would take to completely clarify a suspension of these ribosomes in a high speed centrifuge operating at 10,000 rpm with a tube containing the ribosome suspension in which the maximum distance of travel of particles radially outward is 1 cm and the initial distance from the center of rotation to the particles nearest the center of rotation is 4 cm.

SOLUTION

We can write Equation (5.3.2) as

$$s = \frac{dR}{dt} \frac{1}{\omega^2 R}$$

or

$$\omega^2 s \, dt = \frac{dR}{R}$$

We integrate this equation with the following initial condition:

$$\text{at } t = 0, \quad R = R_0 \text{ (distance from center of rotation to the particles} \\ \text{nearest the center of rotation)}$$

to give

$$\omega^2 st = \ln \frac{R}{R_0}$$

To determine the maximum time required, we evaluate R at the maximum travel of the cells measured from the center of rotation (5 cm):

$$t = \frac{\ln\left(\dfrac{R}{R_0}\right)}{\omega^2 s} = \frac{\ln(5/4)\dfrac{1\,\text{h}}{3600\,\text{s}}}{\left(10,000\,\dfrac{\text{rev}}{\text{min}} \times \dfrac{2\pi\,\text{rad}}{\text{rev}} \times \dfrac{1\,\text{min}}{60\,\text{s}}\right)^2 (70 \times 10^{-13}\,\text{s})} = 8.1\,\text{h}$$

This should not be an unreasonable amount of time to centrifuge the ribosomes. However, since the time varies inversely with the square of the rotation speed, the time can be reduced to 2 h by doubling the speed.

5.3.3 EQUIVALENT TIME

To assess the approximate properties of a particle type to be separated, it is sometimes convenient to calculate an "equivalent time." To do this, we first define a dimensionless acceleration, G, the ratio of the centrifugal to gravitational acceleration for a particular centrifuge:

$$G \equiv \frac{\omega^2 R}{g} \qquad (5.3.4)$$

where R is usually defined as the radius of the centrifuge bowl. Thus, this dimensionless unit is measured in "g's"—multiples of the earth's gravitational acceleration. A rough approximation of the difficulty of a given separation by centrifugation is the product of the dimensionless acceleration and the time required for the separation. This product is called the equivalent time for the separation, and is written as

$$\text{Equivalent time} \equiv Gt = \frac{\omega^2 R}{g} t \qquad (5.3.5)$$

Typical values of equivalent time are as follows: 0.3×10^6 s for eukaryotic cells, 9×10^6 s for protein precipitates, 18×10^6 s for bacteria, and 1100×10^6 s for ribosomes [10].

The equivalent time for the centrifugation of cells or biological particles of unknown sedimentation properties may be estimated in a laboratory centrifuge. Samples are centrifuged for various times until a constant volume of packed cells is reached. The equivalent time Gt is calculated as the product of the G for the particular centrifuge and the time required to reach constant packed cell volume. A centrifuge that has commonly been used for this determination is the Gyro-Tester (Alfa Laval, Inc.).

One approach to scaleup of a centrifugal operation is to assume constant equivalent time:

$$(Gt)_1 = (Gt)_2 \qquad (5.3.6)$$

where the subscripts refer to centrifuges 1 and 2, respectively.

| **EXAMPLE 5.2** | **Scaleup Based on Equivalent Time** If bacterial cell debris has $Gt = 54 \times 10^6$ s [10], how large must the centrifuge bowl be and what centrifuge speed is needed to effect a full sedimentation in a reasonable amount of time? |

Assume that a reasonable amount of time is about 2 h. From Equation (5.3.5) for Gt, we can estimate the centrifuge speed ω if we know the centrifuge bowl size and the time of centrifuging. It is reasonable to have a centrifuge that is 10 cm in diameter. Solving Equation (5.3.5) for ω using these values gives

$$\omega = \left(\frac{Gtg}{Rt}\right)^{1/2} = \left(\frac{54 \times 10^6\,\text{s} \times 9.81\,\frac{\text{m}}{\text{s}^2}}{0.05\,\text{m} \times 2(3600)\,\text{s}}\right)^{1/2}$$

$$= 1213\,\frac{\text{rad}}{\text{s}} \times \frac{1\,\text{rev}}{2\pi\,\text{rad}} \times \frac{60\,\text{s}}{\text{min}} = 11{,}590\,\text{rpm}$$

This speed can be achieved in a production tubular bowl centrifuge (see later: Table 5.5).

5.3.4 SIGMA ANALYSIS

The more commonly used analysis in industry is "sigma analysis," which uses the operation constant Σ to characterize a centrifuge into which feed flows at volumetric flow rate Q. Since the engineer often needs to estimate Q, a convenient relationship is

$$Q = \{v_g\}[\Sigma] \tag{5.3.7}$$

where v_g is the sedimentation velocity at $1 \times g$, namely,

$$v_g = \frac{2a^2(\rho - \rho_0)g}{9\mu} \tag{5.3.8}$$

and Σ represents the geometry and speed of the centrifuge, as derived in the discussion of individual centrifuges in the next section; Σ can also be thought of as the cross-sectional area equivalent of the centrifuge, with units of length squared. Therefore, in Equation (5.3.7) the accolades { } indicate properties of the particle to be separated and of the fluid in which separation is occurring, and the square brackets [ ] indicate properties of the centrifuge.

The sedimentation velocity at $1 \times g$ can be estimated directly by using Equation (5.3.8), or it can be measured in the laboratory. Combining Equations (5.3.8) and (5.2.5), we obtain

$$v_g = \frac{g \ln\left(\dfrac{R}{R_0}\right)}{\omega^2 t} \tag{5.3.9}$$

This is a useful equation for determining v_g in the laboratory. Parameters in Equation (5.3.9) can be measured as follows. The minimum time t to clarify the sample in a laboratory centrifuge at speed ω is determined; R and R_0 are the distances from the center of rotation to the top of the packed solids and to the top of the liquid in the centrifuge tube, respectively.

5.4 Production Centrifuges: Comparison and Engineering Analysis

The common types of production centrifuges are illustrated in Figure 5.2, and a comparison of the advantages and disadvantages of the different centrifuge designs is given in

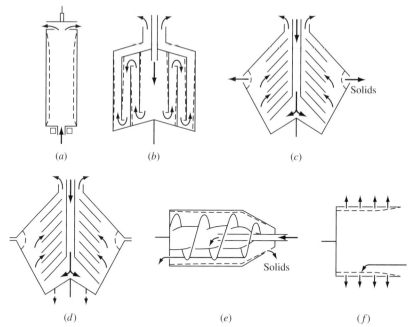

Figure 5.2 Common types of production centrifuge: (*a*) tubular bowl, (*b*) multichamber, (*c*) disk, nozzle, (*d*) disk, intermittent discharge, (*e*) scroll, and (*f*) basket. Arrows indicate the path of the liquid phase; dashed lines show where the solids accumulate.

Table 5.4 [11]. In the tubular centrifuge, solids deposit on the wall of the bowl, and feed continues until the bowl is almost full, at which time the operation is stopped and the solids removed. This type of centrifuge works well for particles of relatively low sedimentation coefficient that must be recovered, such as protein precipitates. The disk centrifuges have a relatively high sedimentation area for their volume and allow for continuous or intermittent solids discharge; they have been successfully used for the centrifugation of cells and cell lysates, where the entire process often must be contained to avoid the escape of aerosols. The scroll (or decanter) and basket centrifuges are typically used for particles that sediment relatively rapidly and can be washed well as packed solids, such as antibiotic crystals.

Of all the centrifuges, the tubular and the disk types are probably the most likely to be found in a bioseparation process involving the recovery of a protein produced by cells. The capabilities of tubular and disk centrifuges are given in Table 5.5 [12]. Note that there is generally a reduction in the maximum *g*-force (dimensionless acceleration) as the diameter of the bowl increases. The tubular bowl and disk centrifuges are analyzed to develop the Σ value that can be used in scaleup.

5.4.1 TUBULAR BOWL CENTRIFUGE

The tubular bowl centrifuge allows fluid to enter at one end of a rotating cylinder and exit at the opposite end, while particles are centrifuged toward the wall of the cylinder and

TABLE 5.4
Comparison of Production Centrifuges[a]

System	Advantages	Disadvantages
Tubular bowl	(a) High centrifugal force (b) Good dewatering (c) Easy to clean (d) Simple dismantling of bowl	(a) Limited solids capacity (b) Foaming unless special skimming or centripetal pump used (c) Recovery of solids difficult

Chamber bowl	(a) Clarification efficiency remains
	    constant until sludge space full
	(b) Large solids holding capacity
	(c) Good dewatering
	(d) Bowl cooling possible
		(a) No solids discharge
		(b) Cleaning more difficult than
		    tubular bowl
		(c) Solids recovery difficult

Disk centrifuge	(a) Solids discharge possible (b) Liquid discharge under pressure eliminates foaming (c) Bowl cooling possible	(a) Poor dewatering (b) Difficult to clean
Scroll or decanter centrifuge	(a) Continuous solids discharge (b) High feed solids concentration	(a) Low centrifugal force (b) Turbulence created by scroll
Basket centrifuge	(a) Solids can be washed well (b) Good dewatering (c) Large solids holding capacity	(a) Not suitable for soft biological solids (b) No solids discharge (c) Recovery of solids difficult

[a] See reference 11.

TABLE 5.5
Capabilities of Tubular and Disk Centrifuges[a]

Type	Bowl diameter (mm)	Speed (rpm)	Maximum dimensionless acceleration (G), $\omega^2 R/g$	Throughput (liters/min)
Tubular bowl	44	50,000	61,400	0.2–1.0
	105	15,000	13,200	0.4–38
	127	15,000	16,000	0.8–75
Disk with nozzle discharge	254	10,000	14,200	40–150
	406	6,250	8,850	100–570
	686	4,200	6,760	150–1500
	762	3,300	4,630	150–1500

[a] See reference 12.

captured both at the wall and by a weir at the exit, as indicated in Figure 5.3. The liquid enters the bowl through an opening in the center of the lower bowl head. The liquid is pushed by centrifugal force toward the periphery of the rotating bowl. Clarified liquid overflows a ring weir in the upper bowl head, the radius of which establishes the depth of the pool of liquid around the periphery of the rotating bowl.

As in most engineering calculations, we wish to determine the flow rate Q. The equations of motion that give the trajectory of sedimented particles are, first, in the radial direction from Equation (5.2.2)

$$\frac{dR}{dt} = \frac{2a^2(\rho - \rho_0)\omega^2 R}{9\mu} \tag{5.4.1}$$

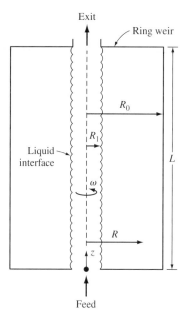

Exit

Ring weir

R_0

R_1

Liquid interface

ω

L

R

z

Feed

Figure 5.3 Cross section of a tubular centrifuge in operation.

then in the axial direction, due only to pumped flow, Q

$$\frac{dz}{dt} = \frac{Q}{A} = \frac{Q}{\pi\left(R_0^2 - R_1^2\right)} \tag{5.4.2}$$

where A is the cross-sectional area for liquid flow in the centrifuge. These equations of motion are combined to give the trajectory equation

$$\frac{\dfrac{dR}{dt}}{\dfrac{dz}{dt}} = \frac{dR}{dz} \tag{5.4.3}$$

Substituting Equations (5.4.1) and (5.4.2) into this ratio, integrating dR between R_0 and R_1, and integrating dz between 0 and L and solving for Q gives

$$Q = \left\{\frac{2a^2(\rho - \rho_0)}{9\mu}\right\}\left[\frac{\pi L\left(R_0^2 - R_1^2\right)\omega^2}{\ln\left(\dfrac{R_0}{R_1}\right)}\right] \tag{5.4.4}$$

In view of Equations (5.3.7) and (5.3.8), the first factor in Equation (5.4.4) can be multiplied by g while the second is divided by g to give, again, Equation (5.3.7) for Σ analysis:

$$Q = \{v_g\}[\Sigma] \tag{5.4.5}$$

where, for a tubular bowl centrifuge,

$$\Sigma = \frac{\pi L\left(R_0^2 - R_1^2\right)\omega^2}{g\ln\left(\dfrac{R_0}{R_1}\right)} \tag{5.4.6}$$

In practice, one uses a benchtop centrifuge to determine v_g, calculates Q from process requirements, uses these two values to determine Σ, and asks a centrifuge vendor or manufacturer to provide such a centrifuge. Conversely, the optimum Q for a known centrifuge with a given Σ can be determined, and Q/Σ can be used to scale the process. This is also true for the continuous disk-stack centrifuge, which is the subject of the next section.

EXAMPLE 5.3

Complete Recovery of Bacterial Cells in a Tubular Bowl Centrifuge   It is desired to achieve complete recovery of bacterial cells from a fermentation broth with a pilot plant scale tubular centrifuge. It has been already determined that the cells are approximately spherical with a radius of 0.5 μm and have a density of 1.10 g/cm³. The speed of the centrifuge is 5000 rpm, the bowl diameter is 10 cm, the bowl length is 100 cm, and the outlet opening of the bowl has a diameter of 4 cm. Estimate the maximum flow rate of the fermentation broth that can be attained.

SOLUTION

The flow rate can be estimated from Equation (5.4.5) by determining the settling velocity under gravity (v_g) and the Σ factor for the centrifuge. We can estimate v_g from Equation (5.3.8) and assuming the viscosity μ is the same as for water (1.0 cp):

$$v_g = \frac{2a^2(\rho - \rho_0)g}{9\mu}$$

$$= \frac{2(0.5 \times 10^{-6}\,\text{m})^2 \times (1.10 - 1.00)\dfrac{\text{g}}{\text{cm}^3} \times 9.81\dfrac{\text{m}}{\text{s}^2} \times \dfrac{10^6\,\text{cm}^3}{\text{m}^3}}{9\left(0.01\dfrac{\text{g}}{\text{cm s}}\right)} = 5.45 \times 10^{-6}\,\text{cm/s}$$

For complete recovery of the cells, we can use Equation (5.4.6) to estimate Σ:

$$\Sigma = \frac{\pi L\left(R_0^2 - R_1^2\right)\omega^2}{g\ln\left(\dfrac{R_0}{R_1}\right)}$$

$$= \frac{\pi(100\,\text{cm}) \times (5^2 - 2^2)\,\text{cm}^2 \times \left(5000\dfrac{\text{rev}}{\text{min}} \times \dfrac{2\pi\,\text{rad}}{\text{rev}}\right)^2}{9.81\dfrac{\text{m}}{\text{s}^2} \times \ln(5/2) \times \dfrac{100\,\text{cm}}{\text{m}} \times \left(\dfrac{60\,\text{s}}{\text{min}}\right)^2} = 2.01 \times 10^6\,\text{cm}^2$$

From Equation (5.4.5):

$$Q = v_g\Sigma = (5.45 \times 10^{-6}\,\text{cm/s}) \times (2.01 \times 10^6\,\text{cm}^2) \times \frac{\text{liter}}{10^3\,\text{cm}^3} \times \frac{60\,\text{s}}{\text{min}} = 0.66\,\frac{\text{liter}}{\text{min}}$$

5.4.2 DISK CENTRIFUGE

A disk centrifuge is a system of rapidly rotating concentric inverted cones placed close together to minimize the time to capture dense particles or liquids while allowing forced flow

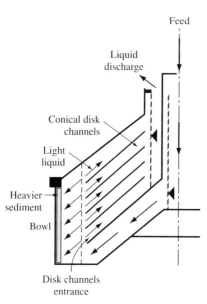

Figure 5.4 Cross-sectional diagram of a disk centrifuge, showing the path of the liquid flow and the collection of solids at the periphery.

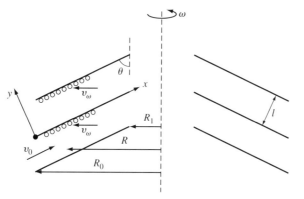

Figure 5.5 Diagram of the zone between two disks and the definition of variables for a disk centrifuge.

to continuously add feed and discharge liquids. In this configuration (Figure 5.4), the feed suspension enters on the axis of rotation and is forced to the bottom of the rotating bowl. Pressure forces the suspension upward. The heavier fluid is forced through holes at the end of each disk channel until it reaches the outer periphery of the bowl. The lighter fluid flows up the disk channels and out of the centrifuge. Dense particles are ejected through a nozzle if it is open; otherwise they collect on the outer wall of the bowl. Depending on whether the nozzle is open or closed, this centrifuge is operating in continuous or batch mode, respectively. The nozzle can be designed to open intermittently.

To determine the maximum feed rate Q, a simplified diagram of a single zone between two disks is used, as shown in Figure 5.5. For convenience, x-y coordinates are placed in a vertical plane that intersects the rotor axis with the x axis parallel to the disk surface. Equations of motion of a suspended particle are then determined with the constraint that every particle must sediment from the lower to the upper wall of a pair of disks. Thus, a suitable relationship must be obtained between v_0, the flow velocity, and v_ω, the sedimentation velocity of the particles. The following assumptions simplify the analysis:

1. Typically, v_0, the flow velocity, is much greater than v_ω.
2. $v_0 = Q/A$, where A decreases as particles move toward the center.
3. The fluid velocity v_0 is a function of y and goes to zero at the surface of the disks.

The equation of motion in the x direction is

$$\frac{dx}{dt} = v_0 - v_\omega \sin\theta \tag{5.4.7}$$

This is simplified by the assumption that v_0 is much greater than $v_\omega \sin\theta$. The average value of v_0, denoted $\langle v_0 \rangle$, is given by the flow rate Q divided by the cross-sectional area A perpendicular to the flow for n disks:

$$\langle v_0 \rangle = \frac{Q}{A} = \frac{Q}{n(2\pi Rl)} \tag{5.4.8}$$

where the R is the radial distance from the center of rotation, which is varying, and l is the spacing between disks, which is fixed. The local value of v_0 can be found by multiplying $\langle v_0 \rangle$ by a function $f(y)$ that gives the velocity variation between the disks:

$$v_0 = \frac{Q}{n(2\pi Rl)} f(y) \tag{5.4.9}$$

Integrating v_0 across the space between the disks gives the average value of v_0:

$$\frac{\int_0^l v_0(y)\,dy}{l} = \frac{Q}{n(2\pi Rl)} \tag{5.4.10}$$

From Equations (5.4.9) and (5.4.10), it is easily deduced that

$$\frac{\int_0^l f(y)\,dy}{l} = 1 \tag{5.4.11}$$

From Equations (5.4.7) and (5.4.9), neglecting $v_\omega \sin\theta$ in comparison to v_0, we have

$$\frac{dx}{dt} = \frac{Q}{n(2\pi Rl)} f(y) \tag{5.4.12}$$

The equation of motion in the y direction is centrifugal particle motion:

$$\frac{dy}{dt} = v_\omega \cos\theta = \frac{2a^2(\rho - \rho_0)\omega^2 R}{9\mu} \cos\theta \tag{5.4.13}$$

where the centrifugal velocity v_ω is obtained from Equation (5.2.2). The slope of the trajectory of particles moving between any pair of disks is determined by combining the equations of motion in the x and y directions:

$$\frac{dy}{dx} = \frac{\dfrac{dy}{dt}}{\dfrac{dx}{dt}} = \frac{4n\pi a^2(\rho - \rho_0)\omega^2 R^2 l \cos\theta}{9\mu Q f(y)} \tag{5.4.14}$$

Since $dx = -dR/\sin\theta$, Equation (5.4.14) can be rearranged to give

$$\frac{Qf(y)\,dy}{l} = -\frac{4n\pi a^2(\rho - \rho_0)\omega^2 R^2 \cot\theta}{9\mu}\,dR \qquad (5.4.15)$$

For the boundary conditions of integration, we focus on the particles that are the most difficult to capture: such a particle entering at $y = 0$ and $R = R_0$ would exit at $y = l$ and $R = R_1$ (see Figure 5.5). After integration of Equation (5.4.15) and using the result of Equation (5.4.11), the result can be rearranged after multiplying and dividing by g to yield

$$Q = \left\{\frac{2a^2(\rho - \rho_0)g}{9\mu}\right\}\left[\frac{2n\pi\omega^2\left(R_0^3 - R_1^3\right)\cot\theta}{3g}\right]$$

$$= \{v_g\}[\Sigma] \qquad (5.4.16)$$

Therefore, in a sensitivity analysis it is seen that the Σ factor depends on the cube of the bowl radius, the cotangent of the disk acute angle, the number of disks in the stack, and, as in the tubular centrifuge, the square of the rotor speed. The disk acute angle θ made by the conical disks is typically between 35 and 45 degrees [13].

5.5 Ultracentrifugation

Ultracentrifuges operate over a range of inertial accelerations of 50,000 to 100,000 $\times g$. These accelerations are so great that it is possible to sediment very small particles and even macromolecules in solution by an ultracentrifuge. In *analytical* centrifugation [14] a sample volume of less than 1.0 ml is centrifuged in an optical cell while the concentration is monitored optically as a function of distance from the center of rotation. In *preparative* centrifugation, samples of up to 50 ml are centrifuged in a batch operation and collected, usually as a function of final distance from the center of rotation. This collection is usually accomplished by carefully puncturing the bottom of the centrifuge tube and collecting the outflow in a series of tubes.

5.5.1 DETERMINATION OF MOLECULAR WEIGHT

One of the most important uses of ultracentrifugation has been the determination of molecular weights by the combined measurement of sedimentation and diffusion coefficients—the "sedimentation–diffusion molecular weight." Proteins are the most common macromolecular compounds that are homogeneous with respect to molecular weight, and it is for these compounds that the sedimentation–diffusion method has been mainly used. The molecular weights of proteins, as well as for viruses, obtained by this method have been found to be in excellent agreement with those obtained by other methods [15]. Today, protein and nucleic acid molecular weights are commonly determined by electrophoresis or chromatography (see Chapter 2).

The development of the dependence of molecular weight on the sedimentation and diffusion coefficients starts with the equation of motion for a sedimenting molecule at steady state [Equation (5.2.1)], which can be written as follows:

$$V(\rho - \rho_0)\omega^2 R - 6\pi\mu av = 0 \qquad (5.5.1)$$

where V is the volume of a single molecule. If we let $V = m\bar{V} = m/\rho$, where $\bar{V}$ is the molecule's specific volume and m is the mass of a single molecule, then Equation (5.5.1) can be rearranged to give

$$m = \frac{6\pi\mu a v}{(1 - \bar{V}\rho_0)\omega^2 R} \tag{5.5.2}$$

Viscosity can be related to the diffusion coefficient by the Stokes–Einstein equation:

$$\frac{\mathscr{D}\mu}{kT} = \frac{1}{6\pi a} \tag{5.5.3}$$

where k is Boltzmann's constant. Substituting the Stokes–Einstein equation into Equation (5.5.2) and making use of the definition of the sedimentation coefficient s [Equation (5.3.2)] and of Boltzmann's constant ($= R/\mathscr{N}$, where R is the universal gas constant and $\mathscr{N}$ is Avogadro's number) leads to

$$M = \frac{sRT}{\mathscr{D}(1 - \bar{V}\rho_0)} \tag{5.5.4}$$

where M is molecular weight and here R is the gas constant. The partial specific volume $\bar{V}$ can be easily determined as the slope of a plot of the specific volume of the solution versus the weight fraction of the substance (see, e.g., the determination of M of chymotrypsinogen by Schwert [16]). For accurate determinations of M using this method, it is necessary that s and $\mathscr{D}$ be extrapolated to zero concentration. The diffusion coefficient of the macromolecule can be measured from concentration profiles at equilibrium during sedimentation with a density gradient, or more commonly, in a separate procedure such as by the free diffusion method [15].

5.6 Flocculation and Sedimentation

After cells have been lysed or bioparticles have been dispersed, it is often useful to hasten the subsequent sedimentation or filtration step by reversibly increasing the size of the particles to be separated. To this end, flocculation is used, and it occurs as the result of adding a suitable chemical called a "flocculant" or by the selection of naturally flocculating cells for fermentation, as in the case of lager yeast. Flocculants can act by forming interparticle molecular "bridges" between particles, in which case the flocculants are usually polymers or oligomers; they can also act by reducing the repulsive forces between cells, usually by reducing the strength of the electrostatic field (for details see Chapter 3, Section 3.4).

Centrifugation is frequently used after the addition of a flocculant to uncleared broth or cell lysates. While the resulting aggregates are often treated as Stokes spheres, they are actually open structures in which internal convection may occur. Various theories of sedimentation of flocs have therefore been proposed, in view of the importance of this motion in the water-processing (sewage treatment) industry [17].

Chapter 3 introduces the subject of flocculation but not the collection of the flocs that are produced. In process engineering the sedimentation of flocs is usually treated empirically. Nevertheless, a number of formal treatments of this problem have been carried out. In the standard Equation (5.2.2) for the sedimentation of spheres, empirical adjustments for sedimentation velocity of flocs can be made on the basis of the reduction of density by the void volume fraction ε of the floc and the reduction of the drag force by the drag reduction factor Ω, which is also a function of the void volume fraction ε and the floc radius a. The resulting Stokes sedimentation velocity of spherical flocs composed of particles of density ρ is then

$$v = \frac{2a^2(1 - \varepsilon)(\rho - \rho_0)\omega^2 R}{9\mu\Omega(\varepsilon, a)} \tag{5.6.1}$$

where a is the radius of the floc. To estimate the drag force reduction due to liquid flow through the void volume of the floc, a dimensionless, normalized diameter $\beta \equiv a/k_f^{1/2}$ is defined, in which k_f is the permeability of the floc for the fluid in which it is settling. The drag force reduction factor has been related to the void volume fraction and the floc radius by a variety of closely related functions, the most widely used of which seems to be [18]:

$$\Omega(\varepsilon, a) = \frac{2\beta^2 \left(1 - \dfrac{\tanh \beta}{\beta}\right)}{2\beta^2 + 3 \left(1 - \dfrac{\tanh \beta}{\beta}\right)} \tag{5.6.2}$$

The determination of β requires knowing the floc permeability k_f, which can be calculated by using the following form of the Carman–Kozeny equation:

$$k_f = \frac{\varepsilon^2}{KS^2(1 - \varepsilon)^2} \tag{5.6.3}$$

where Kozeny's constant $K = 4.8$ [19], and S is Carman's specific surface area, defined as the area of nonporous materials exposed to the liquid per unit volume, as determined from the geometry of the individual particles that form the aggregate.

In addition to the collective behavior of particles in flocs, particles do not behave totally independently when they are suspended at high concentration, such as at the outer rim of a centrifuge or the bottom of a settling tank.

5.7 Sedimentation at Low Accelerations

At low g forces, the sedimentation rate slows down and can sometimes be similar to the rate of transport by diffusion, which is not affected by inertial forces. Here, we examine ways to compare sedimentation and diffusion rates. The case of isothermal settling is evaluated, which can lead to exponential distributions of concentration as a function of height for small particles. Inclined sedimentation and field-flow fractionation, two unit operations that separate suspended particles on the basis of inertial motion at $1 \times g$, are also described.

5.7.1 DIFFUSION, BROWNIAN MOTION

Einstein described diffusion as the consequence of a "random walk" by particles due to their thermal energy kT (k = Boltzmann's constant). The surprisingly simple result was

$$\langle x^2 \rangle = 2 \mathcal{D} t \tag{5.7.1}$$

where $\langle x^2 \rangle$ is the mean square distance traveled by a particle having diffusion coefficient $\mathscr{D}$ in time t. Using the Stokes–Einstein equation [Equation (5.5.3)] for spherical particles of radius a undergoing Brownian movement in a fluid of viscosity μ, we can relate the diffusion coefficient to the thermal energy kT as follows:

$$\mathscr{D} = \frac{kT}{6\pi\mu a} \tag{5.7.2}$$

In a concentration gradient, the net unidirectional flux of particles is proportional to $\mathscr{D}$ and the gradient, dc/dx, of the particle concentration c. Diffusion is not affected by gravity. Diffusion and sedimentation velocities, however, are sometimes similar, and their sum results in gradual settling.

5.7.2 ISOTHERMAL SETTLING

If the temperature T does not change over the height h of an ensemble of particles, then the mean kinetic energy, which is proportional to kT, of all particles is the same at all heights. The potential energy of a particle of mass m is usually expressed as mgh; but if the particles are subject to buoyant forces in the fluid, the potential energy becomes $V(\rho - \rho_0)gh$ for particle volume V. From the Boltzmann distribution rule, the concentration of particles at height h at equilibrium is therefore

$$c(h) = c(0) \exp\left[-\frac{V(\rho - \rho_0)gh}{kT} \right] \tag{5.7.3}$$

This means that concentration is an exponential function of height under isothermal conditions and that large, dense particles with potential energy much greater than kT (from mammalian cells to marbles) will be concentrated at $h = 0$ and that small particles (mainly molecules) will have $c(h) \approx$ constant. However, submicrometer organic particles and certain macromolecules have values of V and ρ that lead to measurable exponential distributions of $c(h)$.

5.7.3 CONVECTIVE MOTION AND PÉCLET ANALYSIS

One way to determine whether $c(h)$ will be distributed as in Equation (5.7.3) is to estimate the value of the Péclet number, Pe, commonly used in the dimensionless analysis of fluid motion relative to diffusive transport. The Péclet number is a ratio of the sedimentation velocity to the characteristic rate of diffusive transport over distance L:

$$\text{Pe} = \frac{v}{\mathscr{D}/L} \tag{5.7.4}$$

If Pe < 0.1, diffusion is dominant; and $c(h)$ is distributed. In this case, Equation (5.7.3) should be used. If Pe > 10, sedimentation dominates.

5.7.4 INCLINED SEDIMENTATION

Rapid removal of high density solids can be achieved at $1 \times g$ by using inclined sedimentation. An inclined settler is shown in Figure 5.6; the dimensions and the flow rates are

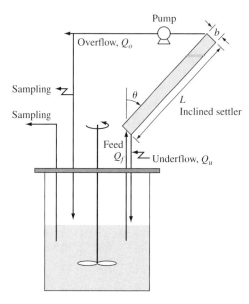

Figure 5.6 Diagram of an inclined settler system indicating variables used in the mass balance equation (5.7.5) and volumetric clearing rate equation (5.7.7).

labeled. In a typical application, feed containing suspended particles is pumped into the settler near or at its lower end at flow rate Q_f, particle-free overflow exits the upper end at flow rate Q_o, and particle-rich suspension leaves in the underflow at rate Q_u. This is an excellent method for harvesting supernatants continuously or batchwise from particle (cell)-laden broth. The material balance relationships are

$$Q_f = Q_u + Q_o \tag{5.7.5}$$

$$c_f Q_f = c_u Q_u + c_o Q_o \tag{5.7.6}$$

where c's designate particle concentrations. It is often desirable that $c_o = 0$. To achieve this, Q_o must equal the volumetric clearing rate $S(v)$. This volumetric clearing rate S is equal to the vertical settling velocity v_g of the cells multiplied by the horizontal projected area of the upward-facing surfaces of the channel onto which the cells may settle, which is given by the following equation [20]:

$$S = v_g w (L \sin\theta + b \cos\theta) \tag{5.7.7}$$

where θ is the angle of inclination of the plates from the vertical, and b, w, and L are the height, width, and length of the rectangular settler, respectively (Figure 5.6).

Inclined settlers are designed so that the path to the completion of sedimentation of a particle is extremely short, only a few millimeters, before the sedimented particles begin to be convected toward the underflow. If the particulate fraction is desired, it can be batch concentrated by continuous recycle of the underflow back to the tank while the overflow bleeds off the supernatant. By the use of appropriate settings, governed and predicted by the foregoing equations, it is also possible to remove small particles in the overflow while retaining larger ones in the underflow, thereby effecting a binary particle classification by size [20].

An important application of inclined settlers is in the removal of unproductive or parasitic cells from bioreactors or cell culture systems; this method was used to remove nonviable hybridoma cells from a cell culture, which resulted in high viable cell concentrations and high monoclonal antibody productivity over a 2-week culture period [20].

Since inclined settlers can be scaled up directly by increasing the area for settling, they potentially could be used at larger than laboratory scale. These settlers operate at much higher capacities than vertical settlers because cells need to settle only a distance of order b (see Figure 5.6) in an inclined settler, compared with a distance of order L in a vertical settler.

5.7.5 FIELD-FLOW FRACTIONATION

Field-flow fractionation (FFF) is designed to separate particles of different sizes on the basis of the hydrodynamics of a very thin, flat, horizontal channel through which a sample suspension is pumped and subjected to the laminar flow velocity gradient in the channel as shown in Figure 5.7. In the case of sedimentation, the driving force toward the lower channel wall is gravity or centrifugal sedimentation. In the latter case, the channel is "wrapped" around the perimeter of a rotating centrifuge. Steep velocity gradients occur at the upper and lower walls of the channel, and these result in a distribution of larger particles toward the center and smaller particles toward the lower wall. The separation of particles by this phenomenon is called "steric" FFF. As particle transport proceeds along the lower wall, the particles bunch up according to their velocity; thus the lower wall is called an "accumulation wall." For continuous operation, a horizontal splitter at the outlet permits the collection of large particles in an upper outlet and small particles in a lower outlet. The governing equation for field-flow fractionation is

$$R = \frac{6\gamma a}{b} \tag{5.7.8}$$

where R is the retention ratio, defined as the ratio of the particle velocity to the mean fluid velocity, a is the particle radius, b is the channel height, and γ is the "steric factor" that determines the particle net velocity and is approximately equal to $(1 + \delta/a)$, where δ is the

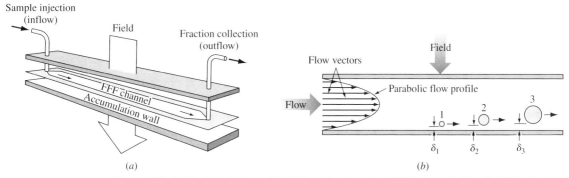

Figure 5.7 (*a*) Exploded view of field-flow fractionation (FFF) channel. The "field" is the driving force for separation, expected to act differentially on particles of different types. The field could be gravitational, electrical, thermal, adsorptive, or steric, to name a few. (*b*) The principle of steric FFF. Larger particles protrude into the higher velocity region of laminar flow, hence are carried farther in a specific amount of time than their smaller counterparts.

distance between the particle and the accumulation wall, as noted in Figure 5.7. This technique has been shown to separate two cell-sized latex populations having sphere diameters of 10 and 15 μm, and also to separate white and red blood cells [21].

5.8 Centrifugal Elutriation

Centrifugal elutriation is similar to inclined sedimentation and field-flow fractionation in that sedimentation takes place in the presence of fluid flow. In centrifugal elutriation, also called counterstreaming centrifugation, the effective length of a sedimentation path is greatly extended by continuously pumping a counterstreaming fluid in the opposite direction to that of sedimentation. To accomplish this, "elutriation rotors" have been designed. Such rotors perform functions similar to those of the tubular bowl and disk-stack centrifuges. The sample suspension is held in the rotor chamber, and the particles remain there as long as the two opposing forces are in balance. By incremental increases in the flow rate of the fluid or by decreases in the centrifugal force, distinct populations of particles (including cells) with relatively homogeneous sizes can be eluted out of the rotor sequentially. The objective of elutriation is thus to collect particles of a specified volume by modifying the velocity of the eluent fluid or the angular velocity of the rotor [22]. Both these variables must be controlled with extreme care if precision of separation is to be achieved. The volume and radius of the largest particle that can escape against the counterflow and be collected in the effluent can be determined from the equation of motion for a spherical particle in an inertial field [Equation (5.2.1)], using the velocity v_0 for the eluting fluid [22]:

$$V = \frac{9\pi \sqrt{2} \left[\dfrac{v_0 \mu}{R(\rho - \rho_0)} \right]^{3/2}}{\omega^3} \tag{5.8.1}$$

$$a = \frac{3 \left[\dfrac{v_0 \mu}{2R(\rho - \rho_0)} \right]^{1/2}}{\omega} \tag{5.8.2}$$

where V and a are the volume and radius of the particle, respectively. The capacity of centrifugal elutriation in the batch mode is about 10^8 particles of about 10 μm diameter.

5.9 Summary

Sedimentation is the movement of particles or macromolecules in an inertial field. Inertial accelerations vary from $1 \times g$ in flocculation tanks to $100,000 \times g$ in ultracentrifuges.

- Sedimentation, like filtration, is used in early stages in downstream bioprocessing mainly for liquid–solid separations. Particles can be separated at large scale in continuous centrifuges, and macromolecules can be separated, either for analysis or collection, at small scale by ultracentrifugation at very high speeds.

- The sedimentation velocity v of a particle depends on the square of its radius a and linearly on the difference between its density ρ and that of the suspending solvent ρ_0:

$$v = \frac{2a^2(\rho - \rho_0)\omega^2 R}{9\mu}$$

  where ω is angular velocity (rad/s), R is the distance of the particle from the center of rotation, and μ is fluid viscosity. The term $\omega^2 R$ is the centrifugal acceleration.

- Corrections related to sedimentation calculations are required when particle concentrations are high enough to hinder settling. Functions of concentration are available for this correction.

- The sedimentation coefficient s, a property of both the particle and the medium, is defined as

$$s \equiv \frac{v}{\omega^2 R}$$

  which leads to

$$s = \frac{2a^2(\rho - \rho_0)}{9\mu}$$

- Equivalent time is the product Gt, where G is defined as

$$G \equiv \frac{\omega^2 R}{g}$$

  One approach to scaleup of centrifugation is to assume constant equivalent time.

- Engineering analyses and scaling calculations on large-scale centrifuges are often performed by means of "sigma analysis," which uses the operation constant Σ to characterize a centrifuge into which feed flows at volumetric flow rate Q. Since the engineer often needs to estimate Q, a convenient relationship is

$$Q = \{v_g\}[\Sigma]$$

  where v_g is the sedimentation velocity at $1 \times g$,

$$v_g = \frac{2a^2(\rho - \rho_0)g}{9\mu}$$

  and Σ represents the geometry and speed of the centrifuge.

- The appropriate centrifuge for a particular liquid–solid separation is chosen on the basis of the objectives of the operation, the solids content and flow rate of the feed stream, and the maximum acceleration at which the device can operate.

- The molecular weight of a macromolecule such as a protein can be determined from ultracentrifuge data, which measure the sedimentation and diffusion coefficients.

- After flocculation, loosely structured, porous particles form, and these sediment more rapidly than their solid counterparts having the same size. Correction factors allow the calculation of the velocity of flocculated particles.

- Diffusion occurs in sedimentation operations, and its significance, specifically at low accelerations, can be ascertained by calculating the Péclet number.

- Two useful sedimentation operations at $1 \times g$ are inclined sedimentation and field-flow fractionation.
- In centrifugal elutriation, a counterstreaming fluid is continuously pumped in the opposite direction to that of sedimentation. This greatly extends the sedimentation path and allows distinct populations of particles (including cells) with relatively homogeneous size to be eluted out of the centrifuge rotor sequentially.

NOMENCLATURE

a radius of particle (μm)

A cross-sectional area (cm^2)

b height (cm)

c concentration (M, or g liter^{-1})

$\mathscr{D}$ diffusion coefficient (cm^2 s^{-1})

g gravitational acceleration (9.8066 m s^{-2})

G multiple of gravitational acceleration ($= \omega^2 R/g$) (dimensionless)

h height (cm)

F_x force in x direction (N)

k Boltzmann's constant (1.3807×10^{-23} J K^{-1})

k_f floc permeability (cm^2)

K Kozeny's constant ($= 4.8$) (dimensionless)

m mass of particle or molecule (g)

M molecular weight (Daltons)

n exponent in Equation (5.2.7) (dimensionless)

n number of disks in a disk centrifige (dimensionless)

$\mathscr{N}$ Avogadro's number (6.0221×10^{23} molecules g-mol^{-1})

l spacing between disks in a centrifuge (cm)

L distance (cm)

L length of an inclined settler (cm)

Pe   Péclet number ($= vL/\mathscr{D}$) (dimensionless)

Q flow rate (cm^3 s^{-1})

R distance from center of rotation (cm)

R retention ratio in field-flow fractionation ($=$ ratio of particle velocity to the mean fluid velocity) (dimensionless)

R gas law constant (8.3145 J g-mol^{-1} K^{-1})

Re   Reynolds number ($2av\rho/\mu$) (dimensionless)

s sedimentation coefficient [Equation (5.3.2)] (s)

S volumetric clearing rate in an inclined settler (cm^3 s^{-1})

t time (s)

T temperature (K)

v sedimentation velocity (cm s^{-1})

v_c sedimentation velocity in a concentrated suspension (cm s^{-1})

v_g sedimentation velocity at $1 \times g$ (cm s^{-1})

v_ω sedimentation velocity at ω angular velocity (cm s^{-1})

v_0 fluid velocity (cm s^{-1})

V volume of a molecule or particle (cm^3)

V_0 volume of cylinder in a gradient mixer [Equation (5.3.1)] (cm^3)

$\bar{V}$ specific volume of a molecule (cm^3 g^{-1})

w width of an inclined settler (cm)

Greek Letters

β normalized diameter of a floc ($= a/k_f^{1/2}$) (dimensionless)

γ steric factor in field-flow fractionation ($\cong 1 + \delta/a$) (dimensionless)

δ distance between particle and accumulation wall in field-flow fractionation (cm)

ε void fraction of particles in a floc (dimensionless)

θ centrifuge disk acute angle (Figure 5.5) (degrees)

θ angle of inclination of an inclined settler from vertical (degrees)

μ viscosity of fluid (g cm^{-1} s^{-1} = poise)

ρ density of particle or molecule (g cm^{-3})

ρ_0 density of medium (g cm^{-3})

Σ operation constant of a centrifuge (cm^2)

ϕ volume fraction of particles (dimensionless)

ω angular velocity (rad s^{-1})

Ω drag reduction factor [Equation (5.6.2)] (dimensionless)

PROBLEMS

5.1   **Sedimentation versus Filtration**   Four particulate materials, A, B, C and D, are suspended in water (density $= 1.00$ g/cm^3) and have the properties given in Table P5.1. Choose between sedimentation and filtration for the separation of the following pairs of particles from one another in mixed suspensions in water. Explain your answer in each case.
(a) A from B
(b) B from C
(c) C from D

5.2   **Strategies for Product Separation**   Yeast cells ($a = 3\,\mu$m) in a fermentor secrete a low molecular weight product at a concentration that produces uniform rod-shaped crystals $2 \times 6\,\mu$m at about 20 times the number concentration (particles/ml) as the cells. Using concise statements, design two possible strategies that take advantage of the particulate nature of the product to separate the product from the broth and from the cells. What additional information about the product crystals would be useful?

5.3   **Isopycnic Sedimentation**   You wish to capture $3\,\mu$m particles in a linear density gradient having a density of 1.12 g/cm^3 at the bottom and 1.00 at the top. You layer a thin particle suspension on the top of the 6 cm column of fluid with a viscosity of 1.0 cp and allow particles to settle at $1\,g$.
(a) How long must you wait for the particles you want (density $= 1.07$ g/cm^3) to sediment to within 0.1 cm of their isopycnic level? Is it

TABLE P5.1

Particle name	Density (g/cm^3)	Radius (μm)
A	1.05	10
B	1.05	12
C	1.01	50
D	1.04	25

possible to determine the time required for particles to sediment to *exactly* their isopycnic level?
(b) If instead of $1\,g$ you use a centrifuge running at 800 rpm, and the top of the fluid is 5 cm from the center of rotation, how long must you centrifuge for the particles to move to within 0.1 cm of their isopycnic level?

5.4   **Time Required for Sedimentation by Gravity**   A certain reagent is added to a suspension of cells $4\,\mu$m in diameter. These cells have a density of 1.08 g/cm^3, and they are suspended in liquid with a density of 1.00 g/cm^3 and viscosity of 1.0 cp. This reagent causes about half of the cells to form fairly solid aggregates, all of which are $90\,\mu$m in diameter and have density midway between that of the liquid and the cells. How much time is required for all the aggregates to sediment to within 1 cm of the bottom of a vessel filled with suspension that is 0.5 m high? Approximately what fraction of the single cells would have sedimented to this depth in the same amount of time? How much time is required for all the single cells to sediment to within 1 cm of the bottom of the vessel?

5.5   **Time Required for Sedimentation in a Centrifuge**   Using the results of Problem 5.4, determine the diameter and speed of a centrifuge required to reduce the total sedimentation time for the aggregates by a factor of 10, assuming you will use containers that are 20 cm high in the centrifuge. Also assume that the center of rotation is 3 cm from the tops of the containers. How much time must the same centrifuge be operated to also sediment all the single cells? For simplicity, assume a swinging-bucket type of centrifuge, in which the axis of the cylindrical vessel is horizontal, hence parallel to the direction of sedimentation.

5.6   **Determination of Sigma for a New Pilot Scale Centrifuge**   You test a new pilot scale centrifuge by doing a breakthrough experiment using yeast as test particles. The yeast were previously found to sediment at $100\,\mu$m/s in a laboratory centrifuge operated at an acceleration of $500 \times g$. The breakthrough flow rate is found to be 10 liters/min. What is the sigma (Σ) of this new centrifuge?

5.7   **Bench Scale Tests for a Tubular Bowl Centrifuge**   You can bench-test a tubular bowl separation by first characterizing the product in a test tube centrifugation. Without actually knowing the size and density of the particles in the suspension, derive an expression for the angular velocity required to capture the solids at a given volumetric flow rate Q in terms of the geometry of the tubular bowl and the quantities you would measure in the test tube centrifugation.

5.8 Recovery of *E. coli* in a Tubular Bowl Centrifuge You are using a tubular bowl centrifuge to recover *E. coli* cells containing an important bioproduct from a fermentation broth. In a preliminary run you find that 50% of the cells are recovered at a flow rate of 5 liters/min and rotation speed of 6000 rpm.
(a) To increase the yield to 95% using the same centrifuge, what must the flow rate be?
(b) How much does the sedimentation velocity change if you double the rotation speed?

5.9 Scaleup of a Disk Centrifuge Based on Laboratory Data Determine the maximum flow rate for the clarification of a suspension of lysed *Escherichia coli* cells by a plant scale disk centrifuge based on laboratory data. The plant centrifuge has a bowl diameter of 25.4 cm and capabilities shown in Table 5.5. For this centrifuge, you also know that $\theta = 42°$, $R_1 = 8$ cm, $R_0 = 20$ cm, and number of disks = 100 (see Figure 5.5).

In a laboratory centrifuge, you determined that it took a minimum of 17 min to clarify the cell lysate at 12,000 rpm. The top of the culture being centrifuged was 32 mm from the center of rotation, and the top of the packed solids was 79 mm from the center of rotation after 17 min.

5.10 Determination of Molecular Weight by Ultracentrifugation A new biopharmaceutical "X" has been discovered. Only crude extracts are available, and the material is known to be a macromolecule. You are given a preparative ultracentrifuge and asked to estimate the molecular weight of the macromolecule. You then do two experiments. In the first one you set up a linear sucrose density gradient and layer a crude sample of X on top of it, using a 5 cm long centrifuge tube (completely filled) and centrifuge to equilibrium for 3 days (72 h) at 25,000 rpm. In the second experiment you place the sample in its dilute buffer (viscosity the same as water) directly into *two* of the same plastic ultracentrifuge tubes and run the centrifuge for 24 h at 25,000 rpm, at which time you stop and remove fractions from one of the two tubes. After an additional 72 h at 25,000 rpm, you stop the centrifuge and remove fractions from the remaining tube. In all three cases, you collect 20 fractions, and each fraction corresponds to a 2.5 mm layer, so fraction 1 came from the bottom of the tube and fraction 20 came from the top. Assume that the top of each centrifuge tube is 3 cm from the center of rotation.

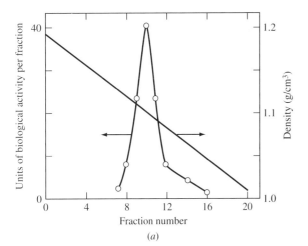
(a)

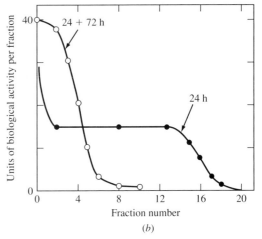
(b)

Figure P5.10 Ultracentrifugation results. (*a*) First experiment: crude extract layered on the top of a tube with a linear sucrose density gradient and centrifuged to equilibrium for 72 h. (*b*) Second experiment: crude extract in its dilute buffer in each of two of the same plastic tubes with one tube centrifuged for 24 h and the other tube for 72 h. Centrifuge speed 25,000 rpm. Fractions, each corresponding to a 2.5 mm layer in the tube, are numbered from the bottom to the top of each tube.

The company biologist then takes the three sets of 20 fractions and tests them on cell cultures. The biologist returns data to you in terms of biological activity units in each fraction, on a scale that is known to be linear with product mass. You then plot the data from the three centrifuge tubes (Figure P5.10).

(a) From the appropriate equations for ultracentrifugation, use the data to estimate the molecular weight of product X. (*Hint:* Use widths of profiles to estimate diffusivity.)

(b) Make a sketch of the method for collecting fractions.

(c) After you completed your material balance calculations on the tubes, which were 8 mm in diameter, the biologist told you that each tube originally contained 20 units of biological activity. Speculate about why the material balance did not close.

5.11 Isothermal Settling Based on the data in Table 5.1, estimate the reduced concentration profile, $c(h)/c(0)$, for the isothermal settling at 1 g of a protein with a sedimentation radius of 0.005 μm at ambient temperature up to a height of 100 cm. Recalculate the profile for a protein with a sedimentation radius of 0.002 μm. Also calculate the molecular weight of each protein. Explain the meaning of the concentration profiles.

5.12 Cost of Centrifugation In a bioprocess for the production of foreveron (a hypothetical youth-giving protein made by Kindergen, Inc.), a centrifuge is used to remove the cells from the culture broth. The centrifuge operates continuously at 20 kW and processes 20 liters of feed culture broth per hour, discharging a liquid supernatant phase containing 92 vol% of the feed liquid and no cells. The cell paste output has a density of 1.08 g/cm^3. The feed is processed batchwise in 40 liters of feed per batch. It takes a busy technician ($20/h) 15 min to start the feed flow and 15 min to collect each batch and deliver the supernatant and sediment to the next steps. Kindergen paid $50,000 for the centrifuge, which has a useful life of 10 years, buys a service contract for $5000/year and replaces the rotor every year for $10,000. The centrifuge is used only for foreveron processing at 2 batches per day, 300 days a year.

Calculate the cost of the centrifugal processing per kilogram of cell paste produced. What contributes the most to the cost of centrifuging?

References

1. Bird, R. B., Stewart, W. E., and Lightfoot, E. N. (2002). *Transport Phenomena,* 2nd ed., Wiley, New York.
2. Kubitschek, H. E. (1987). Buoyant density variation during the cell cycle in microorganisms. *Crit. Rev. Microbiol.,* vol. 14, p. 73.
3. Hart, A., and Edwards, C. (1987). Buoyant density fluctuations during the cell cycle of *Bacillus subtilis. Arch. Microbiol.,* vol. 147, p. 68.
4. Illmer, P. (2001). How big is a bacterium? Approximation to a correct total cell density of *Arthrobacter* sp. *FEMS Microbiol. Lett.,* vol. 196, p. 85.
5. Sheeler, P. (1981). *Centrifugation in Biology and Medicine,* Wiley-Interscience, New York.
6. Richardson, J. F., and Zaki, W. N. (1954). Sedimentation and fluidisation: Part I. *Trans. Inst. Chem. Eng.,* vol. 32, p. 36.
7. Richardson, J. F., and Shabi, F. A. (1960). The determination of concentration distribution in a sedimenting suspension using radioactive solids. *Trans. Inst. Chem. Eng.,* vol. 38, p. 33.
8. Meselson, M., and Stahl, F. W. (1957). The replication of DNA in *E. coli. Proc. Natl. Acad. Sci. USA,* vol. 44, p. 671.
9. Ladisch, M. R. (2001). *Bioseparations Engineering,* Wiley, New York.
10. Belter, P. A., Cussler, E. L., and Hu, W.-S. (1988). *Bioseparations,* Wiley, New York.
11. Bell, D. J., Hoare, M., and Dunnill, P. (1983). The formation of protein precipitates and their centrifugal recovery. In *Advances in Biochemical Engineering/Biotechnology,* vol. 26, A. Fiechter, ed., Springer-Verlag, Berlin, p. 1.

12. Perry, R. H., Green, D. W., and Maloney, J. D., eds. (1997). Liquid–solids operations and equipment (Section 18). *Perry's Chemical Engineers' Handbook,* McGraw-Hill, New York.
13. Leung, W. W. (1998), *Industrial Centrifugation Technology,* McGraw-Hill, New York.
14. Schachman, H. K. (1959). *Ultracentrifucation in Biochemistry,* Academic Press, New York.
15. Tanford, C. (1961). *Physical Chemistry of Macromolecules,* Wiley, New York.
16. Schwert, G. W. (1951). The molecular size and shape of the pancreatic proteases. II. Chymotrypsinogen. *J. Biol. Chem.,* vol. 190, p. 799.
17. Li, D., and Ganczarczyk, J. (1992). Advective transport in activated sludge flocs. *Water Environ. Res.,* vol. 64, p. 236.
18. Neale, G., Epstein, N., and Nader, W. (1973). Creeping flow relative to permeable spheres. *Chem. Eng. Sci.,* vol. 28, p. 1865.
19. Adler, P. M. (1986). Transport processes in fractals. VI. Stokes flow through Sierpinski carpets. *Phys. Fluids,* vol. 29, p. 15.
20. Davis, R. H., Lee, C.-Y., Batt, B. C., and Kompala, D. S. (1991). Cell separations using differential sedimentation in inclined settlers. In *Cell Separation Science and Technology,* D. S. Kompala and P. Todd, eds., ACS Symposium Series 464, American Chemical Society, Washington, DC, p. 113.
21. Giddings, J. C., Barman, B. N., and Liu, M.-K. (1991). Separation of cells by field-flow fractionation. In *Cell Separation Science and Technology,* D. S. Kompala and P. Todd, eds., ACS Symposium Series 464, American Chemical Society, Washington, DC, p. 128.
22. Keng, P. C. (1991). High-capacity separation of homogeneous cell subpopulations by centrifugal elutriation. In *Cell Separation Science and Technology,* D. S. Kompala and P. Todd, eds., ACS Symposium Series 464, American Chemical Society, Washington, DC, p. 103.

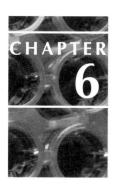

CHAPTER 6

Extraction

Extraction is a process in which two phases come into contact with the objective of transferring a solute or particle from one phase to the other. For the separation and purification of biological products, the phases are most commonly immiscible liquids, and the solute is in soluble form. In certain instances, however, one phase is a liquid and the other phase is a solid; the extraction of caffeine from coffee beans is one example. Although most extractions in biotechnology involve the transfer of soluble bioproducts, organelles and cells have at times been transferred between phases. An organic solvent is often used as the extracting liquid when the solute to be extracted is stable in the organic solvent, typical examples being low molecular weight antibiotics. It is usually not feasible to extract proteins with organic solvents, since proteins are often denatured or degraded as a result of contact with the organic solvent. Proteins can often be successfully extracted by means of two immiscible liquid phases that consist of solutions of two water-soluble but incompatible polymers, or one polymer plus a high concentration of certain salts.

Extraction usually comes early in the purification process for a bioproduct and typically would precede a high resolution step such as chromatography. Extraction is often advantageous because it can bring about a significant reduction in volume and/or can separate the desired product from cells or cell debris. It is desirable to reduce the volume as soon as possible in the process, since large volumes typically lead to large costs.

The extractions of interest in the purification of biotechnological and pharmaceutical products are mainly liquid-to-liquid, and this is the emphasis in this chapter. The basic definitions and principles of extraction are developed first, followed by an explanation of scaleup and design procedures for the extractors most commonly used for bioproducts.

6.1 Instructional Objectives

After completing this chapter, the reader should be able to do the following:

- Define and use key constants such as the partition coefficient, solvent to feed ratio, and extraction factor.
- Explain the factors that affect the partitioning of biomolecules.

- Construct a phase diagram for aqueous two-phase systems and understand their applications to the extraction of proteins.
- Calculate solute concentrations in multistage countercurrent extraction cascades.
- Draw equilibrium and operating lines and use them to calculate equilibrium stages in countercurrent extraction.
- Perform scaling calculations for reciprocating-plate and centrifugal extractors.

6.2 Extraction Principles

Separation of biomolecules in liquid–liquid extraction depends on the partitioning of the biomolecules between the liquid phases. Design of the extraction process depends on the miscibility of the two liquid phases in each other, and the rate of equilibration of the biomolecules between the two phases. Aqueous two-phase extraction is a nondenaturing and nondegrading technique that can be used for a number of biomolecules such as proteins, viruses, cells, and cell organelles; therefore, the analysis of partitioning in aqueous two-phase extraction is emphasized. Both graphical and analytical approaches to extractor stage calculations are discussed.

6.2.1 PHASE SEPARATION AND PARTITIONING EQUILIBRIA

In a single-stage extraction process, one feed stream contacts one extraction solvent stream, and the mixture divides into equilibrium extract and raffinate phases, as shown schematically in Figure 6.1. The distribution of a solute at equilibrium between two liquid phases is defined by the *partition coefficient* as

$$K = \frac{y}{x} \tag{6.2.1}$$

where y and x are the concentrations of the solute in the extract and raffinate phases, respectively. Thus, for the lowest volume of extraction solvent, it is desirable to have K as large as possible. Partition coefficients near unity would require large volumes and many serial extractions for full recovery. Partition coefficients near zero would indicate no extraction at all.

The partition coefficient can depend on many parameters, such as the size of the molecule being extracted, pH, types of solvent, temperature, and concentration and molecular weight of polymers (or salt) in the phases. The literature contains partition coefficients for many biological molecules, especially amino acids and antibiotics, but caution must be exercised if conditions other than those reported are used. An example of the effect of pH on the partition coefficient is shown in Figure 6.2 for the antibiotic penicillin G and

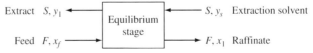

Figure 6.1 Flow diagram of one equilibrium stage, showing stream variables. The two phases are assumed to be immiscible.

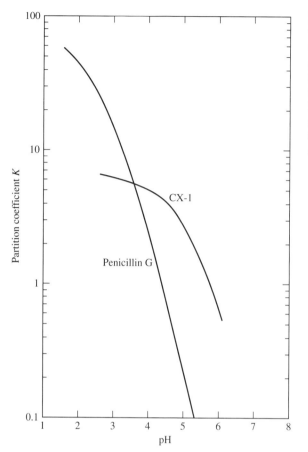

Figure 6.2 Dependence of the partition coefficient on pH for penicillin G and acidic impurity CX-1: an organic solvent was used for extraction from filtered fermentation broth. (Data from M. Souders, G. J. Pierotti, and C. L. Dunn, "The recovery of penicillin by extraction with a pH gradient," *Chem. Eng. Prog. Symp. Ser.* 100, vol. 66, p. 39, 1970.)

an acidic impurity. It can be seen that below pH 4 the extraction of penicillin G into the solvent phase is favored over that of the acidic impurity CX-1.

Models developed to describe the partitioning of biomolecules are useful not only in extending existing partitioning data to new conditions, but also in obtaining reasonable first estimates of partition coefficients. One of the simplest models to describe partitioning between phases is the one developed by Brønsted [1]:

$$K = \exp\left(\frac{M\lambda}{kT}\right) \tag{6.2.2}$$

where M is the molecular weight of the molecule being partitioned, k is the Boltzmann constant, T is absolute temperature, and λ is a constant (lumped parameter) that includes characteristics of both the phase system and the partitioning substance. It can be observed that for the partitioning of penicillin G (Figure 6.2) the Brønsted model corresponds to a steadily decreasing λ from pH 2 to 5.

Aqueous two-phase extraction systems are made by combining two water-soluble polymers or a polymer and a salt in water, above a "critical concentration." These systems separate into two immiscible liquid phases, one of them enriched in one polymer and the

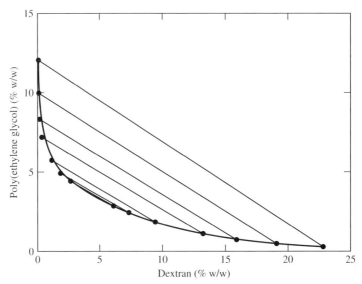

Figure 6.3 Phase diagram for a PEG 6000–dextran D48 system at 20°C. (Data from P.-A. Albertsson, *Partition of Cell Particles and Macromolecules*, 3rd ed., Wiley, New York, 1986.)

other enriched in the other polymer or salt. Soluble or particulate separands, added to the system, partition themselves between the two phases. Aqueous two-phase extraction has become recognized as a nondenaturing and nondegrading technique for the separation of a number of biological entities such as proteins, enzymes, viruses, cells, and cell organelles. This method also has been useful for the removal of undesirable contaminating by-products such as nucleic acids and polysaccharides.

Each two-phase aqueous system can be characterized by a phase diagram, an example of which is shown in Figure 6.3. At concentrations of poly(ethylene glycol) (PEG) and dextran below the curve in Figure 6.3, there is only one liquid phase. On the curve, there are two liquid phases, and tie lines connect the compositions of the phases that are in equilibrium. As observed in Figure 6.3, the phase enriched in PEG can contain almost no dextran. The most common systems use dextran and PEG, or PEG and potassium phosphate, although purified dextran is expensive and alternatives to it have been used [2]. The PEG-rich phase is less dense than either the dextran-rich or salt-rich phase and thus is the lighter or top phase. Typical concentrations of PEG–dextran systems are 10% (w/w) PEG and 15% dextran, or with salt-rich bottom phases, 15% PEG and 15% salt [3].

In general for aqueous two-phase extraction, contaminants such as cells and cell debris partition to the bottom phase or interface. For proteins, the partitioning is affected by many parameters, which are given in Table 6.1. These parameters are difficult to separate and analyze individually, which means that an a priori prediction of partition coefficients is very difficult.

Models have been developed to explain several molecular level mechanisms influencing the partition of proteins in aqueous two-phase extraction. These models can be categorized into three types: lattice models, virial expansions, and scaling–thermodynamic approaches. These models tend to be complicated, requiring the independent measurement

TABLE 6.1
Factors That Affect Protein Partitioning in Two-Phase Aqueous Systems[a]

Protein molecular weight
Protein charge, surface properties
Polymer(s) molecular weight
Phase composition, tie-line length
Salt effects
Affinity ligands attached to polymers

[a]See reference 3.

of several parameters. An example of these models is the virial expansion model developed by King et al. [4]. For a solution with two polymers that is very dilute in protein, such that the concentration of either polymer is much greater than the protein concentration, the protein partition coefficient K_p is given by

$$\ln K_p = \ln \frac{y_p}{x_p} = a_{1p}(x_1 - y_1) + a_{2p}(x_2 - y_2) + \frac{z_p F(\phi_x - \phi_y)}{RT} \qquad (6.2.3)$$

where y_p, x_p = concentration of protein in light and heavy phase, respectively
  y_1, x_1 = concentration of polymer 1 in light and heavy phase, respectively
  y_2, x_2 = concentration of polymer 2 in light and heavy phase, respectively
  a_{1p} = second virial coefficient for interaction between protein and polymer 1
  a_{2p} = second virial coefficient for interaction between protein and polymer 2
  ϕ_y, ϕ_x = electrical potential for light and heavy phase, respectively, relative to some reference
  z_p = net surface charge of protein
  F = Faraday number
  R = gas constant

The virial coefficients can be measured for each type of polymer and salt by membrane osmometry or by low angle laser light scattering. The electrical potentials can be determined as a function of the type of salt and tie-line length. Data are also necessary for the protein surface charge as a function of pH. The use of these methods gave good agreement between predicted and experimentally measured partition coefficients for albumin, α-chymotrypsin, and lysozyme in PEG–dextran systems [4].

The influence of protein size on the phase partitioning has been shown by correlating the partition coefficient with molecular weight for a number of proteins. As indicated in Figure 6.4, partitioning of the protein to the top phase is weakly favored at lower molecular weights, while bottom phase partitioning is strongly favored for the largest proteins. For all these data, the pH of the solution was at the protein's isoelectric point (pI) to minimize any effects of protein charge.

Other factors that can have a large effect on protein partitioning are the type and concentration of the salt in the system and the charge on the protein. For example, Figure 6.5 shows that the partition coefficient for ovalbumin can vary widely depending on the pH, which means that this is an electrostatic effect. This is further supported by the observation

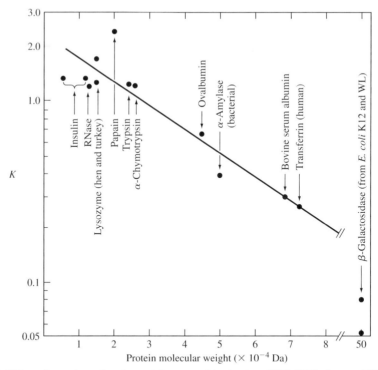

Figure 6.4 Effect of protein molecular weight on partitioning in a PEG 6000–dextran 500 system with pH at the isoelectric point (pI) for all proteins. (Data from S. Saskawa and H. Walter, "Partition behavior of native proteins in aqueous dextran–poly(ethylene glycol)-phase systems," *Biochemistry,* vol. 11, p. 2760, 1972.)

that all the curves in Figure 6.5 intersect at the isoelectric pH of the protein, where there is zero net charge on the molecule. It is also seen that the partition coefficient is greatly affected by whether chloride or sulfate salts are used. Thus, the effect of any change in the salt type in an aqueous two-phase system should be checked experimentally.

The selectivity of the extraction for the protein of interest may be enhanced by the coupling of a biospecific ligand to one of the polymers. PEG is a polymer often selected for conjugation because chemical techniques for modifying PEG are simple, highly efficient, and well known. By coupling a biospecific ligand to PEG and using conditions that favor the partition of contaminants to the bottom phase, the partition coefficient of a desired protein may be greatly increased. The most common strategy is to attach the ligand to the PEG polymer by derivatization of the free hydroxyls at each end of the polymer chain. Reactive intermediates have been formed by conversion of the PEG terminal hydroxyl to halides, sulfonate esters, or epoxide derivatives, which couple to a ligand to form a polymer with a protein binding site on each end [5]. Ligands from the reactive dyes, including Cibacron blue, Procion red, and Procion yellow, have been used for the affinity separation of many proteins. The dye functions as a competitive inhibitor for the substrate, coenzyme, or effector of a variety of proteins, often with an affinity greater than that shown by the competitive molecule [6]. The improvement in the partition coefficient for three enzymes when

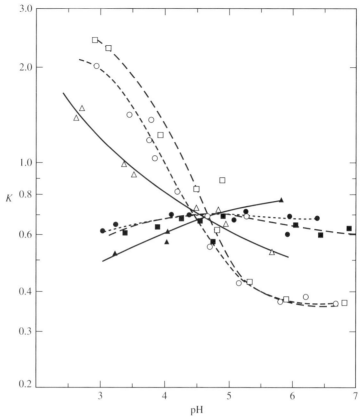

Figure 6.5 Dependence of ovalbumin partitioning on solution pH and salt type in a PEG 6000–dextran 500 system. Open symbols denote chloride salts, solid symbols, sulfates (squares, potassium; circles, sodium; triangles, lithium). (Data from H. Walter, S. Sasakawa, and P.-Å. Albertsson, "Cross-partition of proteins: Effect of ionic composition and concentration," *Biochemistry,* vol. 11, p. 3880, 1972.)

Procion yellow was attached to PEG in the PEG–dextran aqueous two-phase extraction system is shown in Figure 6.6.

After binding of the protein to the ligand-polymer, it is necessary to recover the protein in free form. This has been performed by adding salt to the top phase, rich in ligand-polymer, yielding two phases and consequent partitioning of the protein to the resulting bottom phase [7]. In another approach, a soluble effector was added to a new two-phase system to compete with the bound ligand for the protein's binding site, causing the protein to be released from the ligand-polymer and shift to the bottom phase [8].

6.2.2 COUNTERCURRENT STAGE CALCULATIONS

To achieve high bioproduct recovery in extraction, it is frequently necessary to use more than one extraction stage. Extraction processes are usually set up where the extraction solvent and the feed run countercurrent to each other. Countercurrent flow is preferable to

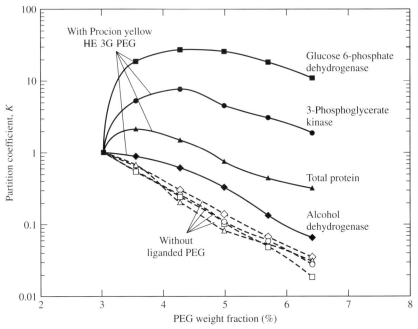

Figure 6.6 Affinity partitioning of enzymes in a PEG 6000–dextran 500 system with and without Procion yellow–PEG. (Data from G. Johansson and M. Andersson, "Parameters determining affinity partitioning of yeast enzymes using polymer-bound triazine dye ligands," *J. Chromatogr.,* vol. 303, p. 39, 1984.)

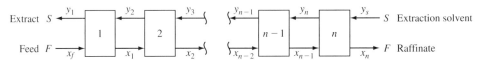

Figure 6.7 Countercurrent extraction cascade with n equilibrium stages, showing stream variables.

cocurrent flow because the solute concentration difference between the raffinate and extraction phases, the driving force for the process, is greater in the countercurrent case. Countercurrent extraction is analyzed both graphically and analytically for equilibrium in each stage of contacting. This analysis is helpful in determining the balance between the number of stages and the ratio of the extraction solvent flow rate to the feed flow rate. This analysis can also be used to determine the extraction efficiency of pilot plant and plant scale extractors.

A countercurrent extraction cascade is shown schematically in Figure 6.7 for n stages, each like that shown in Figure 6.1. The streams leaving each stage are in equilibrium and are numbered according to the stage they are leaving. Feed enters at stage 1 and leaves at stage n, while the extraction solvent is flowing in the opposite direction. Once the feed has entered the cascade, it is called the "raffinate." For these calculations, we make three key assumptions:

1. The two solvents are immiscible or are already in phase equilibrium (i.e., no solvent or polymer is exchanged between phases).

2. The solute concentrations are sufficiently low that the flow rates of the raffinate and extract are constant.

3. Equilibrium is achieved in each stage.

The first assumption depends on the solvents that are chosen. Usually, the extraction solvent is selected to minimize its solubility in the raffinate; otherwise, the loss of extraction solvent in the raffinate can be so great that it must be recovered for the process to be economically viable. The second assumption typically holds for the extraction of bioproducts. Concentrations of bioproducts in fermentation broths are typically less than 10 g/liter (or 1%) and often are less than 1 g/liter (0.1%), which means that the error caused by the second assumption is usually less than the error in the assay of the bioproduct. Should deviations from the second assumption become significant, then F and S should be the flow rates of pure solvent and the concentrations should be ratios of solute to pure solvent.

A material balance on the solute around the feed end of the cascade down to stage $n - 1$ is

$$x_f F + y_n S = x_{n-1} F + y_1 S \tag{6.2.4}$$

where F is the flow rate of feed or raffinate phase, S is the flow rate of extract phase, and the concentrations are defined as in Figure 6.7. This equation can be written in terms of y_n as

$$y_n = \frac{F}{S} x_{n-1} + \frac{y_1 S - x_f F}{S} \tag{6.2.5}$$

From Equation (6.2.5) it is seen that y_n and x_{n-1} are concentrations of passing streams on a line of slope F/S, which is called the *operating line*. The calculation of the number of stages required to reduce the concentration of solute in the feed stream to x_n can be carried out graphically on a McCabe–Thiele-type diagram with y plotted versus x as shown in Figure 6.8. The cascade operating line of slope F/S intersects the x axis at point (x_n, y_s). In drawing point (x_n, y_s), it is assumed here that $y_s = 0$ (since extraction solvent is usually solute free). The *equilibrium curve*, not necessarily a straight line [i.e., the partition coefficient, defined by Equation (6.2.1), may not be constant for the range of concentrations that apply], originates at $x = y = 0$. The stepping off of equilibrium stages starts on the operating line where the feed enters and the extract exits the cascade (x_f, y_1). The concentrations x_1 and y_1 are in equilibrium and are thus on the equilibrium curve. One equilibrium stage is graphed as the horizontal line drawn from (x_f, y_1) that intersects the equilibrium curve at (x_1, y_1); a vertical line is drawn from (x_1, y_1) that intersects the operating line at (x_1, y_2). This procedure is continued until y_s is reached.

For isothermal, dilute solutions, the partition coefficient K is often constant. For this situation, an analytical solution can be developed for the concentration of solute in the raffinate as a function of the concentration in the feed and other process variables. It is convenient to define an *extraction factor* E as the slope of the equilibrium line divided by the slope of the operating line:

$$E = \frac{K}{F/S} = \frac{KS}{F} \tag{6.2.6}$$

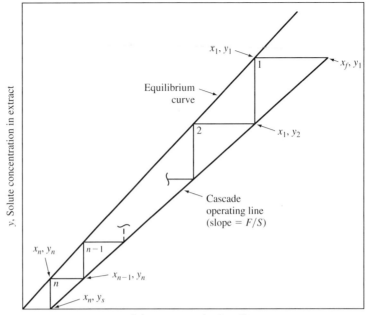

Figure 6.8 Graphical calculation of equilibrium stages for countercurrent extraction.

A material balance on the solute for the nth stage, assuming no solute in the entering extraction solvent ($y_s = 0$), gives

$$x_{n-1}F = x_n F + y_n S \tag{6.2.7}$$

Using the definitions of K and E leads to

$$x_{n-1} = (E + 1)x_n \tag{6.2.8}$$

A material balance on the solute for the $(n - 1)$th stage results in

$$x_{n-2} = (E^2 + E + 1)x_n \tag{6.2.9}$$

Continuing this procedure until stage 1 gives

$$x_f = (E^n + E^{n-1} + \cdots + E^2 + E + 1)x_n \tag{6.2.10}$$

which is mathematically equivalent to

$$x_f = \left(\frac{E^{n+1} - 1}{E - 1} \right) x_n \tag{6.2.11}$$

or

$$x_n = \left(\frac{E - 1}{E^{n+1} - 1} \right) x_f \tag{6.2.12}$$

This equation can be rearranged to give the more commonly used form

$$n = \frac{\ln\left[\dfrac{x_f}{x_n}(E-1) + 1\right]}{\ln E} - 1 \tag{6.2.13}$$

As the number of stages becomes large ($n \to \infty$), we see from Equation (6.2.11) that

$$x_n \to \frac{x_f}{E^n} \to 0 \tag{6.2.14}$$

which we know by intuition—we can extract almost all the solute with a very large number of stages.

For the special case of the extraction factor $E = 1.0$, we can see from Equation (6.2.10) that

$$x_n = \frac{x_f}{n+1} \tag{6.2.15}$$

For values of $E < 1.0$ and a large number of extraction stages ($n \to \infty$), it is clear from Equation (6.2.12) that

$$x_n \to (1 - E)x_f \tag{6.2.16}$$

EXAMPLE 6.1

Separation of a Bioproduct and an Impurity by Countercurrent Extraction A countercurrent extractor with four equilibrium stages is available for separating a desired bioproduct from a contaminating impurity, which is 10% of the weight of the bioproduct in a feed stream. For the extraction solvent being used, which is immiscible with the feed stream, the bioproduct has a partition coefficient K of 10, while the impurity has $K = 1$. For an S/F ratio of 0.2, what will be the ratio of impurity to bioproduct in the extract phase at the outlet of the extractor?

SOLUTION

We can use Equation (6.2.12) to calculate the ratio of bioproduct or impurity in the outlet raffinate to that in the inlet feed:

$$\text{Bioproduct:} \quad E = \frac{KS}{F} = 10(0.2) = 2.0$$

$$\frac{x_n}{x_f} = \frac{E-1}{E^{n+1} - 1} = \frac{2-1}{2^{4+1} - 1} = 0.0323$$

$$\text{Impurity:} \quad E = 1.0(0.2) = 0.2$$

$$\frac{x_n}{x_f} = \frac{0.2 - 1}{0.2^{4+1} - 1} = 0.800$$

To calculate the ratio of impurity to bioproduct in the outlet extract, we use 100 g of bioproduct in the feed as a basis:

$$\text{Ratio of impurity to bioproduct in the extract} = \frac{10 - 0.8(10)}{100 - 0.0323(100)} = 0.021$$

EXAMPLE 6.2

Effect of Solvent Rate in Countercurrent Staged Extraction of an Antibiotic   An antibiotic has been found to have a partition coefficient K of 8.4 at pH 10.0 in a system with aqueous and butyl acetate phases. The pK_a of the antibiotic is 8.5, due to a single tertiary amine group on the molecule. It is desired to recover 95% of this antibiotic by countercurrent staged extraction with butyl acetate from an aqueous phase feed at a pH of 10.0. The flow rate of the feed is 200 liters/min.

Determine the dependence of the number of ideal stages on the flow rate of the extract phase. In what flow rate range would there likely be an economic optimum? Why is it desirable to do the extraction at a pH of 10.0?

SOLUTION

To solve this problem, we need an expression relating the number of ideal countercurrent stages n to the flow rate of extraction solvent S. Substituting into Equation (6.2.13) for E gives

$$n = \frac{\ln\left[\frac{x_f}{x_n}\left(\frac{KS}{F} - 1\right) + 1\right]}{\ln\left(\frac{KS}{F}\right)} - 1$$

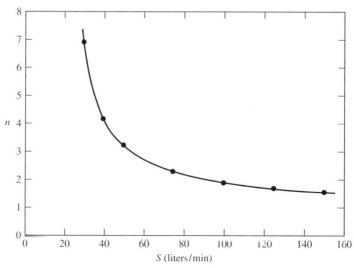

Figure E6.2 Dependence of the number n of theoretical extraction stages on the flow rate S of the butyl acetate extract phase.

The above equation assumes that water and butyl acetate are immiscible. Since the solubility of water in butyl acetate is less than 1%, this is a good assumption. For this problem we know that the ratio $x_n/x_f = 0.05$. By substituting the values for x_n/x_f, K, and F, and various values of S in this equation, we obtain a plot of n versus S (Figure E6.2).

The determination of an exact economic optimum would involve calculating the total annualized cost as a function of the extract flow rate, which would require putting the equipment costs on an annualized basis. Nevertheless, the range of the extraction solvent flow rate where an economic optimum lies can be estimated from the n versus S curve. From this plot, below an extract flow rate of 30 liters/min, the number of ideal stages appears to increase exponentially; while above a flow rate of about 80 liters/min, the change in the number of stages with the flow rate becomes relatively small. Thus, the economic optimum will likely be in the butyl acetate flow rate range of 30–80 liters/min.

Above its pK_a of 8.5, the tertiary amine becomes progressively more unprotonated, while below this pK_a the reverse is true. At a pH of 10.0, this amine is essentially unprotonated, and the antibiotic is thus more readily extracted into the more hydrophobic butyl acetate phase than if it were protonated.

6.3 Scaleup and Design of Extractors

Although many different types of extractors are used at large scale in the chemical industry, only a few are used in the biotechnology field. By far the most commonly used extractors in biotechnology are the mechanically agitated extraction columns and the centrifugal extractors [9]. Of the mechanically agitated extraction columns, the reciprocating-plate extraction columns have been the most widely used. Centrifugal extractors have been widely used in the recovery of antibiotics, and probably the most widely used extractor of this type is the Podbielniak centrifugal extractor. A summary of the general features of these two extractors is given in Table 6.2. Both these extractors will be discussed in detail.

Two concepts that are useful in scaleup and design are *overall stage efficiency* and *height equivalent to a theoretical stage (HETS)*. The column may be treated as a stack of stages, in which the product is at equilibrium between the two phases in each stage as in

TABLE 6.2
Summary of Features of Two Extractors Commonly Used in Biotechnology[a]

Types of extractor	General features
Reciprocating-plate column	High throughput, low HETS; great versatility and flexibility; simplicity in construction; handles liquids containing suspended solids and mixtures with emulsifying tendencies
Centrifugal extractor	Short contacting time for unstable material; limited space required; handles easily emulsified material and systems with little liquid density differences

[a]See reference 10.

Figure 6.7. These variables are

$$\text{Percent stage efficiency} = \frac{\text{number of theoretical stages}}{\text{number of actual stages}} \times 100 \qquad (6.3.1)$$

$$\text{HETS} = \frac{\text{height of extractor}}{\text{number of theoretical stages}} \qquad (6.3.2)$$

The number of theoretical stages is defined by Equation (6.2.13). Design goals are to maximize stage efficiency and minimize HETS.

6.3.1 RECIPROCATING-PLATE EXTRACTION COLUMNS

The most important feature for biotechnology applications of reciprocating-plate extraction columns is their ability to handle liquids with suspended solids and mixtures that emulsify easily. In these columns, first developed by Karr [11], interdispersion is achieved by reciprocating or vibrating plates. The column consists of a stack of perforated plates, which are reciprocated vertically by means of a reciprocating drive mechanism located at the top of the column (Figure 6.9). The amplitude of reciprocation is variable and generally in the range 3 to 50 mm; the reciprocation frequency is variable up to 1000 strokes/min.

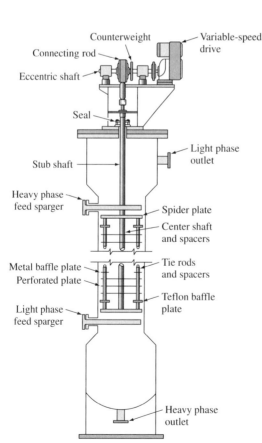

Figure 6.9 Schematic diagram of a 900 mm (36 in.) diameter reciprocating-plate extraction column. (Drawing based on a design by Glitsch Process Systems Inc.)

Research studies on the hydrodynamics and axial mixing in reciprocating-plate extraction columns have shown that the presence of many closely spaced plates leads to conditions that closely approximate uniform isotropic turbulence [12]. The more uniform distribution of energy dissipation in these columns gives high turbulence of interdispersing phases and low axial mixing, and thus gives high mass transfer rates and low HETS. Thus commercial columns are designed with optimal plate spacing built in.

This column obeys the relationship for the number of theoretical stages in a countercurrent staged extractor [Equation (6.2.13)] and is further scaled on the basis of column diameter D and rate of reciprocation of the plates, SPM. Based on performance data obtained for columns having diameters of 25, 76, 305, and 914 mm (1, 3, 12, and 36 in.), the following equations have been developed for the scaleup of reciprocating-plate extractors [13]:

$$\frac{(\text{HETS})_2}{(\text{HETS})_1} = \left(\frac{D_2}{D_1}\right)^{0.38} \tag{6.3.3}$$

$$\frac{(\text{SPM})_2}{(\text{SPM})_1} = \left(\frac{D_1}{D_2}\right)^{0.14} \tag{6.3.4}$$

where subscripts 1 and 2 refer to the small- and large-scale columns, respectively. A successful design of a large-scale reciprocating-plate extractor can be achieved after optimization experiments have been performed on the pilot plant scale. The following procedures are recommended for the scaleup:

1. Data such as x_n/x_f at specified S and F are obtained in a column of 25, 51, or 76 mm (1, 2, or 3 in.) diameter.

2. The optimum performance of the pilot column is determined. The criterion for optimum performance is maximum volumetric efficiency in a column having optimal plate spacing.

3. When scaling up from the pilot scale data, the following parameters are held constant: solvent-to-feed ratio S/F, throughput per column cross-sectional area $(S + F)/A$, plate spacing, and stroke length.

4. The expected minimum HETS in the large diameter column is calculated from Equation (6.3.3). The height of the extractor is calculated from Equation (6.3.2), with the number of theoretical stages estimated from Equation (6.2.13), which is based on the three assumptions introduced in Section 6.2.2.

5. The corresponding reciprocating speed required is calculated from Equation (6.3.4).

6. Suitable baffle plates in design and spacing are provided.

Column volumetric efficiency is defined as

$$\text{Volumetric efficiency} = \frac{\text{total volumetric throughput/column cross-sectional area}}{\text{HETS}} \tag{6.3.5}$$

where total volumetric throughput is defined as the sum of the feed flow rate and the solvent flow rate, $S + F$. This procedure has been used to successfully scale up columns to a diameter of 1.5 m (5 ft) [12] and to a plate stack height of 12.2 m (40 ft) [13].

EXAMPLE 6.3

Scaleup of a Reciprocating-Plate Extraction Column   The operation of a pilot scale reciprocating-plate extraction column has been optimized for the extraction of an antibiotic from whole fermentation broth using amyl acetate as solvent. The antibiotic has a partition coefficient K of 7.5. The optimal operating conditions are as follows: solvent flow rate of 105 ml/min, flow rate of fermentation broth of 70 ml/min, and ratio of antibiotic in raffinate to antibiotic in feed of 0.07. The column was 2.54 cm in diameter, and the height of the extractor (height of the reciprocating plates) was 1.83 m. The agitator speed was 280 strokes/min. What column size and agitator speed are required to give a ratio of antibiotic in the raffinate to antibiotic in the feed of 0.03 and to handle fermentation broth at a rate of 150,000 liters every 12 h? (*Note:* 150,000 liters is a typical volume for a plant antibiotic fermentor.)

SOLUTION

Our first objective is to find the HETS for the optimal pilot plant run. We know for the pilot run that

$$\frac{x_f}{x_n} = \frac{1}{0.07} = 14.29$$

$$\frac{S}{F} = \frac{105}{70} = 1.5$$

We let subscripts 1 and 2 denote the pilot plant and plant scales, respectively. Since the solubility of amyl acetate in water is well below 1%, we make the assumption of immiscible phases. Thus, we can use Equation (6.2.13) to determine the number of theoretical stages

$$n_1 = \frac{\ln\left[\frac{x_f}{x_n}\left(\frac{KS}{F} - 1\right) + 1\right]}{\ln\left(\frac{KS}{F}\right)} - 1 = \frac{\ln[14.29(7.5 \times 1.5 - 1) + 1]}{\ln(7.5 \times 1.5)} - 1 = 1.06$$

and from Equation (6.3.2)

$$(\text{HETS})_1 = \frac{\text{height of plates}}{n_1} = \frac{1.83}{1.06} = 1.73 \text{ m}$$

For the large-scale unit, we can first find the diameter by maintaining a constant total throughput per unit cross-sectional area. For the pilot unit

$$\frac{\text{Total throughput}}{\text{Cross-sectional area}} = \frac{70 + 105}{\pi(2.54)^2/4} = 34.6 \frac{\text{ml}}{\text{min cm}^2}$$

Maintaining S/F constant at large scale gives

$$(\text{Total throughput})_{\text{large scale}} = \frac{(105/70)(150,000) + 150,000}{12} \frac{\text{liters}}{\text{h}} \times \frac{1 \text{ h}}{60 \text{ min}}$$

$$= 520.8 \frac{\text{liters}}{\text{min}}$$

We can now solve for the diameter of the large column:

$$D = \left(\frac{4 \times 520{,}800}{34.6\pi}\right)^{1/2} = 136.3 \, \text{cm}$$

The HETS for the large-scale column can be found by using Equation (6.3.3):

$$(\text{HETS})_2 = (\text{HETS})_1 \left(\frac{D_2}{D_1}\right)^{0.38} = (1.73)\left(\frac{138.5}{2.54}\right)^{0.38} = 7.91 \, \text{m}$$

To find the height of the large-scale column, we need to use the desired concentration of antibiotic in the raffinate to calculate the number of theoretical stages for this column:

$$n_2 = \frac{\ln[(1/0.03)(7.5 \times 1.5 - 1) + 1]}{\ln(7.5 \times 1.5)} - 1 = 1.41$$

From Equation (6.3.2)

$$(\text{Height of plates})_{\text{large column}} = n_2(\text{HETS})_2 = 1.41(7.91) = 11.2 \, \text{m}$$

The agitator speed at large scale can be found by using Equation (6.3.4):

$$(\text{SPM})_2 = (\text{SPM})_1 \left(\frac{D_1}{D_2}\right)^{0.14} = (280)\left(\frac{2.54}{138.5}\right)^{0.14} = 160 \, \text{strokes/min}$$

Thus, the large-scale column is not wider or higher than the largest columns of these types that have been built.

6.3.2 CENTRIFUGAL EXTRACTORS

Centrifugal extractors were first developed for biotechnology processing, and that is still their main application. These extractors are the most popular in antibiotics production, where the large volumes of fermentation broth that must be processed are easily emulsified. Key advantages of centrifugal extractors are their ability to avoid emulsions and to separate liquid phases with small density differences (as small as 0.01 g/cm^3).

Centrifugal extractors consist of several concentric perforated cylinders that rotate around a shaft. The principle of this design is to feed the heavy phase near the axis of the rotating cylinder, while the light phase is fed close to the periphery. Figure 6.10 is a cross-sectional diagram of the widely used Podbielniak centrifugal extractor that indicates the flows of the heavy and light phases. The centrifugal force applied causes the two phases to move radially in a countercurrent mode, such that the heavy phase moves to the periphery and the light phase to the axis of rotation. The capabilities of the various sizes of Podbielniak centrifugal extractors for the extraction of fermentation broth are given in Table 6.3.

The scaleup of Podbielniak centrifugal extractors uses the concept of *equivalent time*, Gt, discussed earlier for centrifuges [see Equation (5.3.5)]. First, optimum performance of the pilot centrifugal extractor is determined. The ratio of feed volume to solvent volume is kept constant upon scaleup. The contact time t for the feed stream in the centrifuge is calculated for the large-scale extractor, assuming that Gt is constant in the scaleup. The feed rate at large scale is calculated from this contact time [14].

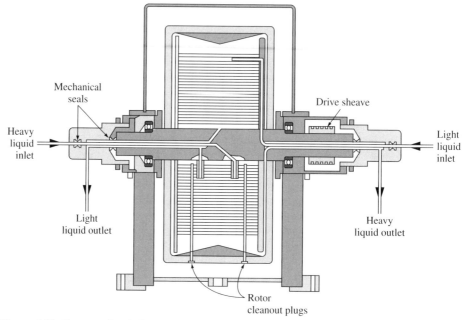

Figure 6.10 Cross-sectional diagram of a Podbielniak centrifugal extractor.

TABLE 6.3
Capabilities of Podbielniak Centrifugal Extractors for Fermentation Broth Extraction

Capacity $(m^3/h)^a$	Maximum speed (rpm)	Maximum centrifugal force $\times g$ $(G,$ dimensionless acceleration)
0.05	10,000	11,400
6.0	3,200	3,600
15.0	2,100	2,300
30.0	2,100	2,300

^{a}Sum of feed and solvent flow rates.

Source: B&P Process Equipment and Systems, 2000.

6.4 Summary

In extraction, two phases come into contact with the goal of transferring a solute or particle from one phase to the other. Liquid–liquid extraction is the type of extraction that is most commonly used for bioproducts. Liquid–liquid extraction aids in product isolation and can be used for partial purification of solutes when used in the multistage mode.

- The key physical constants used in extraction are the partition coefficient $K = y/x$, the ratio of the concentration of a solute in the extract phase to that in the raffinate phase at equilibrium; the solvent-to-feed ratio S/F, the extract phase volume or flow rate divided by the feed phase volume or flow rate; and the extraction factor $E = KS/F$.

- Chemical conditions and properties have predictable effects on the partition coefficient. Both pH and the type of salt or salts present generally have a strong effect on K.

- An equilibrium phase diagram for aqueous two-phase systems consists of a plot of the concentration of a polymer enriched in one of the phases against the concentration of the another polymer or salt enriched in the other phase. Points formed by phases in equilibrium on the equilibrium (binodal) curve are connected by tie lines.

- The extraction of enzymes by affinity partitioning consists of coupling one of the polymers, usually poly(ethylene glycol) (PEG), to an affinity ligand to which the target protein will bind.

- In an n-stage countercurrent extraction cascade, the fraction of solute extracted is calculated from the relationship

$$\frac{x_n}{x_f} = \frac{E - 1}{E^{n+1} - 1}$$

where n is the number of theoretical stages, x_f is the feed concentration, and x_n is the concentration in the exit raffinate.

- An equilibrium line is constructed as a plot of y versus x, and an operating line is constructed by plotting the material balance on the same axes. The combined plot is known as a McCabe–Thiele diagram and is used to calculate the number of equilibrium stages required by starting at $x = x_f$ and reducing x by graphing steps between the two plots until the desired x_n is reached.

- A reciprocating-plate extractor consists of a vertical column through which phases flow counter to each other and in which interdispersion of phases is driven by reciprocating plates. Scaling rules for column height and speed of the reciprocating perforated plates are based on a power of the ratio of the large-scale diameter to the small-scale diameter.

- Centrifugal (Podbielniak) extractors effect phase separation by applying a centrifugal force field as the phases flow countercurrent to each other through perforated cylinders to effect contacting. Scaling rules are those applied to centrifugation.

NOMENCLATURE

a_{ij}	second virial coefficient for interaction between protein i and polymer j (liter mol^{-1})	g	gravitational acceleration (= 9.807 m s^{-2})
A	cross-sectional area (cm^2)	G	multiple of gravitational acceleration (= $\omega^2 R/g$) (dimensionless)
D	column diameter (cm)	HETS	height equivalent to a theoretical stage [Equation (6.3.2)] (cm)
E	extraction factor (= KS/F)		
F	feed flow rate (= raffinate flow rate when solvents are immiscible and solute concentrations are low) (liter h^{-1})	k	Boltzmann's constant (= 1.3807×10^{-23} J K^{-1})
		K	partition coefficient (= y/x)
		M	molecular weight (Daltons)
F	Faraday number (9.6485×10^4 C g-mol^{-1})	R	gas constant (8.3145 J g-mol^{-1} K^{-1})

S extraction solvent flow rate ($=$ extract flow rate when solvents are immiscible and solute concentrations are low) (liter h^{-1})

SPM   rate of reciprocation of plates (strokes min^{-1})

t time (h)

x concentration of solute in raffinate phase or heavy phase (M, or g liter^{-1})

y concentration of solute in extract phase or light phase (M, or g liter^{-1})

z_p net surface charge of protein

Greek Letters

λ lumped parameter in equation for partition coefficient K [Equation (6.2.2)]

ϕ electrical potential (mV)

PROBLEMS

6.1   **Two-Stage Countercurrent Extraction**   In a two-stage countercurrent extraction of a pharmaceutical product, what is the relationship between the feed concentration and the final raffinate concentration in terms of the extraction factor E if both stages are at equilibrium? For a partition coefficient K of 5.0 and a solvent-to-feed ratio (S/F) of 0.5, what will be the ratio of the raffinate concentration to the feed concentration?

6.2   **Purification Factor for Extraction of an Enzyme**   You are preparing an industrial enzyme, alcohol dehydrogenase, from yeast using two-phase aqueous extraction. The crude, clarified extract contains only proteins and has a specific activity of 200 units/g protein. After a single stage of affinity extraction using a Cibacron dye–PEG complex, the extract contains 400,000 units of enzyme activity and 20 g of protein. What is the purification factor for this step?

6.3   **Extraction Conditions for Various Bioproducts**   For the following bioproducts, outline how extraction can be used in the purification process using information obtained from work reported in the literature:
(a) Magnamycin, a macrolide antibiotic from *Streptomyces halstedii*
(b) Glumamycin, a peptide antibiotic from *Streptomyces zaomyceticus*
(c) β-Galactosidase, an enzyme from *Escherichia coli*
(d) Pullulanase, an enzyme from *Klebsiella pneumoniae*

Give as much information as possible about what the process conditions should be (solvents, pH, concentration, etc.).

6.4   **Required Solvent-to-Feed Ratio in Countercurrent Extraction**   In a countercurrent, equilibrium staged extractor with four equilibrium stages, determine the necessary ratio of extract to feed (S/F) to purify bioproduct A to 90% purity from contaminant B. The feed contains two components, 60% A and 40% B, and the partition coefficients are $K_A = 13.0$ and $K_B = 1.0$.

6.5   **Purification of an Antibiotic by Countercurrent Extraction**   A countercurrent extraction unit with three equilibrium stages is used for the separation of a desired antibiotic (partition coefficient $= 6.0$) from a major contaminant (partition coefficient $= 1.0$) in an aqueous feed stream. The feed (or raffinate) and extract phase flow rates are equal. What fraction of each is discarded from the raffinate? Assuming that the feed contains only these two components at an antibiotic-to-contaminant ratio of 3:1, what is the purity of the antibiotic in the exit extract from stage 3? What do you conclude about the effectiveness of this extraction as a means of purification of the antibiotic?

6.6   **Mixer–Settler Extraction System**   You are assigned the task of extracting zithramycin with a yield of at least 90% from a clarified fungal fermentation broth, using a four-stage mixer–settler extraction unit in the pilot plant. (One mixer–settler constitutes one equilibrium stage in the countercurrent extraction cascade shown in Figure 6.7.) The *n*-butanol extraction solvent flows at 20 liters/h, while the clarified broth flows at 30 liters/h. The partition coefficient of zithramycin is adjustable by changing the pH. What minimum value of the partition coefficient should you use for the successful completion of your assigned task?

6.7   **Graphical Equilibrium Stage Calculations for Extraction of a Peptide**   The equilibrium partitioning of a peptide between an aqueous feed phase and an organic solvent extract phase has been found to be nonlinear and can be represented by the following equation:

$$y = 1.47 \ln x + 3.96$$

where y and x are concentrations of the peptide in the extract and aqueous feed (or raffinate) phases, respectively, in grams per liter. It is desired to extract 95% of the peptide from a feed stream having a peptide concentration of 4.0 g/liter. For a feed stream at a flow rate of 5.0 liters/min and an extract stream at a flow rate of 3.3 liters/min, graphically estimate how many equilibrium stages will be required for countercurrent flow of the phases. What is the concentration of the peptide in the exit extract stream? As the concentration of the peptide in the raffinate decreases, does the partitioning of the peptide into the extract become more or less favorable?

6.8 **Scaleup of a Podbielniak Centrifugal Extractor** Pilot plant tests with a Podbielniak centrifugal extractor indicated that excellent recovery (>95%) of a desired bioproduct could be obtained from filtered fermentation broth by extraction with an immiscible organic solvent. The flow rates were the following:

Filtered broth (aqueous) flow rate = 500 ml/min
(continuous phase)

Organic solvent flow rate = 125 ml/min
(dispersed phase)

The pilot plant extractor delivers a centrifugal force of $11,400 \times g$, and the rotating cylinder inside the extractor has a diameter of 20 cm and is 2.5 cm wide.

You have been asked to scale up this extraction by using a larger Podbielniak extractor, which delivers $2300 \times g$ and has a rotating cylinder with a diameter of 91 cm and a width of 91 cm. What flow rates should be used in the larger extractor to achieve the same recovery of bioproduct?

TABLE P6.9

Run number	Flow rates (ml/min)		Antibiotic concentration in raffinate (mg/liter)
	Broth	Chloroform	
1	45	135	2
2	67.5	135	3
3	125	135	30
4	80	120	5
5	100	150	7
6	120	180	9
7	150	225	25

6.9 **Scaleup of Pilot Plant Tests of a Reciprocating-Plate Extraction Column** The pilot plant data in Table P6.9 are for the extraction of an antibiotic from whole fermentation broth using the solvent chloroform in a reciprocating-plate extraction column. The concentration of the antibiotic was 1.4 g/liter in the broth. The column had a diameter of 2.54 cm, and the height of the plates was 3.05 m. The partition coefficient K for the antibiotic is known to be 2.68.

For each pilot run, determine the diameter and height of the plates for the plant column that would be required for processing 50,000 liters of broth in 12 h to give an exit raffinate concentration of antibiotic of 10 mg/liter, assuming a concentration of antibiotic in the feed broth of 1.0 g/liter (a spreadsheet is convenient for these calculations). Without doing a complete economic analysis, in your judgment which scaled-up pilot run appears to be optimum? (Data from A. E. Karr, W. Gebert, and M. Wang, *Can. J. Chem. Eng.*, vol. 58, p. 249, 1980.)

References

1. Brønsted, J. N. Z. (1931). Molecular magnitude and phase distribution, I. *Z. Phys. Chem.*, Abt. A, vol. 155, p. 257.
2. Diamond, A. D., and Hsu, J. T. (1992). Aqueous two-phase systems for biomolecule separation. *Adv. Biochem. Eng. Biotechnol.*, vol. 47, p. 89.
3. Kelley, B. D., and Hatton, T. A. (1993). Protein purification by liquid–liquid extraction. In *Biotechnology*, vol. 3, *Bioprocessing*, G. Stephanopoulos, ed., VCH Publishers, Weinheim, p. 593.
4. King, R. S., Blanch, H. W., and Prausnitz, J. M. (1988). Molecular thermodynamics of aqueous two-phase systems for bioseparations. *AIChE J.*, vol. 34, p. 1585.

5. Harris, J. M., and Yalpani, M. (1985). Polymer-ligands used in affinity partitioning and their synthesis. In *Partition in Aqueous Two-Phase Systems. Theory, Methods, Uses, and Applications,* H. Walter, D. E. Brooks, and D. Fisher, eds., Academic Press, Orlando, FL, p. 589.

6. Stellwagen, E. (1990). Chromatography on immobilized reactive dyes. In *Guide to Protein Purification,* M. P. Deutscher, ed., Academic Press, San Diego, CA, p. 343.

7. Cordes, A., and Kula, M.-R. (1986). Process design for large-scale purification of formate dehydrogenase from *Candida boidinii* by affinity partition. *J. Chromatog.,* vol. 376, p. 375.

8. Kopperschlager, G., and Birkenmeier, G. (1990). Affinity partitioning and extraction of proteins, *Bioseparations,* vol. 1, p. 235.

9. Schugerl, K. (1994). *Solvent Extraction in Biotechnology,* Springer-Verlag, Berlin.

10. Lo, T. C., and Baird, M. H. I. (1997). Extraction, liquid–liquid. In *Encyclopedia of Separation Technology,* vol. 1, D. M. Ruthven, ed., Wiley, New York, p. 760.

11. Karr, A. E. (1959). Performance of a reciprocating plate column, *AIChE J.,* vol. 5, p. 446.

12. Lo, E. C. (1997). Commercial liquid–liquid extraction equipment. In *Handbook of Separation Techniques for Chemical Engineers,* P. A. Schweitzer, ed., McGraw-Hill, New York.

13. Karr, A. E. (1980). Design, scale-up, and applications of the reciprocating plate extraction column. *Sep. Sci. Technol.,* vol. 15, p. 877.

14. Creegan, R. (2000). B&P Process Equipment and Systems, personal communication.

Liquid Chromatography
and Adsorption

Liquid chromatography and adsorption processes are based on the differential affinity of various soluble molecules for specific types of solids. In these processes, equilibrium is approached between a solid phase, often called the resin, or *stationary phase,* and the soluble molecules in the liquid phase. The solid phase is "stationary" because it is often packed in a fixed column. Since the liquid phase is often flowing past the solid phase, it is referred to as the *mobile phase*.

Chromatography is traditionally defined as a sorptive separation process in which a mixture of solutes in a feed solution is introduced at the inlet of a column containing the stationary phase and separated over the length of the column by the action of an elution solvent that is continuously fed to the column after the feed has been introduced. Figure 7.1 shows the method of separating solutes known as *elution chromatography.* By contrast, in pure adsorption processes, where there is no chromatographic effect of different solute bands being separated on the column, the stationary phase is generally uniformly saturated with the target solute before the introduction of an elution solvent which causes the solute to transfer into the mobile phase; this type of operation is sometimes referred to as *on–off chromatography* [1]. Chromatography is usually performed downflow through a stationary phase in a fixed bed, although a recent advance is to operate the column upflow during the feed step to allow the bed to expand with minimal mixing of the adsorbent phase, which can accommodate whole cells and cell fragments in the feed [2].

Besides the traditional elution mode for chromatography, other modes of chromatography are sometimes used, including the *frontal mode* and the *displacement mode*. In frontal chromatography, the target solute does not adsorb to the column, while some impurities are strongly retained by the stationary phase. The feed continues until the front of adsorbed impurities approaches the column outlet. An example of frontal chromatography would be the use of an anion exchanger to adsorb DNA from a protein product stream. A pH is chosen below the isoelectric point of the protein, making it positively charged and unable to bind to the anion exchange column. DNA is adsorbed because it is negatively charged at almost all pH values. This technique allows processing of large amounts of material with a relatively small column, because contaminants are usually present in small amounts relative to the product.

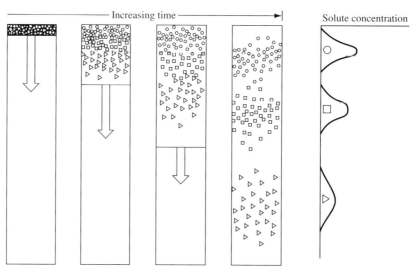

Figure 7.1 Schematic illustration of elution chromatography. Three solutes, fed to a column containing a stationary phase, separate in the column as the elution solvent flows through the column. The three solutes separate as a result of differing affinities for the stationary phase.

In displacement chromatography, the column is loaded with the target solute to a fraction of its capacity and then is eluted by continuously introducing an even more strongly adsorbing substance, known as the displacer. For sufficiently long columns, the solutes in the feed become distributed in a pattern of adjacent pure solute bands, where each upstream solute acts as a displacer for each downstream solute located in the band immediately downstream [3].

Liquid chromatography is a commonly used separation process for biologics and fine and specialty chemicals, while adsorption is widely used for high throughput processes for small biomolecules such as antibiotics, sugars, and amino acids. As a unit operation, chromatography is a specialized version of adsorption-based separations, and many of the concepts taught in this chapter apply to both.

Adsorption is based on a variety of chemistries, some highly customized. At equilibrium, the individual components in the mixture to be separated are distributed between the stationary and mobile phases. This equilibrium can be modified by standard thermodynamic manipulations, such as temperature or composition (chemical activity) of the mobile phase. The most common thermodynamic variable to be modified is the composition of the mobile phase. Salt concentration, organic solvent concentration, and pH are common mobile phase thermodynamic characteristics that can be modified easily.

As in all separation operations, transport processes play a major role. Transport of chemical species from the mobile phase to the stationary phase takes place through a stagnant mass transfer layer at the surface of the particle, and solutes diffuse through pores to a binding site. The kinetics of the binding reaction can sometimes affect the separation. Axial and radial mixing can also be important factors in the separation, especially if the resin bed has not been well packed. Finally, the peripherals to the chromatography or adsorption column, such as the pumps, detectors, injectors, and filters, potentially add dead space and mixing volumes that affect the ultimate quality of the separation.

Adsorption equilibrium is described first in this chapter, followed by the dynamics of adsorption and chromatography. Both fixed-bed adsorbers and agitated-bed adsorbers are included. Various adsorbent types are discussed, as well as particle size, pressure drop, and equipment considerations. Scaleup as applied to both adsorption and chromatography concludes the chapter.

7.1 Instructional Objectives

After completing this chapter, the reader should be able to do the following:

- Write, plot, and use equations for adsorption isotherms.
- Derive the mass balance for fixed-bed adsorption and understand how it can be solved when simplifying assumptions are made.
- Calculate the shock wave velocity of a solute front when local equilibrium and negligible dispersion are assumed.
- Derive the mass balance for a series of agitated-bed adsorber columns.
- Assess chromatographic performance by elution peak analysis.
- Predict the chromatographic separation of solutes under assumptions of a linear isotherm, local equilibrium, and negligible dispersion.
- Explain the origins of band spreading in chromatography and its application to column optimization.
- Select resins based on the adsorption chemistry of the separands.
- Use particle size and pressure drop in the scaling of adsorption and chromatography.
- Use the length of unused bed (LUB) method to scale up a fixed-bed adsorber.
- Use constant resolution scaling rules for elution chromatography.
- Choose equipment for chromatographic separations including columns, packing procedures, pumps, and gradient makers.

7.2 Adsorption Equilibrium

Chromatography and adsorption work by the differential adsorption of species to a resin surface, or ligands, from a complex chemical mixture. The adsorption of a chemical species can be represented by the equilibrium reaction

$$C + S \xrightarrow{K_{eq}} CS \tag{7.2.1}$$

where C is the dissolved chemical species, S is an adsorption site, CS is the chemical bound to the site, and K_{eq} is the equilibrium constant governing the reaction. The equilibrium constant for this adsorption is

$$K_{eq} = \frac{[CS]}{[C][S]} \tag{7.2.2}$$

Consider the three assumptions inherent in this representation of the adsorption reaction. First, it is completely reversible, and the chemical's interaction with the adsorption site causes no alteration in its solution properties or solution state. Second, chemicals bind to sites in a one-to-one fashion, and they bind only to sites. In other words, the binding is "specific," and there is no "nonspecific" binding, or interaction between molecules and the surface. Third, there is only one mode of binding to the site; all binding is equal and is described by a single value for K_{eq}. Despite these limiting assumptions, this "site model" of adsorption still serves as a very accurate description of adsorption in liquid chromatography.

In many cases, the concentration of adsorption sites is very much larger than the concentration of dissolved chemical species ($[S] \gg [C]$), and the equilibrium expression becomes

$$K_{eq} = \frac{[CS]}{[C]} \tag{7.2.3}$$

or,

$$[CS] = K_{eq}[C] \tag{7.2.4}$$

This representation is particularly common in analytical liquid chromatography, where sample volumes are very small and solutes typically dilute. This is known as *linear equilibrium,* because the concentration of the adsorbed species can be expressed as a multiple of the concentration of the dissolved species. There is no limitation to binding set by the inherent ability of the resin to bind the chemical. In other words, no saturation limit of the resin will be reached.

Chromatography in the linear limit is very suitable for analytical chromatography because it emphasizes the difference in K_{eq} between species in order to separate them; these differences can be very small for very similar molecules. The purpose of analytical chromatography is to measure the concentrations of the various components of a mixture. To achieve this goal, "baseline resolution" (zero concentration between eluted separands) is critical, since determining the concentrations of two species that are not completely separated is less accurate. It is also critical that the concentrations of the various solutes do not affect their retention time, which is the case in linear equilibrium.

The linear isotherm approximation is less useful for preparative or industrial scale adsorption and chromatography. The most efficient operation uses all the adsorption sites available. In this case, the concentration of the empty adsorption sites available cannot be ignored. The concentration of unoccupied sites is not readily measurable, but the total number of sites is, so

$$S_{tot} = [CS] + [S] \tag{7.2.5}$$

where S_{tot} is the total site concentration. Combining the general expression for K_{eq}, Equation (7.2.2), with Equation (7.2.5) for S_{tot} gives

$$[CS] = \frac{K_{eq}S_{tot}[C]}{1 + K_{eq}[C]} \tag{7.2.6}$$

the well-known *Langmuir isotherm.* In the limit where $K_{eq}[C]$ is small in comparison to one, the denominator of Equation (7.2.6) goes to one, and the form of the linear adsorption equation is recovered [Equation (7.2.4)]. When $K_{eq}[C]$ is large in comparison to one, $[CS]$ is equal to S_{tot}; that is, the adsorption sites are saturated. The Langmuir isotherm [a plot of Equation (7.2.6)] is "concave downward," having a linear slope in the low concentration

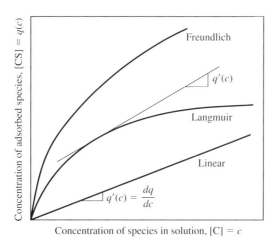

Figure 7.2 Shapes of different equilibrium adsorption isotherms.

limit and a plateau as the resin becomes saturated (Figure 7.2). Equilibrium isotherms that are concave downward are considered to be favorable for adsorption. The Langmuir isotherm has often been used to correlate equilibrium adsorption data for proteins [4].

The consequences of the Langmuir isotherm for preparative and industrial chromatography are many. The adsorption of various species from a complex mixture is now seen as influenced not only by the equilibrium constants (K_{eq}'s), which are species specific, but also by the total binding capacity of the resin (S_{tot}), which is more specific to the resin than to the chemical species. This potentially decreases the differences between species in their chromatographic behavior at high concentrations. Baseline resolution of species is no longer generally attainable, and the concept of resolution begins to lose value in these applications. The goal of large-scale chromatographic separation is also different. Here, we are trying to purify one or possibly two major components. Purification is usually defined as a limit of particular contaminants, for example removal of DNA to less than 10 ppm. Measurement of contaminants is no longer the goal, and separation of contaminants from each other is also not important.

Another equilibrium isotherm that has proven useful is the empirical *Freundlich isotherm* [5]:

$$[CS] = K_{eq}[C]^{1/n} \qquad n > 1 \qquad (7.2.7)$$

With $n > 1$, this isotherm is concave downward (Figure 7.2). This isotherm can be explained as a result of energetic heterogeneity of the surface of the adsorbent. An expression of the Freundlich form results when a set of Langmuir isotherms (with different K_{eq} values) is superimposed. The Freundlich isotherm has been used to describe the adsorption of a wide range of antibiotics, steroids, and hormones [4].

7.3 Adsorption Column Dynamics

A basic understanding of the dynamic behavior of adsorption processes is essential for the design and optimization of large-scale processes. The most important types of operation of adsorbers for biotechnology are fixed-bed columns and agitated-bed columns. The dynamics of each of these systems are analyzed.

7.3.1 FIXED-BED ADSORPTION

The development of a mass balance is the starting point for analyzing fixed-bed adsorption. Simplifications of this mass balance lead to analytical solutions that give insight into the adsorption process.

Mass Balance for Fixed-Bed Adsorption

To understand the dynamics of a fixed-bed adsorption column, a mass balance is performed by considering a disk of cross-sectional area equivalent to that of the column (A), but differential in thickness (Δx). A species to be separated (called the separand) flows in and out of the disk by convection and the combined effects of molecular diffusion and mechanical dispersion. Note that the rate of convection into the disk is the interstitial velocity, that is, the velocity of fluid in the void fraction, ε. This convection rate is the superficial velocity (the flow rate divided by the cross section of the column) divided by the void fraction ($= Q/\varepsilon A$). Within the volume of the disk, separand may accumulate within both the mobile and stationary phases. The mass balance for separand i can be written as follows:

(Rate of separand in) − (rate of separand out) = (rate of accumulation of separand)

$$\tag{7.3.1}$$

$$A\varepsilon\Delta t\left(\frac{v}{\varepsilon}c_i - \mathscr{D}_{\text{eff}}\frac{\partial c_i}{\partial x}\right)\bigg|_{x,t} - A\varepsilon\Delta t\left(\frac{v}{\varepsilon}c_i - \mathscr{D}_{\text{eff}}\frac{\partial c_i}{\partial x}\right)\bigg|_{x+\Delta x,t}$$

$$= A\varepsilon\Delta x(c_i|_{t+\Delta t} - c_i|_t) + A(1-\varepsilon)\Delta x(q_i|_{t+\Delta t} - q_i|_t) \tag{7.3.2}$$

where c_i = concentration of separand i in the mobile phase = $[C]_i$
$\qquad q_i$ = concentration of separand i in the stationary phase averaged over an
$\qquad\qquad$ adsorbent particle = $[CS]_i$
$\qquad \varepsilon$ = void fraction (mobile phase volume/total column volume), commonly
$\qquad\qquad$ 0.3 to 0.4 in fixed beds
$\qquad v$ = mobile phase superficial velocity (flow rate divided by the empty column
$\qquad\qquad$ cross-sectional area, Q/A)
$\quad \mathscr{D}_{\text{eff}}$ = effective dispersivity of the separand in the column
$\qquad t$ = time
$\qquad x$ = longitudinal distance in the column; $x = 0$ at column inlet

Dividing by $A\varepsilon\Delta x\Delta t$, and taking the limit as Δx and Δt go to zero, this mass balance becomes

$$\mathscr{D}_{\text{eff}}\frac{\partial^2 c_i}{\partial x^2} - \frac{v}{\varepsilon}\frac{\partial c_i}{\partial x} = \frac{\partial c_i}{\partial t} + \frac{1-\varepsilon}{\varepsilon}\frac{\partial q_i}{\partial t} \tag{7.3.3}$$

The right-hand side of this equation represents the accumulation rate of solute in any section of the column and recognizes that adsorption is an inherently unsteady state operation. The left-hand side of the equation represents the "in minus out" terms of the mass balance; solute moves through the column via the convection of the mobile phase (second term), or by diffusion or mechanical dispersion down a concentration gradient (first term). The term $\partial q_i/\partial t$ represents the rate of mass transfer of the separand to an average particle in the stationary phase and can be represented in general as

$$\frac{\partial q_i}{\partial t} = f(c_i, c_j, \ldots, q_i, q_j, \ldots) \tag{7.3.4}$$

where the subscripts represent separands $i, j, \ldots$. This rate expression may be a linear driving force expression of the form

$$\frac{\partial q_i}{\partial t} = K_a(c_i - c_i^*) \tag{7.3.5}$$

where K_a is an overall mass transfer coefficient that includes both internal and external mass transfer resistance, and c_i^* is the liquid phase concentration that would exist at equilibrium with q_i.

Assumption of Local Equilibrium and Negligible Dispersion

Since the overall pattern of mass transfer is governed by the form of the equilibrium relationship, the main features of the dynamic behavior can be understood without doing detailed calculations [5]. Thus, this first analysis of the mass balance for adsorption will assume equilibrium locally and ignore the dispersion term for simplicity. This allows us to focus on the velocity, u_i, at which a solute traverses the column. Using an equilibrium isotherm relationship in the form $q_i = f(c_i)$ (Figure 7.2), Equation (7.3.3) becomes

$$\frac{\partial c_i}{\partial t} + \frac{v}{\varepsilon + (1 - \varepsilon)q_i'(c_i)} \frac{\partial c_i}{\partial x} = 0 \tag{7.3.6}$$

where $q_i'(c_i)$ is the slope of the equilibrium isotherm at concentration c_i. If we let

$$u_i = \frac{v}{\varepsilon + (1 - \varepsilon)q_i'(c_i)} \tag{7.3.7}$$

then Equation (7.3.6) becomes

$$\frac{\partial c_i}{\partial t} + u_i \frac{\partial c_i}{\partial x} = 0 \tag{7.3.8}$$

We see that Equation (7.3.8) is identical in form to the equation of continuity for solute i moving at velocity u_i through a column with no packing [6]. Thus, the expression for u_i given by Equation (7.3.7) is the *effective velocity* of component i through the packed bed. When the equilibrium is linear for the solute, $q_i'(c_i) = K_{eq,i}$, so that the effective solute velocity is independent of the concentration and is inversely proportional to $K_{eq,i}$.

For preparative and industrial scale adsorption processes, where it is desired to use high adsorbent loadings, the equilibrium is nonlinear and generally the Langmuir isotherm is applicable. From Figure 7.2 for the Langmuir isotherm, it can be seen that $q_i'(c_i)$ decreases with concentration; according to Equation (7.3.7), a decrease in $q_i'(c_i)$ causes u_i to increase. This would result in the physically unreasonable overhanging concentration profiles shown in Figure 7.3. However, for the solute at the front that advances through the fixed bed, the concentration change is not continuous, and finite differences rather than

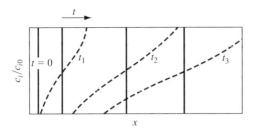

Figure 7.3 Shapes of shock wave concentration profiles (solid lines) and physically unreasonable overhanging concentration profiles for a favorable equilibrium isotherm (dashed lines): c_i, mobile phase concentration of solute; c_{i0}, solute concentration in feed to the adsorber.

differentials must be used for the change in q_i with c_i, as follows:

$$u_{i,\text{sh}} = \frac{v}{\varepsilon + (1 - \varepsilon)\dfrac{\Delta q_i}{\Delta c_i}} \tag{7.3.9}$$

The resulting $u_{i,\text{sh}}$ of the solute front is called the *shock wave velocity*, as the mathematics describing this phenomenon are similar to those describing acoustic waves and ocean waves. The calculation of the shock wave velocity is illustrated later (see Example 7.3).

In real systems, solute dispersion in the column makes perfect step changes in concentration impossible. However, when the equilibrium isotherm is favorable (i.e., concave downward, as for the Langmuir isotherm and the Freundlich isotherm), which is often the case for biological solutes, a solute wave front is *self-sharpening*. This can be explained from an examination of the effect of a favorable isotherm on the solute velocity u_i determined by Equation (7.3.7): low concentrations ahead of the wave result in higher $q_i'(c_i)$ and thus lower u_i, while high concentration within the wave give lower $q_i'(c_i)$ and therefore higher u_i. These effects both work to sharpen the front. On the other hand, low concentrations that trail the front also give higher $q_i'(c_i)$ and thus lower u_i resulting in a broadening tail behind the sharp front. These effects are illustrated later in Example 7.4.

Assumption of a Linear Equilibrium Isotherm and Negligible Dispersion

If the equilibrium isotherm is linear and dispersion is neglected, an analytical solution of the mass balance for adsorption [Equation (7.3.3)] can be obtained. Although numerical solutions of this mass balance can be obtained, analytical solutions are valuable because they provide greater insight into the behavior of the system. The following equations describe adsorption assuming a linear adsorption isotherm, a linear driving force for the mass transfer rate, and negligible dispersion, respectively:

$$q_i = K_{\text{eq},i} c_i \tag{7.3.10}$$

$$\frac{\partial q_i}{\partial t} = K_a(c_i - c_i^*) \tag{7.3.5}$$

$$\frac{\partial c_i}{\partial t} + \frac{1 - \varepsilon}{\varepsilon}\frac{\partial q_i}{\partial t} + \frac{v}{\varepsilon}\frac{\partial c_i}{\partial x} = 0 \tag{7.3.11}$$

The initial and boundary conditions for a column initially free of solute subjected to a step change in the solute concentration at the inlet at time zero are as follows:

$$t < 0 \qquad q_i(0, x) = c_i(0, x) = 0 \tag{7.3.12}$$

$$t \geq 0 \qquad c_i(t, 0) = c_{i0} \tag{7.3.13}$$

For the solution of these equations, it is convenient to write Equations (7.3.11) and (7.3.5) in terms of dimensionless variables:

$$\frac{\partial \phi}{\partial \xi} + \frac{\partial \psi}{\partial \tau} = 0 \tag{7.3.14}$$

$$\frac{\partial \psi}{\partial \tau} = \phi - \psi \tag{7.3.15}$$

where $\phi = c_i/c_{i0}$
$\psi = q_i/q_{i0}$
$q_{i0} = K_{eq,i}c_{i0}$
$\xi = (kK_{eq,i}x/v)(1-\varepsilon)$
$k = K_a/K_{eq,i}$
$\tau = k(t - x\varepsilon/v)$

Equations (7.3.14) and (7.3.15) are exactly analogous to those used to describe heat transfer in a packed bed, and using classical methods the following solution was obtained [7]:

$$\frac{c_i}{c_{i0}} = e^{-\xi}\int_0^\tau e^{-u}I_0(2\sqrt{\xi u})\,du + e^{-(\tau+\xi)}I_0(2\sqrt{\tau\xi}) \tag{7.3.16}$$

where the Bessel function is given by

$$I_0(2\sqrt{\tau\xi}) = \sum_0^\infty \frac{(\tau\xi)^n}{(n!)^2} = 1 + \tau\xi + \frac{\tau^2\xi^2}{4} + \frac{\tau^3\xi^3}{36} + \cdots \tag{7.3.17}$$

The numerical evaluation requires tedious graphical integrations. Representative results of breakthrough curves using the results of graphical integrations [8] are shown in Figure 7.4.

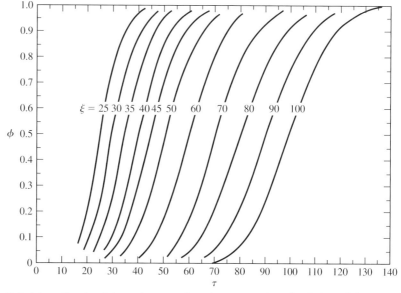

Figure 7.4 Adsorption breakthrough curves based on an analytical solution of the mass balance assuming a linear adsorption isotherm, a linear driving force for the mass transfer rate, and negligible dispersion. (Data from C. C. Furnas, "Heat transfer from a gas stream to a bed of broken solids," *Trans. AIChE,* vol. 24, p. 155, 1930.)

Analytical solutions of the mass transfer equations for adsorption have been obtained with more realistic models, including intraparticle diffusion with and without external film resistance, but the solutions for the breakthrough curve are too cumbersome to be of much practical value [9].

EXAMPLE 7.1

Determination of the Mass Transfer Coefficient from Adsorption Breakthrough Data
A solution containing a biological compound was fed at a superficial velocity of 20 cm/h to a fixed-bed column containing an adsorbent until breakthrough (when $c_i/c_{i0} = 0.1$) was obtained 9.5 h from the start of feeding. The feed concentration was low enough that the equilibrium isotherm was linear, found previously in equilibrium experiments to be

$$q_i = 40c_i^*$$

where q_i and c_i^* have units of milligrams per milliliter. The column was 10 cm long, and the void fraction of the packing was 0.34. From this information, estimate the mass transfer coefficient K_a, assuming a linear driving force for the mass transfer rate,

$$\frac{\partial q_i}{\partial t} = K_a(c_i - c_i^*)$$

SOLUTION

We can solve this problem using the analytical solution of the breakthrough curve in graphical form (Figure 7.4) for the equations that describe mass transfer in adsorption, assuming a linear equilibrium isotherm, a linear driving force for the mass transfer rate, and negligible dispersion. At the column outlet, the parameters in the graphical solution shown in Figure 7.4 are as follows, using the definitions for the variables in Equations (7.3.14) and (7.3.15):

$$\phi = \frac{c_i}{c_{i0}} = 0.1$$

$$\xi = \frac{K_a x}{v}(1 - \varepsilon) = \frac{K_a \times 10\,\text{cm}}{20\,\dfrac{\text{cm}}{\text{h}}}(0.66) = 0.33K_a$$

$$\tau = \frac{K_a}{K_{\text{eq}}}\left(t - \frac{x\varepsilon}{v}\right) = \frac{K_a}{40}\left(9.5\,\text{h} - \frac{10\,\text{cm} \times 0.34}{20\,\dfrac{\text{cm}}{\text{h}}}\right) = 0.233K_a$$

where K_a has units of hours^{-1}. We can rewrite these in terms of K_a as

$$K_a = 3.03\xi = 4.29\tau$$

Finding K_a is a trial-and-error solution. We select values of τ and ξ that give $\phi = 0.1$ in Figure 7.4 and calculate K_a using each of the two foregoing equations. This is continued until the same K_a is obtained for each equation. This procedure leads to

$$K_a \cong 105\,\text{h}^{-1}$$

This value of K_a could now be used to predict breakthrough for other superficial velocities and column lengths.

Constant Pattern Behavior

In preparative and industrial adsorption processes, high feed concentrations, resistance to mass transfer, and axial mixing in the column lead to departures from ideal situations where a linear isotherm, local equilibrium, and negligible dispersion can be assumed. In the case of a favorable equilibrium isotherm (i.e., isotherms that are concave downward, which is the case for the Langmuir isotherm that can often be used for the adsorption of proteins), the concentration profile in the column develops in the initial region; and at some point from the column entrance, the profile propagates without further change in shape—thus the term "constant pattern." This occurs because the dispersion in the column resulting from longitudinal dispersion and mass transfer resistance is opposed by the self-sharpening effect discussed earlier. When the equilibrium isotherm is highly favorable, the distance required to approach the constant pattern limit may be very small, a few centimeters to up to a meter [5]. The constant pattern approximation provides the basis for a very useful and widely utilized method for scaleup design using the concept of the length of unused bed (LUB) (see Section 7.8.1).

7.3.2 AGITATED-BED ADSORPTION

Agitated-bed adsorption processes have been developed to allow removal of a product secreted by the cells without first having to remove the cells. In this type of process, cell culture broth is passed through a series of agitated columns containing an adsorbent, as shown in Figure 7.5. Each column has screens at the inlet and outlet that are designed to retain the adsorbent within the column but allow the broth to pass through. When the concentration of the product in the effluent of the last column in the series reaches a certain value, the flow is stopped, and the lead column is taken out of the train. Periodic countercurrent operation is obtained by advancing each of the remaining columns in the train, placing a regenerated column of adsorbent in the last position, and starting the feed flow again. The lead column taken out of the train is washed with the adsorbent agitated to remove the broth solids, and the product is eluted from the adsorbent, usually in the fixed-bed mode.

This process has advantages over filtration, in that there is no filter aid to dispose of and it is not necessary to wash a filter cake containing the cells, so losses of the product

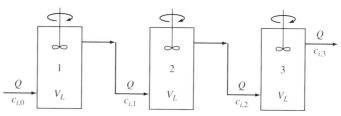

Figure 7.5 Train of agitated-bed adsorption columns for the processing of cell culture broth.

are often less. The equipment for this process is less expensive and easier to maintain than that used for centrifugation. The disadvantage is that expensive solid adsorbents are more easily fouled by the dirtier feed stream and so require harsher or more expensive regeneration procedures and more frequent replacement, compared to adsorbents utilized with streams with fewer impurities. Also, resin attrition can be an issue.

A useful mathematical model for this process has been developed [10]. The continuity equation for the nth column in the train can be written for separand i as

$$\text{(Rate of separand in)} - \text{(rate of separand out)} = \text{(rate of accumulation of separand)} \tag{7.3.18}$$

$$Qc_{i,n-1} - Qc_{i,n} = V_L \frac{dc_{i,n}}{dt} + V_R \frac{dq_{i,n}}{dt} \tag{7.3.19}$$

where Q = volumetric flow rate

$c_{i,n-1}, c_{i,n}$ = separand concentration in feed to and effluent from
        column n, respectively
    V_L = liquid volume in column
    V_R = volume of adsorbent in column
    $q_{i,n}$ = separand concentration in adsorbent phase of
        column n averaged over an adsorbent particle
     t = time

The rate of mass transfer of separand to the adsorbent phase is described by a linear driving force expression similar to Equation (7.3.5):

$$\frac{dq_{i,n}}{dt} = K_a(c_{i,n} - c_{i,n}^*) \tag{7.3.20}$$

where $c_{i,n}^*$ is the separand concentration in the bulk liquid when it is at equilibrium with $q_{i,n}$, and K_a is an overall mass transfer coefficient that can be correlated to experimental data as

$$K_a = A \exp\left(-B \frac{q_i}{q_{i,\text{sat}}}\right) + D \exp\left(-E \frac{q_i}{q_{i,\text{sat}}}\right) \tag{7.3.21}$$

Here A, B, C, and D are constants, and $q_{i,\text{sat}}$ is the adsorbent phase concentration which is in equilibrium with the separand concentration c_{i0} in the feed to the train of mixed columns. In the use of this model for the recovery of an antibiotic, equilibrium was modeled by the Freundlich isotherm written in the form

$$c_{i,n}^* = bq_{i,n}^a \tag{7.3.22}$$

where a and b are constants.

Equations (7.3.19) to (7.3.22) constitute a set of mathematical relationships that govern the performance of each column in the train. These simultaneous equations can be solved by the Runge–Kutta numerical method [11] to predict the effluent and adsorbent concentrations as a function of time. Excellent agreement between predicted and experimental adsorption data for the recovery of the antibiotic novobiocin in a three-stage train has been obtained using this method [10].

7.4 Chromatography Column Dynamics

There are many methods of analysis for the dynamics of chromatography unit operations, too many to cover in detail in this textbook. To speak the language of chromatographers, the most relevant method to know is theoretical plate analysis. To be conversant with engineering methods of analysis, the application of mass balances, rate equations, and equilibria (Chapter 1) are the most relevant. A brief treatment of each of these methods is given here.

7.4.1 Plate Models

Plate models seek to explain the band broadening observed in chromatography by approximating a chromatograph as a series of well-mixed tanks at equilibrium. The terminology comes from analysis of distillation, where plates are sometimes used to hold vapor and liquid in contact to approach equilibrium at various temperatures and compositions. Just as distillation is often performed in a packed column but the concept of a "theoretical equilibrium plate" remains, so has it also come to symbolize resolving power of a chromatography column. A resemblance to the analysis of multistage extraction (Section 6.2) should be noted.

The original analogy was drawn by Martin and Synge [12] in 1941. By dividing the column into a series of imaginary well-mixed tanks at equilibrium and computing the mass balance around each, they deduced when the number of tanks became large, and the initial condition was a solute at a finite concentration in the first tank, that the concentration profile could be described by a Gaussian curve. Properties of a Gaussian curve are shown in Figure 7.6. The volume of the theoretical tanks (also called plates) is found by dividing the

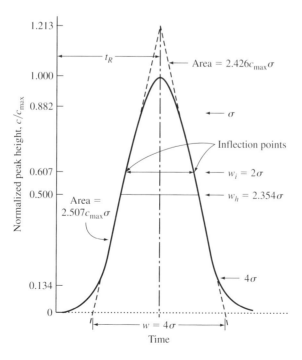

Figure 7.6 Properties of a Gaussian peak c_{max}, maximum peak height; σ, standard deviation; w_i, peak width at inflection points; w_h, peak width at half-height; w, peak width at base (base intercept); t_R, average retention time. (Data from C. Horvath and W. R. Melander, "Theory of chromatography," *Fundamentals and Applications of Chromatographic and Electrophoretic Methods*, Part A, E. Heftmann, ed., p. A41, Elsevier Scientific, Amsterdam, Netherlands, 1983.)

column volume by the number of tanks, and, H, the height of the equivalent theoretical plate (HETP), can be expressed as

$$H = \frac{L}{N} \tag{7.4.1}$$

where L is the length of the column and N is the number of plates. For Gaussian peaks, the plate count (N) can be expressed as the squared average retention time divided by the variance of the peak [13]:

$$N = \frac{t_R^2}{\sigma^2} = \frac{t_R^2}{(w/4)^2} \tag{7.4.2}$$

where w is the peak width at the base (the base intercept as shown in Figure 7.6). The determination of the number of theoretical plates is a primary method by which scientists compare the performance of different columns, packed by different methods or with different resins.

Peak width is used in the definition of *resolution,* R_s, which is a measure of the extent of separation of two peaks in a chromatography:

$$R_s = \frac{t_{R_2} - t_{R_1}}{\frac{1}{2}(w_2 + w_1)} \tag{7.4.3}$$

where t_{R_1}, t_{R_2} = average retention time for separands 1 and 2, respectively
w_1, w_2 = peak width (time) for separands 1 and 2, respectively

Thus, resolution is increased by increasing the difference in the retention times and by decreasing the peak widths.

A low concentration of a moderately bound solute should be used to evaluate a column HETP. Linear response is desired from a molecule that will bind and release from the resin, so that kinetics and mass transfer are measured, as well as mechanical mixing that may be present in the column hydrodynamics. To be certain the solute is in the linear binding range, a variety of concentrations may be tried at first, to measure the effect of concentration on retention time and peak shape. Very often, a water-soluble aromatic compound like benzoic acid is used. A popular series of solutes are the "parabens," methyl parabenzoic acid, ethyl paraben, and propyl paraben. HETP should also be measured under isocratic conditions (i.e., constant eluent buffer composition) to avoid influencing retention time and peak shape with a peak-sharpening gradient.

The undeniable simplicity of this plate model for evaluating the inherent separating power of a chromatography column accounts for its wide use in the field. Its primary use is in comparing multiple packings of the same column geometry with the same resin. Its applicability to the actual separation under study is always under debate, especially when the separation is at high solute concentration and large total solute load, and is eluted with a gradient.

7.4.2 CHROMATOGRAPHY COLUMN MASS BALANCE WITH NEGLIGIBLE DISPERSION

Chromatography separates solutes based on their differential binding to the resin. The separation is effected as the solutes exit the column outlet at different times. How the binding of solutes to resins at equilibrium translates into an elution time difference is found in the mass balance. As the solute is carried by the mobile phase through the resin, each solute partitions between the mobile and stationary phases. The solutes that partition more strongly are retarded with respect to the flowing fluid, and they exit the column later (Figure 7.1). Solutes that do not partition to the stationary phase at all exit the column in the "void volume," which represents the mobile phase volume in the interstices of the packed resin.

The mass balance for chromatography is identical to that for adsorption, Equation (7.3.3). The first analysis of this mass balance for chromatography will ignore the dispersion term for simplicity. This allows us to focus on the peak position, or the rate at which a solute traverses the column. (Another way of doing this is by moment analysis, which is beyond the scope of this text.*) By assuming local equilibrium and neglecting dispersion, we find that the effective velocity of a solute moving through the packed column is the same as that previously found in general for adsorption [Equation (7.3.7)]; this equation is useful in predicting the separation of different solutes, as is illustrated in Example 7.2.

EXAMPLE 7.2

Chromatographic Separation of Two Solutes   Two solutes have linear equilibrium constants of $K_{eq,1} = 7.5$ and $K_{eq,2} = 7.8$, respectively. For a flow rate of 1.5 liter/min, in a column 63 cm in diameter, with a void fraction of 0.33, and local equilibrium, what column length is required to separate the two solutes by 5 min?

SOLUTION

The effective velocity of solute i for negligible dispersion is given by Equation (7.3.7) as

$$u_i = \frac{v}{\varepsilon + (1 - \varepsilon)q_i'(c_i)}$$

For linear equilibrium,

$$q_i = K_{eq,i}c_i$$

$$q_i'(c_i) = K_{eq,i}$$

The superficial velocity is

$$v = \frac{1500 \dfrac{\text{cm}^3}{\text{min}}}{\pi (31.5)^2 \, \text{cm}^2} = 0.481 \frac{\text{cm}}{\text{min}}$$

---

*In moment analysis, the Laplace transform is taken with a Dirac delta function as the inlet condition ($x = 0$), and the derivatives of the transforms with respect to s are evaluated at $s = 0$. The first moment corresponds to the elution time of the peak, the second to its dispersion, the third to the skew, and so forth.

For an excellent reference on moment analysis, see Kucera [14].

For solute 1, the effective velocity is therefore

$$u_1 = \frac{0.481 \frac{cm}{min}}{0.33 + (1 - 0.33)(7.5)} = 0.08982 \frac{cm}{min}$$

This same equation gives 0.08657 cm/min for solute 2. Translating solute velocities into elution times for a constant distance traveled (L),

$$5 \, min = L \left(\frac{1}{u_2} - \frac{1}{u_1} \right)$$

Solving for L gives

$$L = 12.0 \, cm$$

Note that four significant figures are used to calculate u_1 and u_2 to avoid error in calculating L.

Equilibrium relations are usually more complicated than the linear case. For the nonlinear Langmuir equilibrium isotherm [Equation (7.2.6)], we obtain for the first derivative of q_i with respect to c_i,

$$q_i'(c_i) = \frac{K_{eq,i} S_{tot}}{1 + K_{eq,i} c_i} - \frac{K_{eq,i}^2 S_{tot} c_i}{(1 + K_{eq,i} c_i)^2} \tag{7.4.4}$$

As the concentration becomes higher, the value of the derivative becomes smaller and the solute velocity is higher. This makes sense, because to have a high mobile phase concentration, the stationary phase concentration must be near saturation, thus leaving no sites on the resin for the solute in the mobile phase to bind to. This holds true as long as the change in concentration is continuous. However, when a pulse injection is made, the resin immediately preceding the solute front is devoid of bound solute. In this case, the concentration change is not continuous but is a step change, and finite differences must be used for q and c, giving Equation (7.3.9) for the effective velocity of the solute. This velocity is called a shock wave velocity, as discussed for adsorption in Section 7.3.

EXAMPLE 7.3

Calculation of the Shock Wave Velocity for a Nonlinear Isotherm   Solute 1 has a Langmuir isotherm characterized by an S_{tot} of 120 μg/ml, and a K_{eq} of 7.5 ml/mg. Calculate the shock wave velocity for an injection of 1 mg/ml, and column conditions identical to those of Example 7.2.

SOLUTION

To use Equation (7.3.9) for the shock wave velocity, we need to know Δq. For the Langmuir isotherm [Equation (7.2.6)], the resin concentration $q = [CS]$ in equilibrium with concentration $[C] = 1$ mg/ml in the mobile phase is

$$[CS] = \frac{K_{eq} S_{tot}[C]}{1 + K_{eq}[C]} = \frac{\left(7.5 \frac{ml}{mg} \right)\left(0.120 \frac{mg}{ml} \right)\left(1 \frac{mg}{ml} \right)}{1 + \left(7.5 \frac{ml}{mg} \right)\left(1 \frac{mg}{ml} \right)} = 0.106 \frac{mg}{ml}$$

Since the column initially has no solute,

$$\frac{\Delta q_1}{\Delta c_1} = \frac{(0.106 - 0)\dfrac{mg}{ml}}{(1 - 0)\dfrac{mg}{ml}} = 0.106$$

From Equation (7.3.9), we obtain

$$u_{i,sh} = \left(\frac{v}{\varepsilon + (1 - \varepsilon)\dfrac{\Delta q_1}{\Delta c_1}}\right) = \frac{0.481\dfrac{cm}{min}}{0.33 + (1 - \varepsilon)(0.106)} = 1.20\frac{cm}{min}$$

This is over 10 times the solute velocity for the linear isotherm case given in Example 7.2. This is because the concentration of the solute is limited to 0.120 mg/ml on the stationary phase. For the linear case, a 1 mg/ml injection would lead to a 7.5 mg/ml concentration in the stationary phase.

The leading edge of a solute front derived from a Langmuir isotherm is a self-sharpening shock wave, which leads to the constant pattern behavior discussed earlier. The trailing edge, on the other hand, is a continuous diffuse wave. The trailing edge has a continuous concentration from the injection concentration to zero. To predict the elution profile of a peak, both the leading shock wave and the trailing diffuse wave must be calculated. This is done by solving Equation (7.3.7) for the trailing edge with respect to a range of mobile phase concentrations between the injection concentration and zero, and Equation (7.3.9) with respect to the shock front (as in Example 7.3). This is shown in Example 7.4.

EXAMPLE 7.4

Calculation of the Elution Profile   For an injection volume of 5 liters, and other conditions as stated in Examples 7.2 and 7.3, calculate the elution profile for solute 1.

SOLUTION

As determined in Example 7.3, the velocity of the shock front is 1.20 cm/min. The diffuse wave velocity is determined by selecting concentrations between 0 and the injection concentration of 1 mg/ml, and determining the velocity of each of these concentrations trailing the shock front. For example, at a solute concentration of 0.4 mg/ml, we can calculate dq_1/dc_1 from Equation (7.4.4):

$$\frac{dq_1}{dc_1} = \frac{K_{eq,1}S_{tot}}{1 + K_{eq,1}c_1} - \frac{K_{eq,1}^2 S_{tot}c_1}{(1 + K_{eq,1}c_1)^2} = \frac{7.5(0.120)}{1 + 7.5(0.4)} - \frac{(7.5)^2(0.120)(0.4)}{[1 + 7.5(0.4)]^2} = 0.056$$

We can use Equation (7.3.7) to calculate the solute velocity:

$$u_1 = \left(\frac{v}{\varepsilon + (1 - \varepsilon)\dfrac{dq_1}{dc_1}}\right) = \frac{0.481}{0.33 + (1 - 0.33)(0.056)} = 1.31\frac{cm}{min}$$

TABLE E7.4
Diffuse Wave Concentrations Following the Shock Wave for Example 7.3

Solute concentration, c_1 (mg/ml)	Isotherm derivative, dq_1/dc_1	Solute velocity, u_1 (cm/min)
0.8	0.018	1.40
0.6	0.030	1.37
0.5	0.040	1.35
0.4	0.056	1.31
0.23	0.121	1.17
0.1	0.294	0.91
0.01	0.779	0.56
0.005	0.836	0.54
0.001	0.887	0.52

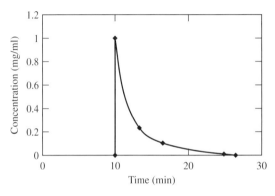

Figure E7.4 The elution profile calculated for a Langmuir isotherm. Conditions are given in the examples in the text.

For a flow rate of 1.5 liters/min, and an injection volume of 5 liters, the diffuse wave trails the shock by 3 min and 20 s [5 liters/(1.5 liters/min)]. Table E7.4 gives the solute velocities for various concentrations.

As can be seen, concentrations above 0.23 mg/ml run faster than the shock front. Since nothing can pass the shock front, these concentrations do not appear in the thermodynamic diffuse wave. They do, however, represent concentrations on the isotherm where the derivative is less than the slope of the chord that connects the shock concentrations. Concentrations of 0.23 mg/ml and less run behind the shock front, and thus constitute the diffuse wave. The shock front elutes from the 12.0 cm bed in 10.0 min. The diffuse wave is delayed by 3.33 min, as mentioned earlier, and then starts at approximately 0.23 mg/ml. The diffuse wave decreases in concentration to 0.1 mg/ml at 16.5 min, to 0.01 mg/ml at 24.8 min, to 0.005 mg/ml at 25.6 min, and to 0.001 mg/ml at 26.4 min. A representative graph is shown in Figure E7.4.

The Langmuir isotherm and other concave downward equilibrium conditions are very common for large-scale liquid chromatography. Analytical chromatography does not generally deal with concentrations large enough to generate a Langmuir isotherm. Ion exchange binding is related to, but not exactly the same as, the Langmuir isotherm. Concave upward isotherms result in peaks that have self-sharpening trailing edges and diffuse leading fronts. Most concave upwards isotherms result from multilayer binding in which the binding becomes more energetically favorable as the layers build up. This is most typical in gas adsorption/desorption applications and is rarely, if ever, found in the liquid chromatography of bioactive molecules.

7.4.3 DISPERSION EFFECTS IN CHROMATOGRAPHY

So far, we have discussed chromatography and adsorption from a purely thermodynamic point of view. Mass transfer and diffusion also play an important role in the development and scaleup of these separation techniques. Diffusion takes place longitudinally in the mobile phase, and also within the pores of the stationary phase. Mass transfer takes place between the mobile and stationary phases. Slow diffusion and mass transfer always decrease the efficiency of the separation, and steps should be taken where possible to increase the rates of each.

Longitudinal dispersion is represented in the complete mass balance for adsorption, Equation (7.3.3). The effective dispersivity $\mathscr{D}_{\text{eff}}$ is a combination of the molecular or binary diffusivity in the mobile phase $\mathscr{D}_{\text{m}}$ and a dispersion coefficient E, which accounts for additional solute transport caused by back-mixing and nonuniformity of the velocity [15]:

$$\mathscr{D}_{\text{eff}} = \mathscr{D}_{\text{m}} + E \tag{7.4.5}$$

For a chromatography process in which the feed solution contains a concentration pulse of the separand, the mass balance including the dispersivity term [Equation (7.3.3)] has been solved by Lapidus and Amundson [16], assuming a linear isotherm, local equilibrium, and a semi-infinite column to give an analytical solution:

$$
\begin{aligned}
c_i(x, t) &= c_{i0}H(x, t) - c_{i0}H(x, t - \bar{t}) \qquad \bar{t} < t < \infty \\
&= c_{i0}H(x, t) \qquad\qquad\qquad\qquad\quad t < \bar{t}
\end{aligned}
\tag{7.4.6}
$$

where the dispersion function is

$$
H(x, t) = \frac{1}{2}\left[1 + \text{erf}\left(\sqrt{\frac{u^2 t}{4\gamma \mathscr{D}_{\text{eff}}}} - x\sqrt{\frac{\gamma}{4t \mathscr{D}_{\text{eff}}}}\right) \right.
$$

$$
\left. + e^{ux/\mathscr{D}_{\text{eff}}}\, \text{erfc}\left(\sqrt{\frac{u^2 t}{4\gamma \mathscr{D}_{\text{eff}}}} + x\sqrt{\frac{\gamma}{4t \mathscr{D}_{\text{eff}}}}\right)\right]
\tag{7.4.7}
$$

in which u is the velocity of fluid through the interstices of the bed ($= v/\varepsilon$), $\gamma = K_{\text{eq}}(1 - \varepsilon)/\varepsilon$, and erf and erfc are the error and complementary error functions, respectively. It is also assumed that the adsorbent is initially free of the separand (clean column) and that the feed concentration is c_{i0} from zero time to time $\bar{t}$, after which it changes back to zero. In this solution, band broadening is dependent on the value of $\mathscr{D}_{\text{eff}}$. While for pure diffusion of a macromolecule in liquid, $\mathscr{D}_{\text{m}}$ is on the order of 10^{-6} to 10^{-7} cm^2/s, the effective dispersivity $\mathscr{D}_{\text{eff}}$ is three to four orders of magnitude higher, indicating the contributions from the mechanical effects are much larger.

An application of Equations (7.4.6) and (7.4.7) is shown in Figure 7.7, which shows the effect of dispersion with respect to mobile phase velocity for a simulated pulse feed. As

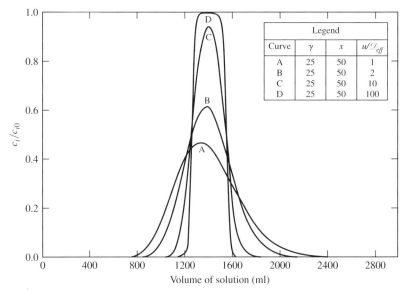

Figure 7.7 Prediction of chromatographic peaks from an analytical solution of the mass balance assuming a linear isotherm, local equilibrium, and a semi-infinite column. In this case, initial adsorbate concentration is zero, and feed solution has a pulse in concentration of c_{i0} of duration equivalent to 300 ml of feed. (Data from L. Lapidus and N. R. Amundson, "Mathematics of adsorption in beds. VI. The effect of longitudinal diffusion in ion exchange and chromatographic columns," *J. Phys. Chem.*, vol. 56, p. 984, 1952.)

the dispersivity increases ($u/\mathscr{D}_{\text{eff}}$ decreases) two orders of magnitude from curve D to curve A, the band widens by a factor of 4, and the maximum concentration decreases to less than 50% of the feed concentration.

In addition to the effect of convective axial dispersion, models have been developed that also consider the effects of intraparticle diffusion and fluid phase mass transfer [17, 18]. The starting point for the formulation of these models is the complete mass balance for adsorption, Equation (7.3.3), and the following equations, which assume a linear equilibrium isotherm and spherical symmetry:

Intraparticle diffusion:

$$\frac{\partial w_{i,p}}{\partial t} = \frac{\mathscr{D}_p}{r^2}\left(r^2 \frac{\partial w_{i,p}}{\partial r}\right) - \frac{\partial w_{i,s}}{\partial t} \tag{7.4.8}$$

Fluid phase mass transfer:

$$\frac{\partial w_{i,s}}{\partial t} = k_a\left(w_{i,p} - \frac{w_{i,s}}{K_{\text{eq}}}\right) \tag{7.4.9}$$

with the matching condition of flux continuity at the particle surface

$$-\mathscr{D}_p \frac{\partial w_{i,p}}{\partial r} = k_c\left(\frac{w_{i,p}}{\varepsilon^*} - c_i\right) \tag{7.4.10}$$

where r = distance from center of spherical resin particle

$w_{i,p}$ = concentration of separand i in the pore fluid based on a unit volume
of stationary phase (i.e., the solid matrix and the pore space)

$w_{i,s}$ = concentration of separand i adsorbed on the internal surfaces of
the stationary phase based on a unit volume of stationary phase

c_i = concentration of solute i in the mobile phase

$\mathscr{D}_p$ = effective separand diffusivity in the pore fluid

k_a = forward rate constant of adsorption

k_c = fluid phase mass transfer coefficient

ε^* = effective volume fraction of the stationary phase that is accessible
to the separand

For the case of isocratic elution, where the concentration of the elution solvent is constant, a lumped-parameter type of solution in a simple form was obtained by Athalye et al. [17]. Their solution relied on using empirical correlations to predict the height equivalent of a theoretical plate (HETP) for the column. The performance of chromatography columns can be predicted using this model based on independent estimates of transport parameters. The general solution of Rasmuson and Neretnieks [18] involves a special numerical integration method, to which the interested reader is referred. The procedure for calculating HETP depends on the approach of van Deemter et al. [19], which is now presented.

While chromatography and adsorption can be modeled accurately, and elution profiles solved exactly on high speed computers by means of finite element analysis, the most practical rule of thumb for estimating the impact of mass transfer and diffusion effects is the van Deemter equation. Van Deemter et al. [19] determined empirically that the bandwidth [measured as the height equivalent of a theoretical plate, or HETP—see Equations (7.4.1) and (7.4.2)] can be related to the linear velocity in the column by the following expression:

$$H = A + \frac{B}{u} + Cu \qquad (7.4.11)$$

The HETP (H) has dimensions of length, and is usually reported in centimeters. The factor A is primarily due to dead volume and other sources of mechanical mixing and is thought to be proportional to resin particle size. Molecular diffusion in the mobile phase is represented by B (cm^2 s^{-1}). If the flow rate is on the order of the liquid phase diffusivity of the solute, the peak broadens. This is almost never the case in liquid chromatography, but this case is relevant for gas chromatography. Finally, mass transfer and intraparticle diffusion are represented by C (s), which is thought to be proportional to the square of the effective liquid film thickness at the surface of the resin particle, divided by the effective diffusivity within the particle. Liquid chromatography is dominated by this mass transfer term. The van Deemter equation is also important because it shows the efficiency of the separation, represented by the lowest possible value of H, as a function of the productivity of the column, represented by the highest possible value for u.

The van Deemter equation can be normalized as

$$H^* = a + \frac{b}{u^*} + cu^* \qquad (7.4.12)$$

where $H^* = H/d_p$ and $u^* = u\,d_p/\mathscr{D}_m$, d_p is the stationary phase particle diameter, and $\mathscr{D}_m$ is the diffusivity of the solute of interest [20]. The reduced velocity u^* is also the

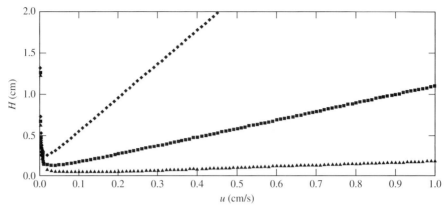

Figure 7.8 Typical van Deemter plot for three particle diameters: 50 μm ($\blacklozenge$), 25 μm ($\blacksquare$), 10 μm ($\blacktriangle$).

Péclet number for chromatography. The reduced plate height, H^*, is rarely below 2, that is, two particle diameters in a chromatography bed constitute a mixing zone. A typical van Deemter plot is shown in Figure 7.8.

When normalized in this manner, the data for many chromatographic applications fall very close to one another [20]. Giddings has given rules of thumb for parameter estimation for a, b, and c [21]. While these rules of thumb are valuable for estimation of peak broadening in the design stage, data for individual applications should always be acquired.

The most efficient separation is achieved with the lowest possible plate height. The definition of efficient separation that is meant here is the separation with the least energy lost to entropy (or thermodynamically speaking, most reversible). More theoretical plates are contained in a given column length when the plate height is minimized. In liquid chromatography, u^* is typically a large number, as diffusivities are on the order of 10^{-5} to 10^{-7} cm^2/s, while linear velocities are typically on the order of 10^{-2} cm/s, and the smallest resin diameter is 10^{-4} cm, so in the best case, u^* is 0.1. If a, b, and c are of the same order of magnitude (b is commonly smallest), then no term in the van Deemter equation dominates. Resins smaller than 1 μm are not commonly available, since the practical limit of pressure drop and resin containment seems to have been reached at this particle size. Larger particle resins are often chosen because they are considerably cheaper, in which case u^* approaches 100 and the cu^* term in the van Deemter equation dominates the plate height. Clearly, small stationary phase particle diameter (1 μm) is desirable for efficient separations.

A comparison of separation efficiency to process cost has been developed for large-scale chromatography for biotechnological applications that gave an optimum particle diameter of 30 μm [22]. This work indicated that particle diameters in the range of 20 to 40 μm make good targets for scaleup of chromatographic processes. Comparisons of these types are subject to the economics of each particular case.

The van Deemter equation has been recast into dimensionless transport parameters by Athalye et al. [17] as follows:

$$\frac{H}{d_p} = \frac{2}{\text{Pe}_g} + \frac{(1-x)}{3(1-\varepsilon)}(\text{Re Sc})\left(\frac{1}{\text{Nu}} + \frac{m}{10} + \frac{m'}{\text{Da}}\right) \qquad (7.4.13)$$

where Pe_g = dispersion Péclet number = $d_p u / \mathscr{D}_{\text{eff}}$

$x = \varepsilon / [\varepsilon + (1 - \varepsilon)\varepsilon^*(1 + K_{eq})]$, or the fraction of separand in the mobile phase at long times

$Re\, Sc$ = (Reynolds number)(Schmidt number) = $(d_p u \varepsilon \rho / \mu)(\mu / \rho \mathscr{D}_m)$ = $d_p u \varepsilon / \mathscr{D}_m$, reduced velocity or diffusion Péclet number

$\mathscr{D}_m$ = separand diffusivity in unbounded solution

Nu = Nusselt number = $d_p k_c / \mathscr{D}_m$

$m = \mathscr{D}_m / \varepsilon^* \mathscr{D}_p$

$m' = \frac{3}{2} m([K_{eq} / (1 + K_{eq})])^2$

Da = Damköhler number = $d_p^2 k_a / 4 \mathscr{D}_p$

The physical significance of each term in Equation (7.4.13) is as follows:

Pe_g term: convective axial dispersion

Nu term: fluid phase mass transfer resistance

m term: intraparticle diffusion resistance

m' term: sorption kinetics resistance

Equation (7.4.13) is valuable because it enables the quantification of each contribution to the plate height, hence a conceptually useful sensitivity analysis.

7.4.4 GRADIENTS AND MODIFIERS

Chemical gradients are used to elute solutes that are bound tightly to the stationary phase. The constituents of the chemical gradient are chosen to compete with the target solute for binding sites on the surface of the stationary phase. This affects the binding equilibrium, such that, at different concentrations of the gradient-forming chemical, different binding profiles are given for the solute of interest.

EXAMPLE 7.5

Equilibrium for a Protein Anion in the Presence of Chloride Ion   Derive the equilibrium for a monovalent protein anion on an ion exchanger in the presence of sodium chloride ion and plot isotherms for various concentrations of chloride ion.

SOLUTION

The equilibrium is described by the following two expressions

$$P^- + S^+ \xrightleftharpoons{K_{eq,1}} S \cdot P$$

$$Cl^- + S^+ \xrightleftharpoons{K_{eq,2}} S \cdot Cl$$

where P^- is the protein anion, S^+ is an unoccupied binding site on the stationary phase, and sodium is a spectator counterion. The equilibrium expressions are

$$K_{eq,1} = \frac{[S \cdot P]}{[P][S]}$$

$$K_{eq,2} = \frac{[S \cdot Cl]}{[Cl][S]}$$

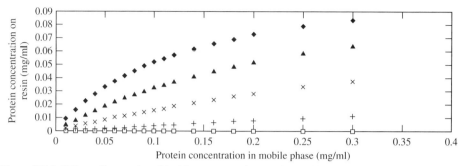

Figure E7.5 Effect of increasing chloride concentration on the equilibrium binding of a monovalent protein. $[NaCl] = 0$ (◆), 0.5 (▲), 2.0 (×), 10 (+), and 100 (□) mg/ml

Mass balances on protein concentration, chloride concentration, and total sites apply.

$$S_{tot} = [S] + [S \cdot P] + [S \cdot Cl]$$

The preceding three equations can be solved for the adsorbed protein concentration to give the "multicomponent" Langmuir isotherm:

$$[S \cdot P] = \frac{K_{eq,1} S_{tot}[P]}{1 + K_{eq,1}[P] + K_{eq,2}[Cl]}$$

where $K_{eq,1}$ is the equilibrium constant for the protein measured in the presence of chloride, and $K_{eq,2}$ is the equilibrium constant for chloride measured in the presence of protein.

As can be seen from the expression derived for $[S \cdot P]$, as the chloride concentration is increased, the concentration of bound protein decreases. A protein that is tightly bound at low ionic strength can be eluted efficiently by an increase in the salt concentration. With $K_{eq,1} = 7.5$, $K_{eq,2} = 2.0$, and $S_{tot} = 120 \, \mu g/ml$, we find the binding isotherms shown in Figure E7.5, where the increasing chloride concentration is seen to decrease binding of the protein. This is the principle of *gradient elution chromatography,* as well as of the regeneration of columns in adsorption processes. Changing the thermodynamic conditions so that the equilibrium is shifted back in favor of the unbound species effects the elution of that species and the regeneration of the resin. The thermodynamic condition may be ionic strength or pH (shifting a multivalent ion to a monovalent or uncharged species), temperature, or solvent composition. Slowly changing the thermodynamic conditions, in a gradient in the mobile phase, can sharpen diffuse waves, as well as cause desorption of the bound species. A step change (step gradient) can regenerate a resin or displace bound constituents.

The rate of change of the gradient with time is important. Most gradients are linear—their rate of change is constant. In reversed-phase chromatography, a gradient of 1%/min is often used; that is, the percentage of organic modifier is increasing smoothly by 1% per minute with a constant flow rate. Another useful but less common gradient is exponential. This provides for a shallower gradient early in the chromatography, and a rapidly increasing gradient late. An exponential gradient is useful when one is separating a wide variety of molecules that range from very loosely bound to extremely tightly bound. Yet another technique is to run a linear gradient to the point of product elution, do an isocratic elution (i.e., constant composition of the elution solvent) through the product zone to maximize resolution, and follow by a gradient to regeneration conditions. Obviously, countless combinations and options exist.

There are two other special cases that arise from the equilibrium isotherm derived in Example 7.5. Ignore, for a moment, the identity of chloride in the equation, for this identity was assigned merely for purpose of example. The first is the situation of very large $K_{eq,1}$. In this case, the isotherm is very steep, approximating a square wave, and the chromatographic mode is called "on–off." The isotherm remains steep until $K_{eq,2} c_2$ is of the order $K_{eq,1} c_1$, where c_1 and c_2 represent protein and salt anion concentration, respectively. Then solute 1 comes "off" the resin. Many proteins follow this case.

In the second case $K_{eq,2}$ is large, and greater than $K_{eq,1}$. In this case, solute 2 "displaces" solute 1 from the resin. Since solute 1 runs ahead of solute 2, and it has nonlinear binding itself, solute 1's leading edge is sharpened by the shock wave effect, and its trailing edge is sharpened by the displacement affected by solute 2. This is the displacement chromatography technique that was discussed earlier.

7.5 Adsorbent Types

Many adsorbent resins have been developed for chromatographic separations. There are two basic resin materials, polymer and silica. Any type of chemistry can be conjugated to either resin material. Typically, however, silica resins have hydrophobic coatings and are used for reversed-phase chromatography. Polymer resins are used in aqueous applications and are conjugated with ion exchange, hydrophobic interaction, or affinity-type ligands.

The selection of the resin material defines many resin physical properties. The resin provides the surface area for the adsorption. Surface area is generally 100 to 1500 m^2/g. The surface area on the outer surface of a 10 μm diameter solid sphere is 1.7 m^2/g, so it follows that most of the surface area is in the internal porosity of the particle. Since this surface area is traditionally accessed by molecular diffusion (one class of resins, called perfusion resins, allows convection through the pores), the path length for this diffusion is important. The path length is defined as the radius of the resin, which is the maximum length a molecule will diffuse to gain access to the internal surface area of the resin. Therefore, both the diameter of the particle and its internal surface area are important for the resin performance.

7.5.1 Silica-Based Resins

Uncoated silica is compatible with water or organic solvent and serves as a good reversible adsorbent for hydrophilic compounds. An organic solvent is often used in the mobile phase, and water is added as the chromatography progresses. Silica is not typically stable at extremes of pH, especially basic pH (exceptions exist). Silica is available with high surface area and small particle size; being very rigid, it does not collapse under high pressures. Uncoated silica used in chromatography came to be known as "normal phase" chromatography. Silica resins are thought to denature some proteins and irreversibly bind others. However, silica-based chromatography is used for the purification of many commercial biotechnology products.

Silica particles coated with long-chain alkanes became widely available in the 1980s. This silica resin has a high affinity for hydrophobic molecules, which increases as the chain length of the bonded alkane increases. The bonding of organic layers to silica has become a science in its own right, and there are many varieties of the same chain length phase. Some phases are polymerized, some are simple monolayers, and some are "end capped," which means that any uncovered surface of the silica is covered with an organic layer after the hydrophobic chain group has bonded. Each variety of resin may have different properties with respect to the mixture to be separated, and in many cases, different lots from the same supplier have different properties as well. The silica left bare around the bonded chains (in resins that are not end capped) often plays a role in the binding, and therefore the percent of the surface covered with each bonding reaction will lead to different properties for a particular resin type or manufacturer's lot. Often, these imperfections in the manufacture of "reversed-phase" chromatography resins are important for the separation that can be achieved, but complicate the strict interpretation of the chromatographic result and can be difficult to reproduce. Reversed-phase chromatography is a very popular laboratory analytical technique and has been scaled up for the production of small molecules and peptides. Concerns about protein denaturation have slowed its introduction into protein manufacture; however, this is changing as the high resolving power of the technique is making it more and more attractive for high value added products.

7.5.2 POLYMER-BASED RESINS

Polymer-based resins are frequently used in industrial applications because of their high stability and low cost. The resins are manufactured by suspension polymerization, in which an emulsion of the polymer is made in an immiscible solvent, and a cross-linking agent is added. The reaction is allowed to proceed to completion, and then the particles are isolated from the suspension, washed, and frequently derivatized. Surfactants are often added in the suspension polymerization to control particle size; in general, however, polymer resins are larger (10–100 μm) than silica-based resins (1–25 μm). Polymer resins are also less rigid and not generally suitable for high pressure (> 4 bar) applications (exceptions exist).

Two synthetic polymers that are commonly used are styrene divinylbenzene and polyacrylamide. Styrene divinylbenzene is very stable at pH extremes, and it has been used primarily as a support for ion exchange chromatography. While the backbone is somewhat hydrophobic, and nonspecific binding does occur, the stability and rigidity of this polymer make it a superior support for chromatography. Polyacrylamide is used less often. It is not used as a polymer solid but as a hydrogel, and its primary use in chromatography has been as a size exclusion gel. The cross-linking in polyacrylamide can be controlled very precisely by the amount of bisacrylamide added in the suspension mixture. The analogy to protein electrophoresis gels (Chapter 2) is apparent here.

Natural polymers such as agarose and dextran are also used in hydrogel form for low pressure chromatography resins. These polymers are naturally hydrophilic and are very compatible with protein and other biomaterials. Agarose can be cross-linked to form a reasonably rigid bead that is capable of tolerating pressures up to 4 bar. Dextran is less rigid and, like polyacrylamide, is used primarily in size exclusion. Dextran gels can be formed with very large pores, capable of including antibody molecules and virus particles. Both dextran and agarose have been derivatized with ion exchange groups, phenol groups, antibodies, dyes, heavy metals, and other biological epitopes that allow very specific binding behavior for the target molecule.

7.5.3 Ion Exchange Resins

Ion exchange resins are those that have been derivatized with an ionic group. The most commonly used ionic groups, listed in order of increasing pK, are sulfoxyl (SO_3^-), carboxyl (COO^-), diethylaminoethyl (DEAE) ($2C_2H_5N^+HC_2H_5$), and quaternary ethylamine (QAE) ($4C_2H_5N^+$). Thus sulfoxyl is the most acidic, and QAE is the most basic. When the acidic ion exchangers are used at pH values above their respective pK values, they carry a negative charge and attract positive counterions. These are called cation exchangers. Conversely the basic ion exchangers are known as anion exchangers.

Sulfoxyl and QAE are a strong acid and a strong base, respectively. Their pK values are close to 1 and 14, meaning that, for all practical purposes, they are fully ionized at all pH values. Carboxyl and DEAE are, respectively, a weak acid and base, with pK values closer to 4 and 10. These resins may be used in a pH gradient to ionize more or fewer of the ionic groups on the resin, thus effecting the separation. When the process calls for a pH shift, but separation of molecules along with that shift is not desirable, "strong" ion exchangers are preferred because their binding capacity is not likely to be altered.

Ion exchange resins are almost always used in aqueous mobile phases, and they are almost always polymer resins. Water is the universal solvent for salts, which dissociate poorly in most organic solvents. The ability to manipulate the pH to effect separation also favors polymer matrices as backbone material. Finally, many common biological contaminant molecules bind nearly irreversibly to ion exchangers (see later). Therefore, to regenerate ion exchange resins, extreme pH values are often used, such as 1.0 M NaOH (pH 14), which would dissolve many common silica-based resins.

Ion exchange resins, once exclusively used to separate small ions, are now heavily used in the separation of proteins, peptides, nucleic acids, small polynucleotides, and other small molecules such as antibiotics. Proteins have distinct isoelectric points (see Chapter 1) based on the content and conformation of their charged amino acids, arginine, lysine, histidine, glutamic acid, and aspartic acid, in decreasing pK of the side chain group. Sometimes tyrosine deprotonates, depending on its microenvironment. Proteins have multiple charges and can be separated based on their number of charges, as well as the sign of the charge and sometimes charge heterogeneity. Nucleic acids, on the other hand, have a charge at each base, such that their charge is proportional to their mass. Since they have a negative charge at each base, they bind very well to anion exchangers, and not at all to cation exchangers. Once bound to an anion exchange column, nucleic acids are very difficult to elute in an aqueous solution that does not simultaneously degrade them.

The selection of the buffer is an important consideration in the use of ion exchangers. It is advisable that the buffering species not interact with the adsorbent, which means that the charged form of the buffer should have the same sign as the charge on the adsorbent. It is also preferable that only simple anions be used in anion exchange (e.g., Cl^-, acetate$^-$) and only simple cations (e.g., Na^+, K^+) in cation exchange [23]. For example, when one is using DEAE–cellulose (an anion exchanger) at pH 8.0, a Tris–chloride buffer might be selected because it has HTris$^+$ (noninteractive) and Tris (neutral) buffering species and Cl^- counterion.

7.5.4 REVERSED-PHASE CHROMATOGRAPHY

Reversed-phase chromatography employs a hydrophobic phase bonded to the surface of the resin. Typically, reversed-phase resins are silica based. Reversed phase is so named because the partitioning of solutes between the mobile phase and stationary phase is opposite to that observed with bare silica. In other words, hydrophobic solutes bind in higher proportions in reversed phase, while hydrophilic solutes bind in higher proportion in "normal phase." Solutes are typically introduced into reversed-phase columns in water, or with minimal amounts of organic solvent, so that most solutes partition to the stationary phase. The organic content of the mobile phase is slowly increased, typically as a percent of acetonitrile, methanol, or isopropanol, thereby decreasing the polarity of the mobile phase.

Hydrophobic phases that are bonded to silica are typically octyl (C_8), octyldecyl (C_{18}), phenyl, and methyl (C_1). The different chain lengths and densities (called "coverage" in the commercial literature) of the different bonded phases obviously lead to more or less hydrophobicity. The entire surface of the silica cannot be fully covered with a monolayer of the desired phase, however, because of steric effects. Bare silica remains exposed, and this bare silica can participate in the separation by interacting with hydrophilic molecules, or hydrophilic domains of large molecules, thereby altering the binding. The strategy employed for covering this exposed silica surface lends almost as much to the specifics of the separation achieved as the chain length of the bonded phase. Polymerized phases represent an attempt to cover the surface by polymerizing the alkyl chains together at their point of attachment to the silica. End-capped resins use a short chain length group, such as methyl or ethyl, to cover the unreacted surface sites. Resins that do not utilize either method are left so intentionally to take advantage of the "mixed-mode" separation that may result. The separation characteristics of these resins are difficult to reproduce precisely, leading to considerable variation in separation from one manufacturer's lot of resin material to another.

Ions do not partition well in hydrophobic phases. It is common, therefore, to choose a counterion for reversed-phase chromatography. For biological mixtures, the counterion is nearly always a strong acid anion, such as trifluoroacetate, acetate, or chloride. The counterion has a strong effect on the separation, by partioning along with the co-ion of interest. The counterion can be used to make solutes more or less hydrophobic and will not affect all solutes equally.

7.5.5 HYDROPHOBIC INTERACTION CHROMATOGRAPHY

Hydrophobic interaction chromatography is typically used for protein separations. It employs derivatized polymer resins, with phenyl, butyl, or octyl ligand groups, typically. Proteins adhere to the hydrophobic surface under high salt conditions and redissolve into the mobile phase as the salt concentration is reduced. Hydrophobic interaction chromatography differs from reversed phase in that the mobile phase is kept aqueous (polar), and the salt concentration is used to effect the partitioning to the surface. The mechanism of partition is related to precipitation as opposed to two-phase partitioning (reversed phase) or ionic interaction (ion exchange). This is due to the use of salts that strip proteins of their solvation water. As the concentration of "lyophilic" salts increases, the probability increases that the protein will aggregate, or nucleate on the surface of the resin [24]. Hydrophobic interaction chromatography is sensitive to pH, salt used, buffer type, and temperature. Each of these must be carefully controlled to achieve reproducible separation but also represent an opportunity for increased selectivity.

7.5.6 AFFINITY CHROMATOGRAPHY

Affinity chromatography takes advantage of biological interactions to effect binding of specific solutes. Antibodies, antigens, or dyes are conjugated to polymer resins for the purpose of binding specific solutes from a mixture. For instance, an antibody could be raised against a target protein. The antibody would be conjugated to a resin, usually via cyanogen bromide activation. The antibody then captures the solute out of the mixture, and the impurities flow through the column. The solute can be recovered by changing the pH, increasing the salt concentration, or adding a displacer, that is, a molecule that also has some affinity for the antibody, or other binding agent (affinity ligand) on the stationary phase.

Other examples of specific interactions that can be used to isolate proteins are enzyme–ligand, enzyme–cofactor, receptor–agonist (antagonist). In each case, one member of any of these pairs may be immobilized to a resin to isolate the desired partner. Affinity chromatography is often coupled with cloning techniques to synthesize the target molecule with an "epitope" or recognition sequence that can be captured by the affinity ligand. Affinity chromatography is used from small-scale research (e.g., high throughput screening) to large-scale purification.

7.5.7 IMMOBILIZED METAL AFFINITY CHROMATOGRAPHY (IMAC)

Some proteins have high affinities for specific metals. This affinity may either be structural, as in the case of metalloproteins, which require metal centers for their biological activity, or based on the content of specific amino acid residues, such as histidine, tryptophan, and cysteine, which have increased affinity for transition metals such as nickel and copper. Techniques have been developed to immobilize metal ions with spacer arms onto polymer resins. These resins are referred to as IMAC resins, and they are used to purify proteins that have one of the two characteristics mentioned above. Genetic engineering has also been used to enable IMAC to be performed, with cloned target biomolecules being fused to polyhistidine tags (Section 1.4).

7.5.8 SIZE EXCLUSION CHROMATOGRAPHY

Sometimes referred to as "gel filtration," size exclusion chromatography (SEC) separates solutes on the basis of their size. There is no derivatization of the polymer gel, and there is no binding between the solutes and the resin. Molecules larger than the largest pores in the gel (the exclusion limit) cannot enter the gel and are eluted first. Smaller molecules enter the gel to varying extents, depending on their size and shape, and thus are retarded on their passage through the bed. In general, SEC resins are hydrophilic polymer gels with a broad distribution of pore sizes. The pore size is dependent on the degree of polymerization of the gel. Size exclusion chromatography is a useful technique, especially for changing buffers, or for removing small molecules from protein solutions. Because of the lack of binding between the solute and the resin, however, the capacity of this technique is low. Size exclusion effects may be in action in all the above-described techniques, since all resins are macroporous.

7.6 Particle Size and Pressure Drop in Fixed Beds

Particle diameter dominates the pressure drop in fixed-bed columns. The pressure drop is given by the Darcy equation [6]:

$$\frac{\Delta p}{L} = \frac{\mu v}{k} \tag{7.6.1}$$

where Δp is the pressure drop over column length L, μ is the viscosity of the mobile phase, v is the superficial velocity, and k is a constant. (In writing Equation [7.6.1], the term $-\rho g$ on the left side in the complete form of the Darcy equation has been omitted because it is very small compared to $\Delta p/L$.) A further correlation, known as the Blake–Kozeny equation [6], gives k as a function of resin particle size and void fraction:

$$k = \frac{d_p^2}{150} \frac{\varepsilon^3}{(1 - \varepsilon)^2} \tag{7.6.2}$$

The Darcy equation applies for rigid particles, such as silicas. For polymer resins that are compressible, k is a function of the linear velocity and column diameter, and a plot of Δp versus v would be nonlinear over a sufficiently wide range of v. Compressible gels get some of their hydrodynamic stability from the walls of the column; therefore the diameter of the column itself plays a role in the pressure drop. These factors need to be investigated for each application. As the stationary phase particle size is decreased, the pressure drop in the column increases as the inverse square. This increased pressure drop then requires additional power in pumping, as well as more specialized requirements for the construction of the column and its seals. The smallest particle sizes create inlet pressures of several hundred bars of pressure (several thousand psi) with a column length of 25 cm, and a velocity v of 0.01 cm/s. Seals and columns that contain these pressures are typically of narrow bore (2.5 mm i.d.), since a small radius of curvature is more effective for holding pressure than a large radius. Large columns packed with small particles are built like pressure vessels, with large lug nuts closely spaced around the flange at the end pieces.

A further complication arising from small particles is containment. The column end pieces must contain the particles without plugging. Plugging can be caused by resin particles, fragments of particles generated through attrition, or other debris in the chromatography feed streams. Particles 1 μm in diameter must be contained by 0.25 μm frits. These frits become easy to foul. The smaller the particle size, the more finely the mobile phase must be filtered and held particle free.

The selection of the resin particle diameter is probably the most critical engineering choice to be made in chromatography design and scaleup. The resolution of the column is favored by small particle size, while the pressure drop is adversely affected. The smaller the particle size, the shorter the column can be made, while maintaining the same separation. To retain binding capacity, the same resin volume must be used. Thus, smaller particles may be packed in short, flat beds, resembling pancakes, while larger particles can be accommodated in long tubes. As the column narrows, linear flow rates increase for the same volumetric flow, thus adversely affecting both pressure drop and resolution. As the columns get shallow and broad, mechanical seals and good flow distribution become more difficult to attain.

7.7 Equipment

Equipment for adsorption and chromatography is similar in that a fixed column or columns are generally involved. The equipment for chromatography, however, tends to be more sophisticated than that for the fixed-bed adsorption process, since the scale of operation is generally smaller and the requirement for resolution is greater. This is reflected in the fact that fixed- and agitated-bed adsorption operations tend to be near the beginning of the purification process, while chromatography is generally at or near the end of the purification. This section describes some of the special design considerations required for adsorption and chromatography equipment.

7.7.1 COLUMNS

Adsorption

Most fixed-bed adsorbers are cylindrical, vertical vessels through which feed, eluent, and regenerant streams are passed down through the bed of resin. There are two types of system to support fixed beds of adsorbents. One is a series of grids and screens, where each higher layer screen has successively smaller openings to prevent adsorbent from passing through and each lower layer has greater strength. In the other type of support, there is a graded system of particles such as ceramic balls. To prevent movement of the adsorbent at the top of the bed, a retention screen is usually installed on the top of the adsorbent and a layer of support balls placed on the top of the screen to hold the screen in place. Because the bed tends to expand and contract during operation of the process, the retention screen must be floating and not attached to the wall of the column. For large columns, it may also be necessary to have the inlet pipe branch into several pipes that are perforated along their length to ensure even distribution of the liquid across the column's cross section [25].

Chromatography

Chromatography columns are also cylindrical, vertical vessels designed to contain resin particles between 2 and $100\,\mu$m in diameter. Fluid is typically pumped downflow through chromatography columns, since air trapped beneath a column can cause voids that are ruinous for resolution. The frit, or in some cases netting, is critical for design of the columns. The frit is held in place by the end fittings on the column, often between gaskets, or incorporated into gaskets. A frit is critical at both the inlet and outlet of the column. The frit serves as a filter of last resort on the column inlet and holds the resin in place in the event the column is back-flushed. Polymer-based resins also swell under different conditions of ionic strength and pH, a phenomenon the inlet frit helps to contain.

The column comes in three basic types, fixed bed height, variable bed height, and axial compression. Fixed beds are used when a process is mature and the columns can be dedicated. The fixed bed is a tube that is fixed directly to end fittings. Threaded fittings are used on the analytical and preparative scales (up to 10 cm diameter); flanged connections are used for larger diameters. Fixed beds are also used for analytical high performance liquid chromatography (HPLC).

Variable or adjustable beds are common in pilot plants and development settings. The variable bed column is typically a tube that fits the outside diameter of the end fittings. The end fittings often consist of two flat plates sandwiching a gasket. When the end pieces are in place, the flat plates are compressed together, squeezing the gasket out against the tube. The top end piece may be placed at any level, depending on the volume of resin required for the process.

Axial compression columns are principally used for large-scale HPLC. When a rigid packing is used for chromatography, such as silica, packing efficiency is critical for high resolution. In this case, it has been found that slurry packing under high flow rate is not adequate. Axial compression columns address this by using a hydraulic piston as the inlet piece that compresses the resin along the length of the column. This not only causes the column to achieve the highest possible packing density, but also holds the resin in place as the action of the pump causes pressure pulses to wash through the column.

7.7.2 Chromatography Column Packing Procedures

Column packing is the most critical step in achieving high column performance in chromatography. There are three objectives in packing the column: to have the resin particles fully wetted, to have the particles fully disassociated from one another, and to achieve the highest possible packing density. To achieve the first objective, resin is slurry packed. This allows the resin to be in full contact with a solvent for an indefinite time prior to packing. It is critical to column capacity that the resin pores be full of solvent so that they are accessible to the process separands. Sometimes, especially at small scale, the resin is actively degassed, by sonication, vacuum, or heat. On larger scales, these methods are often impractical. In these cases, solvents such as ethanol that have high oxygen solubilities can be used to wet the particles.

Ensuring that the particles do not clump or form aggregates that can cause segregation during packing is goal number two. For this reason, solvents such as isopropanol are often added to the slurry. Isopropanol decreases the surface tension sufficiently to allow particles to deaggregate. Other agents, such as buffer salts, are sometimes used to effect particle–particle repulsive forces.

Finally, achieving the highest possible resin density is the ultimate packing goal. There are both chemical and mechanical considerations here. Chemically, particles should be in their least swollen state. If maximally swollen particles are packed, column conditions that cause the particles to subsequently contract will result in a bed that will no longer be packed efficiently, and voids will form. If particles are packed in their most contracted state, conditions that cause them to swell result in a more tightly packed column. High ionic strength is typically used to contract the particles. Attention must be paid to the strength of the seals when swellable resins are used. Resin can push through seals, crack the column, or occlude flow if allowed to swell too much. Highly swellable resins are generally avoided in large-scale processes. Mechanically, the particles should be packed quickly to deprive the slurry of the opportunity to settle and segregate according to particle size. The column should also be packed at or above the highest pressure (or flow rate) conditions expected for the column during normal processing. Finally, the inlet fitting should be placed as quickly and as evenly as possible, so that there are no biases in bed height and so that the pressure developed in the bed is captured mechanically in the seals.

Prior to the packing procedure, the resin is slurried at a packing density much lower than what will ultimately be achieved in the packed column. Therefore, a method is required for providing the extra volume to the column in which to slurry the resin. There are at least two common methods for doing this. The first is to temporarily increase the volume of the column, and the second is to use a secondary vessel. Adjustable-bed columns are easily adapted to the first method. Fixed-bed columns are often fitted with a packing extension tube to use this method. A secondary slurry vessel may also be used. This vessel is typically closed so that it can hold pressure. Then, clean solvent is pumped into the slurry vessel, displacing slurry to the column. Pumping slurries directly through a pump is not recommended because damage to both the resin and the pump may result.

7.7.3 Detectors

Detectors are selected to match the molecule in the application. Almost any detection method can be adapted to chromatography and adsorption, from Fourier transform infrared spectroscopy to antibody conjugation techniques to mass spectroscopy. However, the most common are pH, conductivity, and light absorbance. Conductivity and pH are monitored to check the performance of the gradient, the loading of the column, the regeneration, and reequilibration. Automated chromatography equipment relies on these measurements to proceed to the next steps. Light absorbance, primarily in the ultraviolet range (280, 254, 229, or 214 nm, depending on the application) is used to monitor the effluent for evidence of the target molecule. All three measurements are noninvasive and use well-developed technologies. Other common detection methods in use in large-scale chromatography and adsorption are refractive index, electrochemical detection, and light scattering.

Some applications are sufficiently sensitive to warrant on-line analytical detection. In these applications, a "split stream" is taken, and an analytical instrument, such as an HPLC, is injected with a sample. A purity can then be found for the sample, and a decision made to start, continue, or stop collection of the product. In these cases, the HPLC must be very fast compared with the large-scale chromatography, to permit assessment of purity of the target compound from contaminants and to allow a decision to be made before conditions have changed dramatically in the column effluent. HPLC methods can be developed that run in 1 or 2 min, when specific analytes are sought. In contrast, preparative scale chromatography cycles can last 4 to 8 h, depending on the optimization of the gradient elution sequence. For side stream analysis to be worthwhile, the preparative separation must give no apparent resolution of the target compound from a key contaminant that is present in levels of at least 1%. Also, there must be a requirement for purity of less than approximately 0.1% of this contaminant in the final product. When these conditions exist, and side stream detection must be applied, a high cost–low yield step is in store.

7.7.4 Chromatography System Fluidics

The fluidics of the chromatography system are always important in design and operation. Distribution of fluid across the column cross section is a key aspect in column efficiency and reproducibility. The pumps and tubing that feed the column are often part of this design.

Pumps

Pumps used in chromatography are typically positive displacement pumps. These provide reasonable pressures at reasonable cost. They have low shear, so do not pose a problem for sensitive biomolecules. Positive displacement pumps generally come in two varieties, peristaltic and rotary lobe. Peristaltic pumps use a rotating gear to compress flexible tubing through which the product stream flows. These are very familiar to biochemists and are useful for flow rates up to 10 liters/min. Rotary lobe pumps use two gears whose surfaces align such that at any rotation, they form a seal against each other, as well as against the interior of the pump housing. Rotary lobe pumps, uncommon below 1 liter/min, are almost always used at flow rates above 10 liters/min, where peristaltic pumps generally are not used. Rotary lobe pumps must be fed with positive pressure on the inlet, since they are not self-priming in general.

Gradient Makers

Most laboratory HPLC equipment allows programmed gradients; that is, the user selects a gradient type from a menu, and the gradient is mixed from two reservoirs in a dual-chamber pump. The two reservoirs contain the leanest concentration of salt or solvent and the richest concentration. The gradient is formed by the pump(s) selecting more or less volume from each reservoir as time progresses. The gradient can also be mixed from two tanks using CSTR principles; these principles were illustrated in Figure 5.1 and Equation (5.3.1), which gives the solute concentration at the outlet of the gradient mixer for a linear gradient. A linear gradient results if the rich tank is fed into the lean tank by gravity, while the lean tank (well mixed) feeds the chromatography system. In this case, both the rich tank and the lean tank decrease in volume at half the column flow rate each. It is important that the tanks be level with each other, be at the same height, and have the same cross section. To make an exponential gradient, the volume in the lean tank should remain constant, while the rich tank feeds at the same rate as the column flow rate.

7.8 Scaleup

The scaleup of adsorption and chromatography processes usually involves changes in flow rates and in the dimensions of the column. A safe way to scale up is to keep the column length constant, but this is sometimes not feasible. Because the objectives of adsorption and chromatography are usually different, the scaleup of these two operations is considered separately.

7.8.1 ADSORPTION

The scaleup of a fixed-bed adsorption focuses on the breakthrough curve for a single column, while scaleup of an agitated-bed process is concerned with breakthrough for the last column in a series of columns. The scaleup of mixing becomes important for the mixed-bed process. Scaleup for both types of adsorption process can generally be carried out based on laboratory data.

Fixed-Bed Adsorption

Several approaches can be used to scale up fixed-bed adsorption processes [26]. Two useful approaches are discussed here: (1) the length of unused bed (LUB) concept, which allows scaleup based on data from laboratory columns, keeping the particle size and superficial velocity constant, and (2) the computer simulation method, which also requires laboratory experimental data and constant particle size but does not require the superficial velocity to be constant.

LUB Method   In discussing the LUB method of scaleup, it is necessary to define the *break-point time* (t_b) and the *ideal adsorption time* (t^*) on a breakthrough curve, which are indicated in Figure 7.9. The break-point time is usually taken at the relative concentration, (c_i/c_{i0}, where c_{i0} is the feed concentration) of 0.05 or 0.10 [27]. Since only the fluid last exiting the column has this concentration, the average fraction of the solute removed from the start of feeding to the break-point time is usually 0.99 or higher. The ideal adsorption time is the time for breakthrough that would occur if the solute were in perfect equilibrium with the bed of adsorbent, which would give a vertical breakthrough curve. For a symmetrical breakthrough curve, the ideal adsorption time is the time at which $c_i/c_{i0} = 0.5$. At the ideal adsorption time for a bed initially free of the solute to be adsorbed, based on a unit area of bed cross section,

$$vc_{i0}t^* = L\rho_b q_{i,\text{sat}} \tag{7.8.1}$$

where L = bed length

$q_{i,\text{sat}}$ = average adsorbent phase concentration of solute i in equilibrium
with feed concentration c_{i0}, based on the adsorbent weight
(weight of solute i per weight of adsorbent)

v = superficial velocity

c_{i0} = concentration of solute i in the feed

ρ_b = bulk density of the adsorbent

The ideal adsorption time is therefore given by

$$t^* = \frac{L\rho_b q_{i,\text{sat}}}{vc_{i,0}} \tag{7.8.2}$$

The amount of the solute adsorbed at the break point can be determined by integrating the breakthrough curve up to time t_b, as indicated in Figure 7.10. The width of the breakthrough curve defines the width of the *mass transfer zone* in the bed.

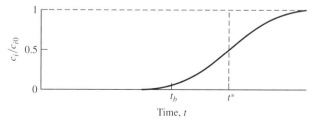

Figure 7.9 Breakthrough curve for a fixed-bed adsorber, showing the break-point time t_b, chosen at 5% of the solute feed concentration, and the ideal adsorption time t^*. The ratio of mobile phase concentration of solute to solute concentration in feed to the adsorber is c_i/c_{i0}.

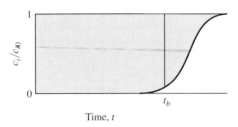

For adsorption where the equilibrium isotherm is favorable, which is true for the Langmuir isotherm that can often be used for protein adsorption, the concentration profile in the mass transfer zone takes on a characteristic shape that does not change as the zone propagates through the bed [5]. At the break-point time, the adsorbent between the inlet of the bed and the beginning of the mass transfer zone is completely saturated (in equilibrium with the solute in the feed). The adsorbent in the mass transfer zone goes from being completely saturated to being almost free of solute, and the adsorbent could be assumed to be on the average about half-saturated. This would be equivalent to half the adsorbent in the mass transfer zone being saturated and the other half being unused. The scaleup principle is that the length of the unused bed in the mass transfer zone does not change as the bed length is changed [28].

The length of unused bed (LUB) can be determined directly from the breakthrough curve obtained experimentally. For this method, LUB is defined as [5]:

$$\text{LUB} = \left(1 - \frac{q_{i,t_b}}{q_{i,\text{sat}}}\right)L = \left(1 - \frac{t_b}{t^*}\right)L \tag{7.8.3}$$

where t_b and t^* are stoichiometric times determined by integration of the breakthrough curve:

$$t^* = \int_0^\infty \left(1 - \frac{c_i}{c_{i0}}\right)dt \tag{7.8.4}$$

$$t_b = \int_0^{t_b} \left(1 - \frac{c_i}{c_{i0}}\right)dt \tag{7.8.5}$$

In scaleup calculations, the length of column required can easily be found by adding the LUB to the length calculated by assuming local equilibrium, with a shock wave concentration front [5].

EXAMPLE 7.6

Scaleup of the Fixed-Bed Adsorption of a Pharmaceutical Product   The breakthrough data given in Table E7.6 were obtained for the adsorption of a pharmaceutical product in a laboratory column (5 cm diameter × 15 cm high) at a feed flow rate of 400 ml/h and feed concentration of 0.75 U/liter, where U is units of biological activity of the pharmaceutical product. It is desired to scale up the process to operate in a column 30 cm high. What break-point time can be expected in the 30 cm high column?

TABLE E7.6

t (h)	c_i (U/liter)
20.5	0.01
26.7	0.20
32.0	0.39
36.0	0.53

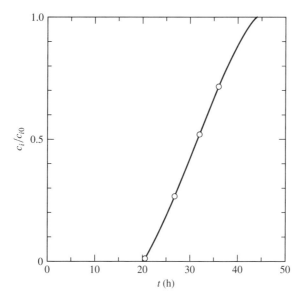

Figure E7.6 Breakthrough curve for the fixed-bed adsorption of a pharmaceutical product in Example 7.6.

SOLUTION

We use the LUB method, which involves graphical integrations of the breakthrough curve to determine the amount of solute adsorbed at various times. The breakthrough curve with concentration in dimensionless form is therefore plotted, with the curve extended to $c_i/c_{i0} = 1.0$ assuming that the curve is symmetrical about $c_i/c_{i0} = 0.5$ (Figure E7.6). By doing graphical integrations of this curve, we find the following:

$$\text{Total solute adsorbed at saturation } (c_i/c_{i0} = 1.00) = Qc_{i0} \int_0^{44.0} \left(1 - \frac{c_i}{c_{i0}}\right) dt$$

$$= Qc_{i0}(31.5\,\text{h})$$

$$\text{Total solute adsorbed at the break point } (c_i/c_{i0} = 0.05) = Qc_{i0} \int_0^{21.5} \left(1 - \frac{c_i}{c_{i0}}\right) dt$$

$$= Qc_{i0}(21.2\,\text{h})$$

where Q is the volumetric flow rate. Thus, t^* and t_b are 31.5 and 21.2 h, respectively, for this column. From Equation (7.8.3)

$$\frac{\text{LUB}}{L} = 1 - \frac{Qc_{i0}(21.2\,\text{h})}{Qc_{i0}(31.5\,\text{h})} = 1 - 0.673 = 0.327$$

$$\text{LUB} = 0.327(15\,\text{cm}) = 4.90\,\text{cm}$$

The adsorbent loading at saturation per total bed volume is

$$\rho_b q_{i,\text{sat}} = \frac{v c_{i0} t^*}{L} = \frac{400\,\dfrac{\text{cm}^3}{\text{h}} \times 0.00075\,\dfrac{\text{U}}{\text{cm}^3} \times 31.5\,\text{h}}{\dfrac{3.14 \times 5^2\,\text{cm}^2 \times 15\,\text{cm}}{4}} = 0.0321\,\frac{\text{U}}{\text{cm}^3}$$

For the column 30 cm high, the LUB does not change, so that

$$\frac{\text{LUB}}{L} = \frac{4.90}{30} = 0.163$$

The superficial velocity stays the same and is

$$v = \frac{400\,\dfrac{\text{cm}^3}{\text{h}}}{\dfrac{3.14 \times 5^2\,\text{cm}^2}{4}} = 20.4\,\frac{\text{cm}}{\text{h}}$$

From Equation (7.8.2),

$$t^* = \frac{L\rho_b q_{i,\text{sat}}}{v c_{i0}} = \frac{30\,\text{cm} \times 32.1\,\dfrac{\text{U}}{\text{liter}}}{20.4\,\dfrac{\text{cm}}{\text{h}} \times 0.75\,\dfrac{\text{U}}{\text{liter}}} = 62.9\,\text{h}$$

Therefore, from Equation (7.8.3),

$$t_b = t^*\left(1 - \frac{\text{LUB}}{L}\right) = (62.9\,\text{h})(1 - 0.163) = 52.6\,\text{h}$$

Computer Simulation Method   In the design of a production scale adsorption process, it is often desirable to simulate the process on a computer based on laboratory or pilot plant data. An example of such a simulation for design and scaleup was carried out by Chen et al. [29] for the fixed-bed adsorption of a valuable pharmaceutical product. Their model neglected dispersion [Equation (7.3.11)] and used a linear driving force expression for the mass transfer rate [Equation (7.3.5)] and a combination of linear and Freundlich isotherms for the equilibrium relations. From a laboratory column, the following relationships for the mass transfer coefficient K_a were obtained:

$$K_a = 1640 - 3700\frac{q_i}{q_{i,\text{sat}}} \qquad \text{for } \frac{q_i}{q_{i,\text{sat}}} \le 0.4 \qquad (7.8.6)$$

$$K_a = 264 - 260\frac{q_i}{q_{i,\text{sat}}} \qquad \text{for } 1 \ge \frac{q_i}{q_{i,\text{sat}}} > 0.4 \qquad (7.8.7)$$

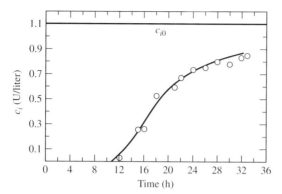

Figure 7.11 Comparison of model predictions (curve) and experimental data (open points) for the fixed-bed adsorption of a pharmaceutical product for a plant scale column 40 in. in diameter at a column height/column diameter ratio of 3.0. (Data from J. W. Chen, J. A. Buege, F. L. Cunningham, and J. I. Northam, "Scale-up of a column adsorption process by computer simulation," *Ind. Eng. Chem. Process Des. Dev.*, vol. 7, p. 1849, 1968.)

The laboratory column experiments indicated that pore diffusion in the adsorbent was the controlling mechanism for mass transfer. The equilibrium isotherm relationships were

$$c_i^* = 0.00006q_i \qquad \text{for } q_i \leq 20 \qquad (7.8.8)$$

$$c_i^* = 2.74 \times 10^{-8}q_i^{3.57} \qquad \text{for } q_i > 20 \qquad (7.8.9)$$

Equations (7.3.5) and (7.3.11) were solved on a computer by the method of finite differences, subject to the mass transfer and equilibrium isotherm relationships for a step change in the concentration in the feed from 0 to c_{i0} and no solute on the adsorbent at time 0. This model was used to predict the operation of 24 and 40 in. columns, and excellent agreement was obtained between the model and experimental data for c_i versus time. Model predictions and experimental data are shown in Figure 7.11. Computer simulation was used to optimize process variables such as flow rate, temperature, cycle time, concentration of product in the feed stream, and amount of adsorbent. Such a computer model of the adsorption process also enables the ready evaluation of changes in the process variables after the plant process is put into operation.

Agitated-Bed Adsorption

The scaleup of a series of agitated columns containing adsorbent (Figure 7.5), operated in the periodic countercurrent mode, follows directly from the mathematical model presented for this process [Equations (7.3.19)–(7.3.22)] [10]. Besides the increases in the flow rate and volume that occur upon scaleup, the mixing patterns may also change. A change in the mixing during scaleup was demonstrated in a mixed-bed adsorption process applied to a fermentation broth containing an antibiotic [30]. For a change in column volume by a factor of 29 from the laboratory column 2 in. in diameter to the pilot plant column with a 6 in. diameter, the mixing changed from perfect mixing in the laboratory column to mixing characterized by the following equation in the pilot plant column, based on the following age distribution function that was determined:

$$1 - \frac{c_i}{c_{i0}} = \exp\left[-\frac{V_L}{V_B}\left(\frac{t}{\tau} - \frac{V_P}{V_L}\right)\right] \qquad (7.8.10)$$

where c_{i0}, c_i = concentration of a nonadsorbing component i in the feed or effluent, respectively

V_L = total volume of liquid in the column

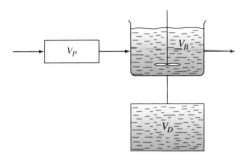

Figure 7.12 Mixing model based on the age distribution function determined experimentally for a pilot scale (6 in. diameter) agitated-bed adsorption column. Stagnant volume $V_D = V_L - V_B - V_P$. The total volume of liquid in the column is V_L.

V_B = volume of back-mixed liquid in the column
V_P = volume of liquid in plug flow in the column
τ = column residence time (V_L/Q, where Q = feed flow rate)
t = time from the start of a step change in the concentration of the nonadsorbing component from 0 to c_{i0}

Equation (7.8.10) corresponds to a mixing model consisting of a plug flow region in series with a perfectly mixed region connected to a stagnant region (Figure 7.12) [31]. A model for nonideal mixing can be incorporated into the mathematical simulation and design of the large-scale process.

7.8.2 CHROMATOGRAPHY

Chromatography scaleup algorithms typically account for changes in bed height and diameter, linear and volumetric flow rate, and particle size. A general approach to scaleup is based on keeping the resolution constant. Yamamoto et al. [32, 33] have developed the following proportionality for resolution, R_s, of proteins in linear gradient elution ion exchange chromatography and hydrophobic interaction chromatography:

$$R_s \propto \left(\frac{\mathscr{D}_m L}{g(V - V_0)u d_p^2} \right)^{1/2} \tag{7.8.11}$$

where $\mathscr{D}_m$ = diffusion coefficient of the protein in solution
L = column length
g = slope of the gradient (change in concentration of gradient per volume of gradient)
V = column volume
V_0 = column void volume
u = interstitial fluid velocity
d_p = particle diameter

This equation has been found to be valid over a wide range of experimental conditions [32, 33]. To remove the volume terms from the expression for resolution, we make the definitions

$$\bar{Q} = \frac{Q}{V} = \frac{\varepsilon u A}{V} \tag{7.8.12}$$

$$G = Vg \tag{7.8.13}$$

where Q is the inlet flow rate, ε is the column void fraction, and A is the column cross-sectional area. These substitutions lead to

$$R_s \propto \left(\frac{\mathscr{D}_m \varepsilon}{G(1-\varepsilon)\bar{Q}d_p^2} \right)^{1/2} \qquad (7.8.14)$$

Thus, for scaleup with constant resolution from scale 1 to scale 2 for the same product and the same column void fraction, the scaleup equation is

$$G_1 \bar{Q}_1 d_{p1}^2 = G_2 \bar{Q}_2 d_{p2}^2 \qquad (7.8.15)$$

Thus, as the particle size increases on scaleup, the flow rate relative to the column volume must decrease and/or the gradient slope must decrease to maintain constant resolution, which seems correct intuitively.

In practice, the gradient and the stationary phase size and chemistry are not changed upon scaleup. This is because it is easy to develop lab scale processes that use the same resin and same gradient that can be used at the commercial process scale. Therefore, in practice, only the ratio between column volume and flow rate need be addressed:

$$\frac{Q_1}{V_1} = \frac{Q_2}{V_2} \qquad (7.8.16)$$

Scaleup from a well-designed process development or preparatory column is reasonably straightforward. When the bed height can be maintained on scaleup, the mobile phase linear velocity remains the same, and the column is simply scaled by diameter. Scaling from a column 1 cm in diameter to column having a diameter of 10 cm would constitute a 100-fold increase in scale. This is the most conservative way to scale up a column but is not a necessary constraint.

EXAMPLE 7.7

Scaleup of a Protein Chromatography   A column 20 cm long, with an internal diameter of 5 cm, gives sufficient purification to merit scaleup. The column produces 3.2 g of purified protein per cycle, and a cycle takes 6 h, from equilibration through regeneration. You want a throughput of 10 g/h. What are the new column's dimensions if linear velocity is held constant?

SOLUTION

As just discussed, for scaleup when the linear velocity is held constant, the column diameter is increased, and the column height is maintained the same. If the linear flow rate is held constant, then the cycle time cannot be altered. Thus, the scaled up column must produce 6 h/cycle $\times$ 10 g/h = 60 g/cycle. Since the flow rate is proportional to the throughput of protein,

$$\frac{Q_2}{Q_1} = \frac{60\,\text{g/cycle}}{3.2\,\text{g/cycle}} = 18.75$$

From Equation (7.8.16), the scaleup relationship when the gradient and the particle size are not changed upon scaleup, and since $L_1 = L_2$,

$$\frac{Q_2}{Q_1} = \frac{V_2}{V_1} = \frac{\pi \left(\dfrac{D_2}{2}\right)^2 L_2}{\pi \left(\dfrac{D_1}{2}\right)^2 L_1} = \left(\frac{D_2}{D_1}\right)^2 = 18.75$$

where D_1 and D_2 are the column diameters for columns 2 and 1, respectively. Since $D_1 = 5.0\,\text{cm}$, we obtain

$$D_2 = (18.75)^{0.5} D_1 = 21.6\,\text{cm}$$

It is not always necessary to scale according to bed diameter. The flow rate may be normalized against the volume of the empty column, to give units of time^{-1}. This normalized flow rate is held constant from one scale to another, as shown in Equation (7.8.16). The case given in Example 7.7 can be shown to be a special case of this more general rule of thumb. Considerable research and practice indicates that this technique is equally effective and less restrictive compared to that just mentioned. In this case, column bed height may be increased or decreased, depending on the requirements of pressure drop and mechanical seals. A shallower bed gives a lower linear velocity and a wider diameter. A deeper bed gives a higher linear velocity (and higher pressure drop) and a proportionally narrower bed diameter. Therefore, bed height may need to be scaled on the basis of pressure drop constraints.

EXAMPLE 7.8

Scaleup of Protein Chromatography Using Standard Column Sizes   Consider the case given in Example 7.7. Available standard column diameters are 20 and 25 cm. What flow rates and bed depths would apply to each of these columns?

SOLUTION

The column volumes for both columns are still 18.75 times that used in the laboratory column. Thus,

$$V_2 = 18.75 V_1 = 18.75 \left(\frac{\pi D_1^2}{4} L\right) = 18.75 \left(\frac{\pi 5^2}{4} \times 20\right) = 7359\,\text{cm}^3$$

For a column 20 cm in diameter,

$$L = \frac{V_2}{\dfrac{\pi D_2^2}{4}} = \frac{7359\,\text{cm}^3}{\dfrac{\pi}{4}(20)^2} = 23.4\,\text{cm}$$

and for column the 25 cm in diameter,

$$L = \frac{V_2}{\dfrac{\pi D_2^2}{4}} = \frac{7359\,\text{cm}^3}{\dfrac{\pi}{4}(25)^2} = 15.0\,\text{cm}$$

In both cases, the gradient is also expressed in column volumes. The total gradient volume, that is, the total volume of eluent used to go from the leanest mobile phase condition to the richest, is expressed in terms of column volumes, and this is held constant on scaleup.

EXAMPLE 7.9

Scaleup of Elution Buffer Volumes in Protein Chromatography   An NaCl gradient is used to elute the product in Example 7.7. The gradient increases from 100 mM NaCl to 250 mM in 3 h. To form this gradient a 100 mM NaCl solution is combined with a 250 mM NaCl solution in a gradient mixer, in which the 100 mM NaCl solution is mixed (see Figure 5.1). The flow rate is 40 ml/min. What volumes of low salt and high salt eluent should be made for the scaleup column with an internal diameter of 20 cm?

SOLUTION

The small-scale volume of each gradient buffer is 3.6 liters (40 ml/min × 180 min/2). The scaleup factor of 18.75 still applies, so that

Volume of each buffer at large scale = (18.75)(3.6 liters) = 67.5 liters

Suppose that in development, the product elutes in 60% of the gradient volume. The gradient may be truncated to save time and buffer volume. From Equation (7.8.13), the gradient can be adjusted according to:

$$G = \frac{c_{hi} - c_{low}}{V_g / V} = \text{constant} \qquad (7.8.17)$$

where c_{hi} and c_{low} are the concentrations of the high strength and low strength buffers, respectively, and V_g is total gradient volume. Keeping the gradient slope constant is essential for maintaining resolution in chromatography.

Finally, pressure drop must be considered. Pressure drop is given by Darcy's equation [Equation (7.6.1)]. As columns increase in diameter, the pressure they can hold generally decreases (or the structural mass is increased, to provide the strength to maintain a seal). This is because as the radius of curvature increases, the force against the pressure decreases, which weakens the seal (think, e.g., about the stability of large bubbles in a glass of beer, compared with the smaller ones). It is possible to design the scaled-up column from the accessible pressure drop of the column.

EXAMPLE 7.10

Consideration of Pressure Drop in Column Scaling   Determine the minimum diameter possible for the columns analyzed in Example 7.8 and subject to the flow rate given in Example 7.9, when the column pressure should not exceed 300 kPa (28.8 psig), the maximum solution viscosity is 1.1 cp, and the void fraction in the column is 0.35. The resin particle size is 100 μm.

SOLUTION

Pressure drop can be calculated using Darcy's law [Equation (7.6.1)] by knowing μ, v, k, and L. We can calculate k from the Blake–Kozeny equation (7.6.2):

$$k = \frac{(100\,\mu\mathrm{m})^2}{150} \frac{0.35^3}{(1 - 0.35)^2} = 6.77\,\mu\mathrm{m}^2$$

The column volume must be 7359 ml (from Example 7.8), and the flow rate is 40 ml/min (from Example 7.9) times the scaleup factor of 18.75 (= 750 ml/min). We calculate the pressure drop for the column 20 cm in diameter and 23.4 cm in length, which would give the higher pressure drop of the two standard column sizes (20 and 25 cm):

$$\text{Cross-sectional area} = \frac{\pi (20\,\mathrm{cm})^2}{4} = 314\,\mathrm{cm}^2$$

$$v = \frac{750\,\dfrac{\mathrm{cm}^3}{\mathrm{min}}}{314\,\mathrm{cm}^2} = 2.39\,\frac{\mathrm{cm}}{\mathrm{min}}$$

From Equation (7.6.1) the pressure drop is

$$\Delta p = \frac{\mu v L}{k} = \frac{0.011\,\dfrac{\mathrm{g}}{\mathrm{cm\,s}} \times 2.39\,\dfrac{\mathrm{cm}}{\mathrm{min}} \times 23.4\,\mathrm{cm} \times \dfrac{1\,\mathrm{min}}{60\,\mathrm{s}} \times 10^{-4}\,\dfrac{\mathrm{kPa\,cm\,s}^2}{\mathrm{g}}}{6.77 \times 10^{-8}\,\mathrm{cm}^2}$$

$$= 15.1\,\mathrm{kPa}$$

Thus, the standard 20 cm diameter column would operate at well below the maximum allowable pressure.

7.9 Summary

The differential affinity of various solute molecules for solids of specific types is the basis for liquid chromatography and adsorption separation processes. Liquid chromatography is traditionally defined as the process where solutes are separated based on the speed at which they move through a column packed with adsorbent, also called the resin or stationary phase. By contrast, in pure adsorption processes the stationary phase is generally uniformly saturated with the target solute before an eluting solvent is introduced to cause the solute to transfer into the liquid or mobile phase.

- Different adsorption equilibrium constants make it possible to separate solutes from one another; equilibrium constants are controlled by solvent composition. In the reaction $C + S \leftrightarrow CS$ at equilibrium, the adsorption equilibrium constant is given by $K_{eq} = [CS]/[C][S]$, where [S] is the concentration of vacant sites in the solid phase. If the total number of sites S_{tot} is not much greater than [CS], then sites are nearly saturated, and combining the site balance with the equilibrium relationship gives the Langmuir adsorption isotherm equation, which shows site saturation when [CS] is plotted versus [C]. The Langmuir isotherm is often used to describe the equilibrium data for proteins.

- A special case occurs when S_{tot} is much greater than [CS]: without saturation, the plot is a linear adsorption isotherm, or $[CS] = K_{eq}[C]$. The linear isotherm is applicable to analytical chromatography.

- Another equilibrium isotherm that has proven useful is the Freundlich isotherm, which states that [CS] is proportional to $[C]^{1/n}$, with $n > 1$. This isotherm has been used to describe the adsorption of a number of antibiotics, steroids, and hormones.

- As a solution containing a species to be separated, called a separand, is flowed into a packed bed, sites become occupied until they are saturated; then separand flows past this saturation zone to find new sites in a transition zone until the transition zone reaches the end of the bed, resulting in outflow of separand known as breakthrough. The analysis of breakthrough curves serves as the basis for scaling and optimizing separand isolation by packed-bed adsorption.

- The complete mass balance for the separand in a fixed-bed adsorber is given by

$$\mathscr{D}_{\text{eff}}\frac{\partial^2 c_i}{dx^2} - \frac{v}{\varepsilon}\frac{\partial c_i}{\partial x} = \frac{\partial c_i}{\partial t} + \frac{1-\varepsilon}{\varepsilon}\frac{\partial q_i}{\partial t}$$

where c_i is the concentration of separand in the mobile phase, q_i is the concentration of separand in the adsorbent phase averaged over a particle, $\mathscr{D}_{\text{eff}}$ is the effective dispersivity of the separand, x is the distance from the bed inlet, v is the superficial velocity, ε is the void fraction of the bed packing, and t is time. This mass balance also applies to chromatography.

- The mass transfer rate of separand into the adsorbent particle is often expressed in terms of a linear driving force as

$$\frac{\partial q_i}{\partial t} = K_a(c_i - c_i^*)$$

where K_a is an overall mass transfer coefficient, and c_i^* is the liquid phase concentration at equilibrium with q_i.

- By assuming local equilibrium and negligible dispersion, the mass balance can be solved to give the effective velocity of separand i as

$$u_i = \frac{v}{\varepsilon + (1-\varepsilon)q_i'(c_i)}$$

where $q_i'(c_i)$ is the slope of the equilibrium isotherm at concentration c_i. When the equilibrium isotherm is nonlinear and favorable (e.g., a Langmuir isotherm) and there is a discontinuous change in concentration in the bed, $q_i'(c_i)$ becomes $\Delta q_i/\Delta c_i$, where the differences are taken across the discontinuity. The resulting velocity of the solute front is called the shock wave velocity.

- The mass balance for adsorption with the linear driving force expression for the mass transfer rate, a linear equilibrium isotherm, and negligible dispersion has been solved analytically to give an expression for the breakthrough curve.

- Chromatographic performance is assessed by elution peak analysis using a plate model, in which the number of theoretical plates is proportional to $t_R^2/(w/4)^2$, where t_R is average retention time and w is peak width. The resolution of two solutes is directly proportional to the difference in average retention times and inversely proportional to the sum of the peak widths.

- Equilibria for protein–salt ion exchange chromatography are calculated by using two equilibrium constants, one for protein and one for salt ion in the multicomponent Langmuir equation. This gives the protein adsorption isotherm as a function of salt concentration, which can be varied continuously in gradient elution chromatography. Gradient elution is thus distinguished from isocratic elution, in which no solute concentrations are varied.

- Band spreading in chromatography can be attributed to mechanical dispersion in the mobile phase, and molecular diffusion in the mobile and adsorbent phases. These factors govern the choice of particle size, pressure drop, and flow velocity. The impact of mass transfer and diffusion effects can be estimated by using the van Deemter equation.

- Adsorbents are selected on the basis of the adsorption chemistry of the separands, such as ion exchange, hydrophobic interaction, affinity reactions, metal affinity, and size exclusion.

- For a given mobile phase flow velocity, the required pressure drop is inversely proportional to the square of the particle diameter and directly proportional to column length and the viscosity and superficial velocity of the mobile phase (Darcy's law). Small particle size, desired for high resolution, can lead to unreasonable pressure requirements.

- Numerous design considerations are required for adsorption and chromatography equipment.

- A common method to scale up fixed-bed adsorbers is the length of unused bed (LUB) method, which is dependent on the equilibrium isotherm being favorable (concave downward) and on the particle size and superficial velocity remaining constant.

- The scaleup of a series of agitated columns containing adsorbent follows directly from the mass balance for each column, linear driving force expression for the mass transfer rate, and equilibrium isotherm, which must be solved by numerical methods. For this system, scaleup of the mixing is an important consideration.

- The constant resolution scaling rules for gradient elution chromatography, in their simplest form, specify that the product of gradient slope, normalized flow rate, and particle diameter squared should be held constant. Since normalized flow rate includes column area, an efficient way to scale up is to increase column diameter.

NOMENCLATURE

A	constant in the van Deemter equation [Equation (7.4.11)] (cm)	c	concentration of a solute in the mobile phase (M, or g liter^{-1})
A	cross-sectional area of column (cm^2)	c^*	equilibrium concentration of solute in the mobile phase (M, or g liter^{-1})
B	constant in the van Deemter equation [Equation (7.4.11)] (cm^2 s^{-1})	c_0	feed concentration of a solute in the mobile phase (M, or g liter^{-1})

C	chemical species in the mobile phase	N	number of theoretical plates
C	constant in the van Deemter equation [Equation (7.4.11)] (s)	Nu	Nusselt number ($= d_p k_c / \mathscr{D}_m$) (dimensionless)
CS	chemical species adsorbed to an adsorption site	Δp	pressure drop (kPa)
d_p	stationary phase particle diameter (cm)	Pe_g	dispersion Péclet number ($= d_p u / \mathscr{D}_{\mathrm{eff}}$) (dimensionless)
D	column diameter (cm)	q	concentration of a solute in the stationary phase averaged over an adsorbent particle (M, or g liter^{-1})
Da	Damköhler number ($= d_p^2 k_a / 4 \mathscr{D}_p$) (dimensionless)		
$\mathscr{D}_m$	molecular diffusivity of separand in unbounded solution (cm^2 s^{-1})	q_{sat}	concentration of solute on the stationary phase at equilibrium with feed concentration c_0, averaged over an adsorbent particle (M, g liter^{-1}, or g g^{-1})
$\mathscr{D}_{\mathrm{eff}}$	effective dispersivity in the column (cm^2 s^{-1})		
$\mathscr{D}_p$	molecular diffusivity of separand in the pore fluid (cm^2 s^{-1})	Q	flow rate (ml min^{-1})
		$\bar{Q}$	(flow rate)/(column volume) (min^{-1})
E	dispersion coefficient of separand, as a result of back mixing in the fixed-bed and nonuniformity of the velocity (cm^2 s^{-1})	r	radial distance in the stationary phase (cm)
		R_s	chromatographic resolution (dimensionless)
g	slope of the elution gradient (M liter^{-1}, or g liter^{-2})	Re	Reynolds number ($= d_p u \varepsilon \rho / \mu$) [Equation (7.4.13)] (dimensionless)
g	gravitational acceleration (9.0866 m sec^{-2})	S_{tot}	total concentration of adsorbent sites (mg ml^{-1})
G	normalized slope of the gradient (M, or g liter^{-1})	Sc	Schmidt number ($= \mu / \rho \mathscr{D}_m$) (dimensionless)
H	height equivalent to a theoretical plate (HETP) (cm)	t	time (s)
$H(x, t)$	dispersion function (dimensionless)	t_R	retention time (s)
H^*	normalized height equivalent of a theoretical plate (H / d_p)	t^*	ideal adsorption time (s)
		u	velocity of fluid through the interstices of the bed (cm s^{-1})
I_0	Bessel function (dimensionless)		
k	constant in the Darcy equation [Equation (7.6.1)] (cm^2)	u_i	solute i effective velocity (cm s^{-1})
k	K_a / K_{eq} (s^{-1})	$u_{i,\mathrm{sh}}$	solute i shock wave velocity (cm s^{-1})
k_a	forward rate constant of adsorption (s^{-1})	v	mobile phase superficial velocity (cm s^{-1})
k_c	fluid phase mass transfer coefficient (cm s^{-1})	V	volume of fixed-bed column (ml)
K_a	overall mass transfer coefficient (s^{-1})	V_B	volume of back-mixed liquid in column (ml)
K_{eq}	equilibrium constant (units vary)	V_L	liquid volume in column (ml)
L	length of the column (cm)	V_P	volume of liquid in plug flow in column (ml)
m	$\mathscr{D}_m / \varepsilon^* \mathscr{D}_p$ (dimensionless)	V_R	volume of adsorbent in column (ml)
m'	$\frac{3}{2} m [K_{\mathrm{eq}} / (1 + K_{\mathrm{eq}})]^2$ (dimensionless)	V_0	column void volume (ml)

w	peak width at base (base intercept) (s)
w_p	concentration of separand in the pore fluid based on a unit volume of stationary phase (i.e., the solid matrix and the pore space) (M, or g liter^{-1})
w_s	concentration of separand i adsorbed on the internal surfaces of the stationary phase based on a unit volume of stationary phase (i.e., the solid matrix and the pore space) (M, or g liter^{-1})
x	longitudinal distance in the column (cm)
x	fraction of separand in the mobile phase at long times $[\varepsilon/\varepsilon + (1-\varepsilon)\varepsilon^*(1+K_{eq})]$

Greek Letters

γ	$1 + K_{eq}(1-\varepsilon)/\varepsilon$
ε	column void fraction (dimensionless)
ε^*	effective volume fraction of the stationary phase that is accessible to the separand (dimensionless)
ξ	$(kK_{eq}x/v)(1-\varepsilon)$ (dimensionless)
μ	viscosity of mobile phase (g cm^{-1} s^{-1})
ρ	density of fluid (g cm^{-3})
ρ_c	bulk density of adsorbent (g cm^{-3})
σ	standard deviation for a peak (s)
τ	$k(t - x\varepsilon/v)$ (dimensionless)
τ	column residence time (V_L/Q) (s)
ϕ	c/c_0 (dimensionless)
ψ	q/q_0 (dimensionless)

Subscripts

i, j	components i, j
$n-1, n$	columns $n-1, n$ in a series of columns

PROBLEMS

7.1   Freundlich versus Langmuir Isotherm Apply both the Freundlich isotherm and the Langmuir isotherm to the set of data shown in Table P7.1. Determine the applicable constants for the two different isotherms. Which is better, and why?

7.2   Langmuir Isotherm For the Langmuir isotherm [Equation (7.2.6)], determine the relationship

TABLE P7.1

q (mg/g)	c (mg/ml)
2.00×10^{-2}	3.20×10^{-9}
8.00×10^{-2}	3.28×10^{-6}
1.00×10^{-1}	1.00×10^{-5}
5.00×10^{-1}	3.12×10^{-2}
7.00×10^{-1}	1.68×10^{-1}
1.00	1.00

between the equilibrium constant K_i and the mobile phase concentration [C] when the adsorbent is half-saturated.

7.3   Three Binding Solutes Derive the isotherms for three binding solutes that all compete for the same sites on the resin. If $K_{eq2} < K_{eq1} < K_{eq3}$, qualitatively describe the effects of the following situations:

(a) c_1 is very small compared to c_2 and c_3.
(b) $K_{eq3} \gg K_{eq1}$
(c) c_2 is low initially, and increases throughout the elution process.

7.4   Solute Binding to Two Binding Sites Derive an isotherm for a single solute molecule that binds simultaneously to two binding sites. An example of this would be a divalent cation binding to a carboxymethyl-derivatized resin. Use this expression to generate an example isotherm.

7.5   Dispersion versus Molecular Diffusivity What is the difference between the effective dispersivity $\mathcal{D}_{eff}$ and molecular diffusivity $\mathcal{D}_m$? What are the units for each?

7.6   Prediction of the Break-Point Time in Fixed-Bed Adsorption At low concentrations, the equilibrium for the antibiotic novobiocin and Dowex 21K anion exchange resin is linear,

$$q_i = 125c_i^*$$

for q_i and c_i^* in units of milligrams per milliliter. For the range of concentrations where this isotherm is valid, the mass transfer coefficient K_a averages 82 h^{-1}. Assuming a linear isotherm, estimate the break-point time (where $c_i/c_{i0} = 0.05$) in a fixed-bed adsorber with a bed length of 20 cm and superficial velocity of 40 cm/h. (Data from P. A. Belter, F. L. Cunningham, and J. W. Chen, "Development of a recovery process for novobiocin," *Biotechnol. Bioeng.*, vol. 15, p. 533, 1973.)

7.7 **Separation of Two Solutes Having Linear Equilibrium** Two solutes have a linear equilibrium with the stationary phase. Their equilibrium constants are 6.5 and 6.6, respectively. The task is to separate 20 ml of the mixture on a column 5 cm in diameter. The flow rate is 10 ml/min. Determine the minimum column length required to just separate the two compounds, with $\varepsilon = 0.3$.

7.8 **Calculation of the Shock Wave Velocity** A solute has a Langmuir isotherm characterized by an S_{tot} of 120 μg/ml and a K_{eq} of 60 ml/mg. Calculate the shock wave velocity for an injection of 1 mg/ml and column conditions identical to those presented in Problem 7.7.

7.9 **Using the LUB Method to Scale Up Fixed-Bed Adsorption** Breakthrough data (see Table P7.9) have been obtained for a fixed-bed adsorption process with a weakly anionic adsorbent using a feed of filtered fermentation broth containing the antibiotic cephalosporin at a concentration of 4.3 g/liter. The bed is 1 m long × 3 cm diameter, and the superficial velocity is 2 m/h. It is desired to scale up this process to a bed length of 3 m using the same superficial velocity. Use the LUB method to estimate the break-point time, defined here as occurring when $c_i/c_{i0} = 0.1$, for the large column. (Data from P. A. Belter, E. L. Cussler, W.-S. Hu, *Bioseparations*, p. 174, Wiley-Interscience, New York, 1988.)

7.10 **Removal of a Low Level Contaminant** One mode in which adsorbents are used is to strip a low level contaminant like DNA from a product solution. No other solution components are negatively charged at the operating pH for the column.

(a) If an anion exchanger is used for this purpose, determine the total bed capacity for DNA given these parameters: bed volume = 2 liters, $S_{tot} =$ 2 mg/ml, $K_{eq} = 2$ ml/mg, and DNA concentration = 5 μg/ml.

(b) Determine the bed capacity for the same conditions except that the DNA concentration is 100 μg/ml.

7.11 **Chromatography Scaleup** It is desired to scale up the throughput by a factor of 150 for a linear gradient ion exchange chromatography of a product protein from the laboratory to the plant. The conditions for the laboratory chromatography are the following: 1.0 cm bed diameter (ID) × 20 cm bed height, 20 μm particle size, and 30 cm/h superficial velocity. The particle size of the same type of ion exchange resin available for the plant operation is 40 μm. Two columns are available in the plant: one column 14.0 cm diameter (ID) × 50 cm high, and another column 18.0 cm diameter (ID) × 50 cm high. To keep the resolution for the chromatography constant in the plant, which column should be used? For this column, what should be the resin bed height and the superficial velocity and what do you estimate the pressure drop to be? The viscosity of the mobile phase is 1.0 cp, and the void fraction for resin in the plant column is 0.33.

7.12 **Design of a Protein Purification Process (Mini-Case Study)** Propose a purification process for the protein described. Assume a bacterial process. Design of the process will involve estimation of protein properties, for which a spreadsheet is provided at the textbook website (http://www.biosep.ou.edu). (Protein adapted from H. Zou, T. J. McGarry, T. Bernal, and M. W. Kirschner, "Identification of a vertebrate sister-chromatid separation inhibitor involved in transformation and tumorigenesis", *Science*, vol. 285, pp. 418–422, 1999.)

Sequence

1 MATLIYVDKENGEPGTRVVAKDGLKLG-
SGPSIKALDGRSQVSTPRFGKTFD

52 APPACLPKATRKALGTVNRATEKSVKT-
|---------S-S---------|
KGPLKQKQPSCFSAKKMTEKTCVKAKS

106 SVPASDDAYPEIEKFFPFNPLD-
FESFDLPEEHQIAHLPLSGVPLMILDEER

157 ELEKLFQLGPPSPVKMPSPPWESNL-
LQSPSSILSTLDVELPPVCCDIDI

TABLE P7.9

t (h)	c_i (g/liter)
4.7	0.2
6.5	0.4
7.3	1.0
7.8	1.8
8.1	2.7
8.7	3.8
9.3	4.2
10.3	4.3

Research Production Process

A PCR-amplified gene was cloned into a modified pET28a vector to express recombinant protein in *E. coli*. Protein more than 90% pure was obtained by affinity purification with Ni-nitrilotriacetic acid (NTA) beads followed by Resource Q column chromatography.

In addition to the traditional process and host-cell-related impurities, the following substance related impurities are present in the process:

Impurity 1: Des-Met (N-terminal methionine deleted for the desired protein)

Impurity 2: C-terminal cleavage product (cleavage between lysine 9 and glutamic acid 10)

1 ENGEPGTRVVAKDGLKLGSGPSIKA-
                   LDGRSQVSTPRFGKTFD

43APPACLPKATRKALGTVNRATEKSVK-
            |---------S-S---------|
    TKGPLKQKQPSCFSAKKMTEKTCVKAKS

97 SVPASDDAYPEIEKFFPFNPLDFESFDL-
          PEEHQIAHLPLSGVPLMILDEER

148 ELEKLFQLGPPSPVKMPSPPWESNL-
         LQSPSSILSTLDVELPPVCCDIDI

Impurity 3: N-terminal cleavage product (cleavage between lysine 9 and glutamic acid 10)

1   MATLIYVDK

Impurity 4: Misfolded isoform

1 MATLIYVDKENGEPGTRVVAKDGL-
   KLGSGPSIKALDGRSQVSTPRFGKTFD

   |----------------------------S-S-------------
52 APPACLPKATRKALGTVNRATEKSVK-

--------------------------|
TKGPLKQKQPSCFSAKKMTEKTCVKAKS

106 SVPASDDAYPEIEKFFPFNPLDFES-
      FDLPEEHQIAHLPLSGVPLMILDEER

157 ELEKLFQLGPPSPVKMPSPPWES-
      NLLQSPSSILSTLDVELPPVCCDIDI

Impurity 5: Covalent dimer between cysteine 56 on adjacent molecules

Information about chromatography adsorbents can be found through links to various companies at the textbook website (www.biosep.ou.edu). There is also a link to a spreadsheet to approximate protein charge as a function of pH at the textbook website.

7.13 **Specification of Equipment for Chromatography Steps (Mini-Case Study)** Use the purification steps selected in Problem 7.12, to choose equipment and specify column scale for the chromatography steps, given the following considerations:

- Fermentation titer: The *E. coli* process produces 500 liters of product at a concentration of 2.5 g/liter. One fermentation cycle is 48 h, and the success rate is 90% (failure means that the entire fermentation batch had to be dumped); 50% of the recovered protein is the target product.
- There will be a need for 3000 g for clinical trial material for *E. coli* product, produced over three or more batches.
- There are 3 months available to campaign the clinical material, starting after the Christmas shutdown.
- Each purification step gives a yield of 90%, and there is an additional 10% penalty every third step.
- The protein degrades with the following zero-order kinetics (independent of concentration):

$$\text{rate } (\%/\text{day}) = -50e^{-1000/T} \qquad T \text{ in K}$$

The degradation product is a deamidation of asparagine 11 and must be purified away if it exceeds 0.1%.

General Plant Resources

- There is one 500-liter fermentor available.
- Fermentation products can be delivered to the first separation step free from solids.
- There are five columns available in the manufacturing area (see Table P7.13).

TABLE P7.13

Number of columns	Diameter (cm)	Minimum/maximum length (cm)	Maximum pressure (bar)
1	100	5/20	3
1	63	7/30	3
2	25	12/50	5
1	15	30/100	100

- One chromatography cycle takes 6 h starting from storage or cleaning solutions, 4.5 h when one is resuming from the elution step.
- A water for injection (WFI) still is available that produces 200 liters/h.
- There are two 4°C cold rooms.
- Two crossflow filter housings are available, one capable of holding 50 ft^2 of membrane, the other capable of holding 200 ft^2.
- Any tank you size can be made available.

Factors to Consider

- Compatibility with current processes (what's currently used in manufacturing?)
- Cleaning and storage requirements (ease of accommodation in the plant, handling issues, etc.)
- Preferred vendor (experience, reliability, delivery, price, validation package?)
- Future scalability

References

1. Simpson, J. M. (1994). Conventional chromatography. In *Protein Purification Process Engineering,* R. G. Harrison, ed., Dekker, New York, p. 209.
2. Shiloach, J., and Kennedy, R. M. (2000). Expanded-bed adsorption process for protein capture. In *Handbook of Separations*, S. Ahuja, ed., Academic Press, San Diego, CA, p. 431.
3. Shukla, A. A., and Cramer, S. M. (2000). Bioseparations by displacement chromatography. In *Handbook of Separations,* S. Ahuja, ed., Academic Press, San Diego, CA, p. 379.
4. Belter, P. A., Cussler, E. L., and Hu, W.-H. (1988). *Bioseparations,* Wiley-Interscience, New York.
5. Ruthven, D. M. (1991). Adsorption. In *Kirk–Othmer Encyclopedia of Chemical Technology,* vol. 1, 4th ed., J. I. Kroschwitz, ed., Wiley-Interscience, New York, p. 493.
6. Bird, R. B., Stewart, W. E., and Lightfoot, E. N. (2002). *Transport Phenomena,* 2nd ed., Wiley, New York.
7. Klinkenberg, A. (1954). Heat transfer in cross-flow heat exchangers and packed beds. *Ind. Eng. Chem.,* vol. 46, p. 2285.
8. Furnas, C. C. (1930). Heat transfer from a gas stream to a bed of broken solids. *Trans. AIChE,* vol. 24, p. 142.
9. Ruthven, D. M. (1984). *Principles of Adsorption and Adsorption Processes,* Wiley-Interscience, New York.
10. Belter, P. A., Cunningham, F. L., and Chen, L. W. (1973). Development of a recovery process for novobiocin. *Biotechnol. Bioeng.,* vol. 15, p. 533.
11. Chapra, S. C., and Canale, R. C. (1998). *Numerical Methods for Engineers,* 3rd ed., WCB/McGraw-Hill, Boston.
12. Martin, A. J. P., and Synge, R. L. M. (1941). A new form of chromatogram employing two liquid phases. I. A theory of chromatography. *Biochem. J.,* vol. 35, p. 1358.
13. Horvath, C., and Melander, W. R. (1983). Theory of chromatography. In *Fundamentals and Applications of Chromatographic and Electrophoretic Methods, Part A,* E. Heftmann, ed., Elsevier, Amsterdam, Netherlands, p. A27.
14. Kucera, E. (1965). Contribution to the theory of chromatography: Linear nonequilibrium elution chromatography. *J. Chromatog.,* vol. 19, p. 237.
15. Lightfoot, E. N., Sanchez-Palma, R. J., and Edwards, D. O. (1962). Chromatography and allied fixed-bed separations processes. In *Interscience Library Chemical Engineering Processing,* H. M. Schoen, ed., Interscience, New York, p. 99.
16. Lapidus, L., and Amundson, N. R. (1952). Mathematics of adsorption in beds. VI. The effect of longitudinal diffusion in ion exchange and chromatographic columns. *J. Phys. Chem.,* vol. 56, p. 984.

17. Athalye, A. M., Gibbs, S. J., Lightfoot, E. N. (1992). Predictability of chromatographic separations. Study of size-exclusion media with narrow particle size distributions. *J. Chromatog.,* vol. 589, p. 71.

18. Rasmuson, A., and Neretnieks, I. (1980). Exact solution of a model for diffusion in particles and longitudinal dispersion in packed beds. *AIChE J.,* vol. 26, p. 686.

19. Van Deemter, J. J., Zuiderweg, F. J., and Klinkenberg, A. (1956). Longitudinal diffusion and resistance to mass transfer as causes of non-ideality in chromatography. *Chem. Eng. Sci.,* vol. 5, p. 271.

20. Knox, J., and Scott, H. (1983). B and C terms in the van Deemter equation for liquid chromatography. *J. Chromatog.,* vol. 282, p. 297.

21. Giddings, J. C. (1965). *Dynamics of Chromatography,* Pt. 1, *Principles and Theory,* Dekker, New York.

22. Peskin, A. P., and Rudge, S. R. (1992). Optimization of large-scale chromatography for biotechnological applications. *Applied Biochem. Biotechnol.,* vol. 34/35, p. 49.

23. Scopes, R. K. (1982). *Protein Purification,* Springer-Verlag, New York.

24. Roettger, B. F., Myers, J. A., Ladisch, M. R., and Regnier, F. E. (1989). Adsorption phenomena in hydrophobic interaction chromatography. *Biotechnol. Prog.,* vol. 5, p. 79.

25. Keller, G. E., Anderson, R. A., and Yon, C. M. (1987). Adsorption. In *Handbook of Separation Process Technology,* R. W. Rousseau, ed., Wiley-Interscience, New York, p. 644.

26. Thomas, W. J., and Crittenden, B. (1998). *Adsorption Technology and Design,* Butterworth-Heinemann, Oxford.

27. McCabe, W. L., Smith, J. C., and Harriott, P. (1993). *Unit Operations of Chemical Engineering,* McGraw-Hill, New York.

28. Collins, J. J. (1967). The LUB/equilibrium section concept for fixed-bed adsorption. *AIChE Symp. Ser.* 74, vol. 63, p. 31.

29. Chen, J. W., Buege, J. A., Cunningham, F. L., and Northam, J. I. (1968). Scale-up of a column adsorption process by computer simulation. *Ind. Eng. Chem. Process Des. Dev.,* vol. 7, p. 1849.

30. Harrison, R. G. (1981). Antibiotic adsorption process using a non-ionic macroporous resin, Presented at AIChE National Meeting, Houston, 1981.

31. Levenspiel, O. (1962). *Chemical Reaction Engineering,* Wiley, New York.

32. Yamamoto, S., Nomura, M., and Sano, Y. (1987). Resolution of proteins in linear gradient elution ion-exchange and hydrophobic interaction chromatography. *J. Chromatog.,* vol. 409, p. 101.

33. Yamamoto, S., Nakanishi, K., and Matsuno, R. (1988). *Ion-Exchange Chromatography of Proteins,* Dekker, New York.

Precipitation

Precipitation, which is the process of coming out of solution as a solid, is an important method in the purification of proteins that usually comes early in the purification process. Precipitation is frequently used in the commercial separation of proteins. The primary advantages of precipitation are that it is relatively inexpensive, can be carried out with simple equipment, can be done continuously, and leads to a form of the protein that is often stable in long-term storage. Since precipitation is quite tolerant of various impurities, including nucleic acids and lipids, it is used early in many bioseparation processes.

The goal of precipitation is often concentration to reduce volume, although significant purification can sometimes be achieved. For example, all the protein in a stream might be precipitated and redissolved in a smaller volume, or a fractional precipitation might be carried out to precipitate the protein of interest and leave many of the contaminating proteins in the mother liquor.

In this chapter the focus is first upon protein solubility, which is the basis of separations by precipitation. Then we discuss the basic concepts of particle formation and breakage and the distribution of precipitate particle sizes. The specific methods that can be used to precipitate proteins are treated next. The chapter concludes with methodology to use for the design of precipitation systems.

8.1 Instructional Objectives

After completing this chapter, the reader should be able to do the following:

- Explain the various factors that influence protein solubility.
- Use the Cohn equation to predict solution equilibria (precipitation recoveries).
- Identify the distinct steps in the development of a precipitate.
- Calculate mixing times in an agitated precipitator, the kinetics of diffusion-limited growth of particles, and the kinetics of particle–particle aggregation.
- Perform particle balances as a function of particle size in a continuous flow stirred tank reactor (CSTR).

- Explain the methods used to cause precipitation.
- Outline the advantages and disadvantages of the three basic types of precipitation reactor: the batch reactor, the CSTR, and the tubular reactor.
- Implement simple scaling rules for a precipitation reactor.

8.2 Protein Solubility

Since proteins precipitate because they become insoluble, a knowledge of the factors that influence protein solubility is important in understanding why proteins precipitate. The most important factors affecting the solubility of proteins are structure and size, protein charge, and the solvent. Explanations follow for each of these factors.

8.2.1 STRUCTURE AND SIZE

In the native state, a protein in an aqueous environment assumes a structure that minimizes the contact of the hydrophobic amino acid residues with the water solvent molecules and maximizes the contact of the polar and charged residues with the water. The major forces acting to stabilize a protein in its native state are hydrogen bonding, van der Waals interactions, and solvophobic interactions. In aqueous solution, these forces tend to push the hydrophobic residues into the interior of the protein and the polar and charged residues on the protein's surface. For example, one study of 36 globular proteins has shown that 95% of the ionizable groups are solvent accessible [1]. In other studies of 69 proteins, the average solvent-(water-) accessible atomic surface was found to be 57% nonpolar, 25% polar, and 19% charged) [2, 3]. Thus, in spite of the forces operating to force hydrophobic residues to the protein's interior, the surface of proteins usually contains a significant fraction of nonpolar atoms.

The forces acting on a protein lead to the achievement of a minimum Gibbs free energy. For a protein in its native configuration, the net Gibbs free energy is on the order of only 10 to 20 kcal/mol. This is a relatively small net free energy, which means that the native structure is only marginally stable and can be destabilized by relatively small environmental changes [4].

Water molecules bind to the surface of the protein molecule because of association of charged and polar groups and immobilization by nonpolar groups. For example, a study of the hydration of human serum albumin found two layers of water around the protein [5]. In the layer next to the protein, the water molecules are almost totally oriented, with the hydrogen atoms adjacent to and facing the albumin surface, while the oxygen atoms face away from the protein surface. In the second layer of water molecules, most of the water molecules (70%) are nonoriented. These hydration layers are thought to promote solubility of the protein by maintaining a distance between the surfaces of protein molecules. This phenomenon is illustrated in Figure 8.1.

The size of a protein becomes important with respect to solubility when the protein is excluded from part of the solvent. This can happen when nonionic polymers that are added to the solution result in steric exclusion of protein molecules from the volume of solution occupied by the polymer. Juckes [6] developed a model for this phenomenon based on the

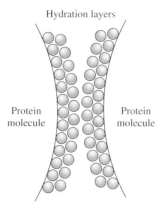

Hydration layers

Protein molecule

Protein molecule

Figure 8.1 Schematic diagram of the limit of approach of two protein molecules to each other because of the hydration layers on each molecular surface.

Ab

Ag

Figure 8.2 Schematic representation of antibody–antigen (Ab–Ag) interaction.

protein molecule being in the form of a solid sphere and the polymer molecule in the form of a rod, which gave the following equation for S, the solubility of the protein:

$$\ln S = \beta' - K'c_p \tag{8.2.1}$$

where

$$K' = \frac{\bar{V}}{2.303}\left(\frac{r_s + r_r}{r_r}\right)^3 \tag{8.2.2}$$

Here, r_s and r_r are the radius of the protein solute and polymer rod, respectively, $\bar{V}$ is the partial specific volume of the polymer, c_p is the polymer concentration, and β' is a constant. Based on this model, we can expect the lowest protein solubility for large proteins.

Molecular size is the predominant factor in a type of precipitation known as affinity precipitation. When affinity groups or antibodies to a specific biomolecule (antigen) are added to a solution, the antibody–antigen interaction can form large multimolecular complexes as shown in Figure 8.2. Such complexes are usually insoluble and cause selective precipitation of the antigen. The average size of the complex agglomerates is maximized when there is a stoichiometric ratio of antibody and antigen. If either is present in great excess, only bimolecular complexes will be formed; and there may be no precipitation, or low recovery, even if a precipitate is formed.

8.2.2 CHARGE

The net charge of a protein has a direct bearing upon the protein's solubility. The solubility of a protein increases as its net charge increases, a result of greater interaction with dipolar

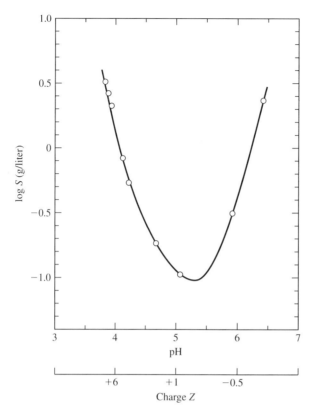

Figure 8.3 The solubility (S) of insulin in 0.1 N NaCl as a function of pH. The charge Z is the average protonic charge per 12,000 g of insulin at the pH values indicated. (Data from C. Tanford, *Physical Chemistry of Macromolecules*, p. 242, Wiley, New York, 1961.)

water molecules. A repulsive reaction between protein molecules of like charge further increases solubility.

A simple way to vary the charge on a protein is by changing the pH of the solution. The pH of the solution in which a protein has zero net charge is called the isoelectric pH or isoelectric point. The solubility of a protein is, in general, at its minimum at the isoelectric point. A typical example is shown in Figure 8.3. Nonuniform charge distribution, however, results in a dipole moment on the molecule, which leads to an increase in solubility and a move in the minimum solubility away from the isoelectric point. The effect of the dipole moment is discussed further in the following section.

The net charge of a protein is determined by the following factors: the total number of ionizable residues, the accessibility of the ionizable residues to the solvent, the dissociation constants (or pK_a values) of the ionizable groups, and the pH of the solution [7]. Besides the chemical makeup of the ionizable groups, factors that can influence the pK_a values are the chemical nature of the neighboring groups (e.g., inductive effects), the temperature, the chemical nature of the solvent as partially reflected by its dielectric constant, and the ionic strength of the solvent.

8.2.3 SOLVENT

The solvent affects the solubility of proteins primarily through two parameters, hydrophobicity and ionic strength. The first of these parameters has been well studied through observations of single-phase solutions of water and monohydric alcohols. Although these solutions can cause protein denaturation at room temperature, denaturation can be avoided at sufficiently low temperatures. Studies of monohydric alcohols have shown that denaturing efficiency is as follows:

$$methanol < ethanol < propanol < butanol$$

This led to the conclusion that the alcohols with longer alkyl chains are binding more effectively to apolar groups on the protein, weakening intraprotein hydrophobic interactions and thus leading to denaturation [8]. It is thought that when the temperature is low, the monohydric alcohols compete for the water of hydration on the protein and cause the protein molecules to approach more closely, so that van der Waals interactions lead to aggregation.

The ionic strength of the solvent can have both solubilizing and precipitating effects. The solubilizing effects are referred to as salting in, while the precipitating actions are called salting out. The salting-in effect has been observed for several proteins, including the class of proteins named euglobulins. These proteins are insoluble in the absence of salt at their isoelectric points but become soluble when salt is added. In contrast, members of another class of proteins, the albumins, are very soluble in water as well as in high concentrations of salt. It is believed that the solubilities of the euglobulins and the albumins differ because the euglobulins have a much higher dipole moment than the albumins [9].

A theoretical treatment of the interactions between ions and dipoles developed in 1943 by Kirkwood [10] accounts for salting-in effects by considering the solute size, solute shape, solute dipole moment, solvent dielectric constant, solution ionic strength, and temperature. One of Kirkwood's models was for a spherical dipolar ion with a point dipole moment u located at the center of the sphere. The equation derived to describe the interactions is as follows:

$$\ln\left(\frac{S_p}{S_0}\right) = K_i I - K_s I \tag{8.2.3}$$

where S_p is the solubility of the dipolar ion at ionic strength I, S_0 is the solubility of the dipolar ion in the absence of salt, K_i is the salting-in constant, and K_s is the salting-out constant. Ionic strength is defined by

$$I = \frac{1}{2}\sum_i c_i z_i^2 \tag{8.2.4}$$

where c_i is the molar concentration of any ion and z_i is its charge. The salting-in and salting-out constants can be related to other variables as follows:

$$K_i \propto \left(\frac{u}{\varepsilon T}\right)^2 \tag{8.2.5}$$

$$K_s \propto \frac{V_e}{\varepsilon T} \tag{8.2.6}$$

where ε is the dielectric constant of the solvent, T is temperature, and V_e is the excluded volume of the dipolar ion. Equations (8.2.3) and (8.2.5) confirm the observed strong relationship between the solubility of dipolar proteins and the size of their dipole moment u, as well as the greater salting-in effect observed with proteins with high dipole moments. Also, it can be seen that the salting-in term increases more than the salting-out term as the dielectric constant decreases. The dielectric constant decreases as the polarity of the solvent decreases. Therefore, the salting-in effect tends to predominate in relatively nonpolar solvents, while the salting-out effect is more dominant in aqueous solvents.

At high ionic strength, the salting-out effect becomes predominant and can be described empirically by the Cohn equation [11]:

$$\ln S = \beta - K'_s I \tag{8.2.7}$$

where S is the solubility of the protein and K'_s is a salting-out constant characteristic of the specific protein and salt that is independent of temperature and pH above the isoelectric point. The constant β, the hypothetical solubility of the protein at zero ionic strength, depends only on temperature and pH for a given protein and is a minimum at the isoelectric point. It is interesting that the Cohn equation is identical in form to the equation that describes the precipitation of proteins by the addition of nonionic polymers, Equation (8.2.1). In addition, the Kirkwood equation for the solubility of dipolar ions [Equation (8.2.3)] can be arranged to give

$$\ln S_p = \ln S_0 - (K_s - K_i)I \tag{8.2.8}$$

which is also identical in form to the Cohn equation, with

$$\beta = \ln S_0 \tag{8.2.9}$$

$$K'_s = K_s - K_i \tag{8.2.10}$$

Both salting in and salting out are illustrated in Figure 8.4 for hemoglobin with ammonium sulfate or sodium sulfate being added. From zero ionic strength, the solubility of the protein increases to a maximum as salt is added and then continuously decreases as even more salt is added.

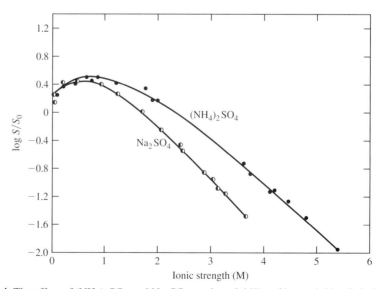

Figure 8.4 The effect of $(NH_4)_2SO_4$ and Na_2SO_4 on the solubility of hemoglobin: S_0 is the solubility in pure water, and S is the solubility in the salt solution. (Data from C. Tanford, *Physical Chemistry of Macromolecules*, p. 244, Wiley, New York, 1961.)

EXAMPLE 8.1

Salting Out of a Protein with Ammonium Sulfate   Data were obtained on the precipitation of a protein by the addition of ammonium sulfate. The initial concentration of the protein was 15 g/liter. At ammonium sulfate concentrations of 0.5 and 1.0 M, the concentrations of the protein remaining in the mother liquor at equilibrium were 13.5 and 5.0 g/liter, respectively. From this information, estimate the ammonium sulfate concentration to give 95% recovery of the protein as precipitate.

SOLUTION

We can use the Cohn equation [Equation (8.2.7)], to solve this problem if we can determine the constants in the equation. Since ionic strength is directly proportional to concentration c for a given salt [Equation (8.2.4)], we can rewrite the Cohn equation as

$$\ln S = \beta - K_s'' c$$

Substituting the experimental data into this equation gives

$$\ln 13.5 = \beta - 0.5 K_s''$$
$$\ln 5.0 = \beta - 1.0 K_s''$$

Solving these equations for the constants yields

$$\beta = 3.60$$
$$K_s'' = 1.99 \, \text{M}^{-1}$$

For 95% recovery, the protein solubility in the mother liquor at equilibrium is 5% of the initial protein concentration. At this solubility, from the Cohn equation

$$c = \frac{\beta - \ln S}{K_s''} = \frac{3.60 - \ln(0.05 \times 15)}{1.99} = 1.95 \, \text{M}$$

8.3 Precipitate Formation Phenomena

By studying the phenomena of precipitate formation, we can maximize our control over the characteristics of the final protein precipitate. Important characteristics of protein precipitates are the particle size distribution, density, and mechanical strength. Normally, it is desired to avoid having a large fraction of particles in the small size range. For proteins, particle sizes near or below 1 μm are considered to be small. Protein precipitates that consist largely of particles with small particle sizes can be difficult to filter or centrifuge. Low particle densities also can lead to filtration or centrifugation problems and can give excessive bulk volumes of the final dried precipitate. Particles with low mechanical strength can give problems with excessive attrition when the dry particles are moved. Low strength can also be interpreted as gel formation, which leads to major problems in filtration and centrifugation.

Precipitates form by a series of steps that occur in sequence, which are the following: (1) initial mixing, (2) nucleation, (3) growth governed by diffusion, and (4) growth governed by fluid motion. There is often some overlap between these steps. The final size of the precipitate particles during step 4 is subject to the limits imposed by particle breakage during mixing. The completion of the growth by fluid motion step can be followed by an "aging" step, where the particles are mixed until reaching a stable size.

8.3.1 INITIAL MIXING

Initial mixing is the mixing required to achieve homogeneity after the addition of a component to cause precipitation. It is important to bring precipitant and product molecules into collision as soon as possible. This is a problem in "micromixing," which is also important in fermentation. This subject was studied by the Russian statistician Kolmogoroff in the form of the homogeneous isotropic turbulence model, which assumes that mixing between randomly dispersed eddies is instantaneous and the mixing within eddies is diffusion limited. It is therefore important to know the mean length of eddies, also known as the "Kolmogoroff length," here designated l_e. It can be calculated from [12]

$$l_e = \left(\frac{\rho v^3}{P/V} \right)^{1/4} \tag{8.3.1}$$

where ρ is the liquid density, v is the liquid kinematic viscosity, and P/V is the agitator power input per unit volume of liquid.

It is necessary to mix until all molecules have diffused across all eddies. This time can be estimated from the Einstein diffusion relationship (see Section 5.7.1):

$$t = \frac{\delta^2}{2 \mathscr{D}} \tag{8.3.2}$$

where δ is the diffusion distance and $\mathscr{D}$ is the diffusion coefficient for the molecule being mixed. For spherical eddies of diameter l_e, this becomes

$$t = \frac{l_e^2}{8 \mathscr{D}} \tag{8.3.3}$$

Thus, precipitation is initiated in a well-stirred tank for a period of time determined on the basis of isotropic turbulence.

8.3.2 NUCLEATION

Nucleation is the generation of particles of ultramicroscopic size. For particles of a given solute to form, the solution must be *supersaturated* with respect to the solute. In a supersaturated solution the concentration of the solute in solution is greater than the normal equilibrium solubility of the solute. The difference between the actual concentration in solution and the equilibrium solubility is called the degree of supersaturation, or just supersaturation. The rate of nucleation increases exponentially up to the maximum level of supersaturation, or supersaturation limit, which is illustrated in Figure 8.5. It should be noted that the rate of nucleation increases to a very high value at the supersaturation limit.

High supersaturations generally have negative consequences in carrying out precipitations. When the supersaturation is high, the precipitate tends to be in the form of a colloid, a gel, or a highly solvated precipitate. To obtain precipitate particles having desirable characteristics, the supersaturation should be kept relatively low.

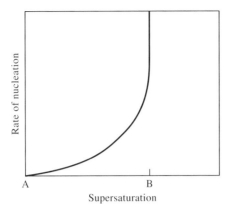

Figure 8.5 Nucleation rate as a function of degree of supersaturation. The normal equilibrium solubility is at A, and the supersaturation limit is at B.

8.3.3 GROWTH GOVERNED BY DIFFUSION

The growth of precipitate particles is limited by diffusion immediately after nucleation and until the particles grow to a limiting particle size defined by the fluid motion, which generally ranges from 0.1 to 10 μm for high and low shear fields, respectively [13]. In a dispersion of particles of uniform size that are growing as dissolved solute diffuses to the particles, the initial rate of decrease of particle number concentration (N) can be described by a second-order rate equation that was first derived by Smoluchowski [14]:

$$-\frac{dN}{dt} = K_A N^2 \tag{8.3.4}$$

Here N is the number of mono-sized particles at any given time t. The constant K_A is determined by the diffusivity $\mathscr{D}$ and diameter L_{mol}, of the molecules that are adding to the particles as follows:

$$K_A = 8\pi \mathscr{D} L_{\text{mol}} \tag{8.3.5}$$

Integrating Equation 8.3.4 gives

$$N = \frac{1}{K_A t + 1/N_0} \tag{8.3.6}$$

For convenience, N_0 is taken as the initial number concentration of dissolved solute molecules. The Stokes–Einstein equation can be used to estimate the diameter of globular proteins, which can be modeled as spheres:

$$L_{\text{mol}} = \frac{kT}{3\pi \mathscr{D} \mu} \tag{8.3.7}$$

where k is the Boltzmann constant, T is the absolute temperature, and μ is the liquid viscosity.

Equation (8.3.6) can be rewritten as

$$\frac{N}{N_0} = \frac{1}{1 + N_0 K_A t} \tag{8.3.8}$$

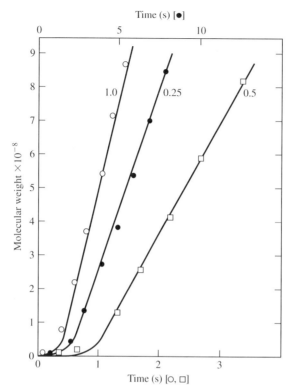

Figure 8.6 Molecular weight–time plots for three concentrations of α_3-casein (concentrations as indicated on graph in kilograms per cubic meter), aggregating in the presence of 0.008 M $CaCl_2$. Molecular weight was determined from light-scattering and turbidity measurements. (Data from D. J. Bell, M. Hoare, and P. Dunnill, "The formation of protein precipitates and their centrifugal recovery," *Adv. Biochem. Eng.,* vol. 26, p. 20, Springer-Verlag, Berlin, 1983.)

With M as the molecular weight of particles at time t and M_0 as the molecular weight of the solute,

$$\frac{M}{M_0} = \frac{N_0}{N} \tag{8.3.9}$$

so that

$$M = M_0(1 + K_A N_0 t) \tag{8.3.10}$$

This equation has been verified experimentally by measuring the molecular weight of precipitating α-casein. The data plotted in Figure 8.6 indicate good agreement with Equation (8.3.10) after an initial lag time.

EXAMPLE 8.2

Calculation of Concentration of Nuclei in a Protein Precipitation   We wish to precipitate the protein α_2-macroglobulin contained in 100 liters of aqueous solution at 20°C in a tank at a concentration of 0.2 g/liter. α_2-Macroglobulin is a globular protein with a molecular weight of 820,000 and a diffusion coefficient of 2.41×10^{-7} cm^2/s at 20°C. (Data from *Handbook of Biochemistry and Molecular Biology,* vol. III, G. D. Fasman, ed., CRC Press, Cleveland, 1976.) The precipitate particles have a density of 1.3 g/cm^3. The solution is stirred with a 75 W (0.1 hp) motor. Calculate the concentration of nuclei at the end of the "initial mixing" period.

SOLUTION

During the initial stirring, diffusion-limited molecular collisions occur, and molecules must travel by diffusion across an eddy in order to meet. The time required for this is determined by combining the Einstein diffusion equation for spherical eddies [Equation (8.3.3)] and the equation for the Kolmogoroff eddy length l_e, or the average diameter of an unstirred zone [Equation (8.3.1)]. Assume the properties of water since the solution is dilute, so that $\rho = 1.0 \, \text{g/cm}^3$ and $\nu = $ kinematic viscosity $= 0.01 \, \text{cm}^2/\text{s}$. First, convert power per unit volume to appropriate units:

$$\frac{P}{V} = \left(\frac{75 \, \text{W}}{100 \, \text{liters}} \right) \left(\frac{10^7 \, \text{g cm}^2 \, \text{s}^{-3}}{\text{W}} \right) \left(\frac{1 \, \text{liter}}{1000 \, \text{cm}^3} \right)$$

$$= 7.5 \times 10^3 \, \frac{\text{g cm}^2 \, \text{s}^{-3}}{\text{cm}^3}$$

Therefore,

$$t = \frac{\left(\dfrac{\rho \nu^3}{P/V} \right)^{1/2}}{8 \mathscr{D}} = \frac{\left(\dfrac{1.00(0.01)^3}{7.5 \times 10^3} \right)^{1/2}}{8(2.41 \times 10^{-7})} = 6.0 \, \text{s}$$

Since diffusion is limiting during this 6 s period, the number concentration of nuclei N can be determined from the integrated form of the second-order rate equation for the growth of particles limited by diffusion [Equation (8.3.6)]:

$$N = \frac{1}{K_A t + 1/N_0}$$

where $K_A = 8\pi \mathscr{D} L_{\text{mol}}$. Since globular proteins are approximately spherical, we can estimate the diameter L_{mol} of the molecule from the Stokes–Einstein equation [Equation (8.3.7)]:

$$L_{\text{mol}} = \frac{kT}{3\pi \mu \mathscr{D}} = \frac{\left(1.38 \times 10^{-23} \dfrac{\text{J}}{\text{K}} \right)(293 \, \text{K}) \left(\dfrac{10^7 \, \text{g cm}^2 \, \text{s}^{-3}}{\text{J s}^{-1}} \right)}{3\pi \left(0.01 \dfrac{\text{g}}{\text{cm s}} \right) \left(2.41 \times 10^{-7} \dfrac{\text{cm}^2}{\text{s}} \right)} = 1.78 \times 10^{-6} \, \text{cm}$$

This value of L_{mol} allows us to calculate K_A:

$$K_A = 8\pi (2.41 \times 10^{-7} \, \text{cm}^2 \, \text{s}^{-1})(1.78 \times 10^{-6} \, \text{cm}) = 1.08 \times 10^{-11} \, \text{cm}^3 \, \text{s}^{-1}$$

N_0 is the number concentration of protein molecules that are dissolved before starting precipitation, so that

$$N_0 = \left(\frac{0.2 \, \text{g}}{\text{liter}} \right) \left(\frac{\text{liters}}{1000 \, \text{cm}^3} \right) \left(\frac{\text{mol}}{820,000 \, \text{g}} \right) \left(6.02 \times 10^{23} \, \frac{\text{molecules}}{\text{mol}} \right)$$

$$= 1.47 \times 10^{14} \, \frac{\text{molecules}}{\text{cm}^3}$$

Substituting into Equation (8.3.6) for N, we obtain after the initial 6 s mixing period

$$N = [(1.08 \times 10^{-11}\,\text{cm}^3\,\text{s}^{-1})(6.0\,\text{s}) + (1.47 \times 10^{14}\,\text{cm}^{-3})^{-1}]^{-1}$$

$$= [(6480 + 0.7) \times 10^{-14}\,\text{cm}^3]^{-1} = 1.54 \times 10^{10}\,\frac{\text{particles}}{\text{cm}^3}$$

In this calculation we note that the term involving N_0 is very small, and thus the initial number of molecules is not important to the calculation.

EXAMPLE 8.3

Diffusion-Limited Growth of Particles   For the protein precipitation in Example 8.2, calculate the time for the particles to reach a size of $1.0\,\mu\text{m}$, assuming that growth is governed by diffusion only up to this particle size. Also calculate the number concentration of the $1.0\,\mu\text{m}$ particles.

SOLUTION

Assuming spherical particles, we can calculate the number of molecules per $1.0\,\mu\text{m}$ particle from the volume per particle, the particle density, and the molecular weight of the protein. This enables us to calculate the ratio of the molecular weight of the particle to that of an individual molecule, M/M_0, and from this ratio we can calculate the time to reach this particle size by rearranging Equation (8.3.10):

$$t = \frac{\dfrac{M}{M_0} - 1}{K_A N_0}$$

$$\frac{\text{Volume}}{\text{Particle}} = \frac{\pi (10^{-4}\,\text{cm})^3}{6} = 5.23 \times 10^{-13}\,\text{cm}^3$$

$$\frac{\text{Mass}}{\text{Particle}} = (5.23 \times 10^{-13}\,\text{cm}^3)\left(\frac{1.3\,\text{g}}{\text{cm}^3}\right) = 6.80 \times 10^{-13}\,\text{g}$$

$$\frac{\text{Molecules}}{\text{Particle}} = (6.80 \times 10^{-13}\,\text{g})\left(\frac{\text{mol}}{820{,}000\,\text{g}}\right)\left(\frac{6.02 \times 10^{23}\,\text{molecules}}{\text{mol}}\right)$$

$$= 4.99 \times 10^5\,\text{molecules}$$

Therefore,

$$\frac{M}{M_0} = 4.99 \times 10^5$$

Substituting into the equation for time using the values of K_A and N_0 calculated in Example 8.2, we obtain

$$t = \frac{(4.99 \times 10^5)\left(\dfrac{1\,\text{min}}{60\,\text{s}}\right) - 1}{(1.08 \times 10^{-11}\,\text{cm}^3\,\text{s}^{-1})(1.47 \times 10^{14}\,\text{cm}^{-3})} = 5.24\,\text{min}$$

We can easily calculate N from Equation (8.3.9):

$$N = \frac{N_0 M_0}{M} = \frac{1.47 \times 10^{14}}{4.99 \times 10^5} = 2.95 \times 10^8 \ \frac{\text{particles}}{\text{cm}^3}$$

8.3.4 GROWTH GOVERNED BY FLUID MOTION

Growth of particles is governed by fluid motion after the particles have reached a critical size, typically $1\,\mu$m in diameter [12]. In this growth regime, particles tend to grow by colliding and then sticking together. This is a flocculation process, governed by the factors discussed in detail in Chapter 3. Flocculation is enhanced when electrostatic repulsion between particles is reduced in comparison to the attractive van der Waals force. This can be accomplished by raising the ionic strength and lowering the temperature, to reduce the thickness of the electrical double layer, or Debye length, around particles.

For particles of uniform size in a suspension, the initial rate of decrease of particle number concentration (N) due to collisions can be described by a second-order rate equation:

$$-\frac{dN}{dt} = \frac{2}{3}\alpha L^3 \gamma N^2 \tag{8.3.11}$$

which was also derived by Smoluchowski [14]. Here α is the collision effectiveness factor (fraction of collisions that result in permanent aggregates), L is the diameter of the particles, and γ is the shear rate (velocity gradient). Assuming that the volume fraction of the particles $(\phi = \pi L^3 N/6)$ is constant during particle growth governed by fluid motion, Equation (8.3.11) becomes

$$-\frac{dN}{dt} = \frac{4}{\pi}\alpha \phi \gamma N \tag{8.3.12}$$

Integrating Equation (8.3.12) yields

$$\frac{N}{N_0} = \exp\left(\frac{-4\alpha\phi\gamma t}{\pi}\right) \tag{8.3.13}$$

where N_0 is now the particle number concentration at the time $[t = 0$ in Equation (8.3.13)] at which particle growth starts to be governed by fluid motion. For turbulent flow, the average shear rate $\bar{\gamma}$ can be estimated by the following equation developed by Camp and Stein [15]:

$$\bar{\gamma} = \left(\frac{P/V}{\rho v}\right)^{1/2} \tag{8.3.14}$$

where P/V is power dissipated per unit volume, and ρ and v are the density and kinematic viscosity of the liquid, respectively.

EXAMPLE 8.4

Growth of Particles Limited by Fluid Motion   Calculate the time required for the 1 μm particles in Example 8.3, to reach a size of 20 μm when growth is limited by fluid motion and assuming that the flow is turbulent.

SOLUTION

Here we use Equation (8.3.13) for particles growing by colliding and sticking together. Solving Equation (8.3.13) for time, we obtain

$$t = \frac{\pi \ln \dfrac{N_0}{N}}{4\alpha\phi\gamma}$$

From Example 8.3, N_0 is the particle concentration of $1\,\mu m$ particles ($= 2.95 \times 10^8$ particles/cm^3). We can find the particle concentration N of $20\,\mu m$ particles from the volume of these particles, assuming spherical particles and 100% of the protein mass is $20\,\mu m$ particles, so that

$$N = \frac{\dfrac{c_{i0}}{\rho_p}}{\dfrac{\pi L^3}{6}} = \frac{\left(\dfrac{0.2\,\text{g}}{\text{liter}}\right)\left(\dfrac{\text{liters}}{10^3\,\text{cm}^3}\right)\left(\dfrac{\text{cm}^3}{1.3\,\text{g}}\right)}{\dfrac{\pi(2 \times 10^{-3}\,\text{cm})^3}{6}} = 3.67 \times 10^4\,\frac{\text{particles}}{\text{cm}^3}$$

where c_{i0} is the feed concentration of the protein before starting the precipitation, and ρ_p is the density of protein particles. The volume fraction ϕ of particles is assumed to be constant, so we will calculate it for the $20\,\mu m$ particles:

$$\phi = \left[\frac{\pi(2 \times 10^{-3})^3}{6}\,\frac{\text{cm}^3}{\text{particle}}\right]\left[3.67 \times 10^4\,\frac{\text{particles}}{\text{cm}^3}\right] = 0.000154$$

The shear rate γ is estimated from the equation of Camp and Stein for the average shear rate in turbulent flow [Equation (8.3.14)], where we use the P/V calculated in Example 8.2:

$$\bar{\gamma} = \left(\frac{P/V}{\rho\nu}\right)^{1/2} = \left(\frac{7.5 \times 10^3}{(1.0)(0.01)}\right)^{1/2} = 866\,\text{s}^{-1}$$

Using a conservative value of the collision effectiveness factor α of 0.05 [12], we obtain

$$t = \frac{\pi \ln\left(\dfrac{2.95 \times 10^8}{3.67 \times 10^4}\right)}{4(0.05)(0.000154)(866)} = 1059\,\text{s} = 17.6\,\text{min}$$

8.3.5 PRECIPITATE BREAKAGE

When precipitate particles grow large enough by colliding and sticking together, they become susceptible to breakage during collisions. The rate of precipitate breakage has been shown to depend on the shear rate and particle concentration. In a study of soy protein precipitate particles, for example, particle breakup dominated at sizes greater than $16\,\mu m$, and breakup became negligible at low particle volume fractions (0.0002) [16].

A model that has successfully described the breakup of protein precipitates is the displacement model, which depicts the rate of aggregate size change as a function of displacement from an equilibrium aggregate diameter, L_e [17]:

$$\frac{dL}{dt} = k(L_e - L)^n \tag{8.3.15}$$

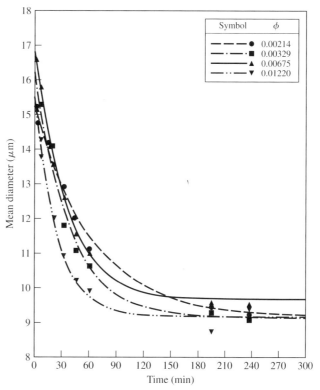

Figure 8.7 Volume mean aggregate diameter as a function of time for soy precipitate particles exposed to shear rate of $1340\,s^{-1}$ at different particle volume fractions (ϕ). Lines are drawn for the displacement model. Points are experimental data. (Data from D. L. Brown and C. E. Glatz, "Aggregate breakage in protein precipitation," *Chem. Eng. Sci.*, vol. 42, p. 1831, 1987.)

where the rate constant k would be expected to depend on the volume fraction of particles ϕ and the shear rate γ. This model with $n = 1$ (first order) fits data well for the mean diameter of soy protein particles at constant shear and various particle concentrations (Figure 8.7). The equilibrium diameter L_e has been shown to depend on the shear rate. For soy protein precipitate in laminar Couette shear [17],

$$\bar{L}_{e,v} \propto \gamma^{-0.14}, \qquad 2000\,s^{-1} \leq \gamma \leq 80,000\,s^{-1} \tag{8.3.16}$$

and for casein precipitated by salting out in a continuous stirred tank reactor [17],

$$\bar{L}_{e,v} \propto \bar{\gamma}^{-0.21}, \qquad 12\,s^{-1} \leq \bar{\gamma} \leq 154\,s^{-1} \tag{8.3.17}$$

where $\bar{L}_{e,v}$ is the equilibrium particle size at the volume mean of the particle size distribution. Particle breakage is also discussed in the analysis of the particle size distribution in a continuous flow stirred tank reactor (Section 8.4).

8.3.6 PRECIPITATE AGING

As indicated in Figure 8.7, protein precipitate particles reach a stable size after a certain length of time in a shear field. The time period for reaching this stable size is called the "aging" time. The strength of protein particles has been correlated with the product of the

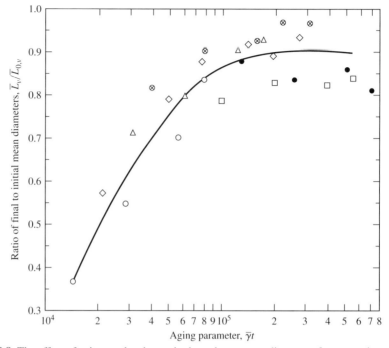

Figure 8.8 The effect of aging on the change in the volume mean diameter of soy protein precipitate particles exposed to capillary shear. Average capillary rate of shear, $1.7 \times 10^4 \, \text{s}^{-1}$; average time of exposure in capillary, 0.065 s; protein concentration, 30 kg m^{-3}; initial mean diameter $\bar{L}_{0,v}(\mu\text{m})$ prior to exposure to capillary shear: ○, 53.5; △, 23.4; ◇, 19.5; ⊗, 15.2; □, 10.2; ●, 8.8. (Data from D. J. Bell and P. Dunnill, "Shear disruption of soya protein precipitate particles and the effect of aging in a stirred tank," *Biotechnol. Bioeng.*, vol. 24, p. 1271, 1982.)

mean shear rate and the aging time, $\bar{\gamma}t$, which is known as the *Camp number*. As indicated in Figure 8.8 for soy protein particles, the mean particle size becomes approximately constant after reaching a Camp number of 10^5. Aging of precipitates helps the particles withstand processing in pumps and centrifuge feed zones without further size reduction.

8.4 Particle Size Distribution in a Continuous Flow Stirred Tank Reactor

The equations describing precipitate formation and breakage in the preceding section are convenient to use when the size of the precipitate is changing with time. In addition to being applicable to the batch reactor, these equations can also be used to describe precipitate growth in the tubular reactor (see Section 8.6). However, for one important type of reactor—the continuous flow stirred tank reactor (CSTR)—operation is at steady state. Because precipitated protein leaves the CSTR at the same time the feed protein solution enters, the product stream leaving the CSTR will have a *range of particle sizes,* which does not vary with time for operation at steady state. The distribution of particle sizes can be characterized by performing a population balance for precipitate particles in the CSTR [18]. Before writing the population balance, it is first necessary to define the population density distribution function $n(L)$ and the linear growth rate G:

$$n(L) \equiv \frac{dN}{dL} \tag{8.4.1}$$

$$G \equiv \frac{dL}{dt} \tag{8.4.2}$$

Thus, if we know $n(L)$, we can take the integral of $n(L)dL$ to find the total number of particles within two given particle diameters.

Making a balance on the number ΔN of precipitate particles within a given size range ΔL gives

(Accumulation of particles) = (particles growing into range)

− (particles growing out of range)

− (particles of given range flowing out)

− (particles in range broken up)   (8.4.3)

For steady state operation there is no accumulation, so that the balance on particles in the L to $L + \Delta L$ size range is

$$0 = \left(V\frac{dN}{dt} \right)_{\text{growth},L} - \left(V\frac{dN}{dt} \right)_{\text{growth},L+\Delta L} - Q\Delta N - VB\Delta L \tag{8.4.4}$$

where Q is the volumetric flow rate, V is the volume of the CSTR, and B is the volumetric breakage or death rate of particles in the size range L to $L + \Delta L$, such that the net rate of disappearance of particles due to breakage per volume is $B\Delta L$. Using the definitions of n and G, this equation becomes

$$(VnG)|_L - (VnG)|_{L+\Delta L} - Qn\Delta L - VB\Delta L = 0 \tag{8.4.5}$$

Dividing by ΔL and taking the limit as $\Delta L \to 0$, we obtain for constant reactor volume V

$$V\frac{d(nG)}{dL} + Qn + VB = 0 \tag{8.4.6}$$

or

$$\frac{d(nG)}{dL} + \frac{n}{\tau} + B = 0 \tag{8.4.7}$$

where τ is the mean residence time ($= V/Q$).

The growth of protein precipitate particles in a CSTR has been successfully described by Petenate and Glatz [19], who used the following expression for the linear growth rate:

$$G = \frac{dL}{dt} = \left(\frac{A}{4\pi} \right) \gamma\phi_1 L = K_0 L \tag{8.4.8}$$

where A is a constant and ϕ_1 is the volume fraction of submicrometer-sized "primary" particles. In this model, aggregates above a critical diameter L_0 grow because small primary

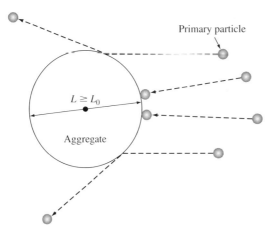

Figure 8.9 Schematic drawing of the growth of a precipitate aggregate by collisions of the aggregate with small primary particles: L_0 is the critical diameter above which particles grow by colliding with and sticking to small primary particles.

particles collide with aggregates and stick (see Figure 8.9). Primary particles typically have diameters of about $0.2\,\mu$m for protein precipitation in a CSTR [19].

The larger a precipitate aggregate becomes, the more susceptible it is to being broken up by the local shear stresses encountered in turbulent flow. For the conditions under which local shear stresses are dominant, the following equation for the volumetric breakage or death rate has been successfully used by Petenate and Glatz [19] to describe the breakup of protein aggregates:

$$B = K\mu\gamma^2 nL^3 = K_1 nL^3 \tag{8.4.9}$$

where K is a constant and μ is the viscosity.

A useful parameter to know for the precipitation in a CSTR is the mass-averaged particle size. For spherical particles, the total mass concentration M_T in terms of the mass-averaged diameter $\bar{L}_m$ is

$$M_T = \rho_p \pi \left(\frac{\bar{L}_m^3}{6}\right) N \tag{8.4.10}$$

where ρ_p is the density of individual particles and N is the total particle number concentration. From $n(L)$, we can determine M_T and N for spherical particles equal or greater than L_0 in size as follows:

$$M_T|_{L\geq L_0} = \frac{\rho_p \pi}{6} \int_{L_0}^{\infty} L^3 n(L)\, dL \tag{8.4.11}$$

$$N|_{L\geq L_0} = \int_{L_0}^{\infty} n(L)\, dL \tag{8.4.12}$$

Combining Equations (8.4.10), (8.4.11), and (8.4.12) gives the mass-averaged particle diameter for $L \geq L_0$:

$$\bar{L}_m|_{L\geq L_0} = \left(\frac{\int_{L_0}^{\infty} L^3 n(L)\, dL}{\int_{L_0}^{\infty} n(L)\, dL}\right)^{1/3} \tag{8.4.13}$$

EXAMPLE 8.5

Dependence of Population Density on Particle Size and Residence Time in a CSTR
For the precipitation of protein in a CSTR, determine the dependence of the population density distribution n on the particle diameter L and residence time τ. Consider both particle growth by aggregation and particle breakup.

SOLUTION

Use the expressions for the linear growth rate and particle breakup [Equations (8.4.8) and (8.4.9)] in the population balance equation [Equation (8.4.7)]. This gives

$$\frac{d(K_0 L n)}{dL} + \frac{n}{\tau} + K_1 n L^3 = 0$$

Differentiating and rearranging, we have

$$\frac{dn}{n} = -dL\left(\frac{1 + \dfrac{1}{K_0 \tau}}{L} + \frac{K_1 L^2}{K_0}\right)$$

With lower limits of integration of n_0 (the population density function at $L = L_0$) and critical diameter L_0, we obtain

$$\frac{n}{n_0} = \left(\frac{L_0}{L}\right)^{1+1/K_0\tau} \exp\left[\frac{-K_1}{3K_0}\left(L^3 - L_0^3\right)\right]$$

8.5 Methods of Precipitation

Methods that have been developed to precipitate proteins are based on a knowledge of the solubility of proteins. Based on our discussion in Section 8.2, the most obvious methods that emerge are pH adjustment to the isoelectric point of the protein (called isoelectric precipitation), addition of organic solvents, salting out, and addition of nonionic polymers.

Isoelectric precipitation is based on the fact that the solubility of a given protein is generally at a minimum at the isoelectric point (pI) of the protein (see Figure 8.3). This is a convenient method to use when fractionating a protein mixture. For this situation the pH should be adjusted above the highest pI or below the lowest pI of all the proteins present. The pH is then changed to the nearest pI, where precipitate is allowed to form and is then removed. This procedure is repeated for as many proteins as one desires to precipitate. Local extremes of pH should be avoided during pH adjustment to minimize protein denaturation (see Chapter 1, Section 1.4.7). There are two advantages of isoelectric precipitation when acids are added to cause precipitation: mineral acids are cheap, and several acids (e.g., phosphoric, hydrochloric, sulfuric) are acceptable in protein food products. This method, however, will not work for all proteins; for example, gelatin, which is a very hydrophilic protein, does not precipitate at its isoelectric point in solvents having low ionic strength [12].

Several organic solvents have been used to precipitate proteins, including alcohols, acetone, and ether. Alcohols, however, have been the most widely used in industry. One of

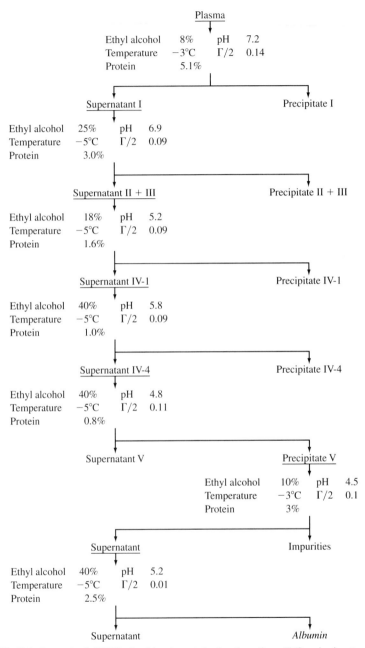

Figure 8.10 Cohn's method (1946) for blood protein fractionation: $\Gamma/2$ = ionic strength. (Data from L. E. Strong, "Blood fractionation," in *Encyclopedia of Chemical Technology,* vol. 2, R. E. Kirk and D. F. Othmer, eds., p. 566, Interscience Encyclopedia, New York, 1948.)

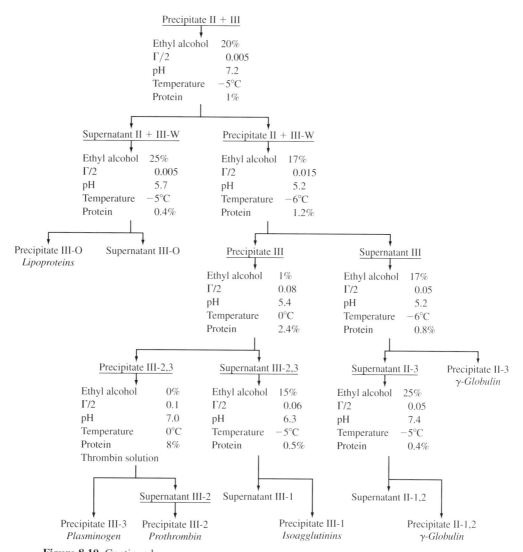

Figure 8.10 Continued.

the most important processes utilizing alcohol to precipitate proteins is the Cohn process to purify therapeutic proteins from human plasma. This process uses ethanol at temperatures below 0°C to minimize denaturation by the organic solvent. The variables that are manipulated in the Cohn process are pH, ionic strength, and ethanol concentration. Ionic strength is kept low, which leads to a salting-in effect (see Section 8.2). This salting-in effect is enhanced when ethanol is added. Cohn's methods for the preparation of albumin, plasminogen, prothrombin, isoagglutinins, and γ-globulin starting with blood plasma are illustrated in Figure 8.10.

In the salting out of proteins, salt is dissolved in the solution containing the proteins. The protein solubility decreases as the salt ionic strength rises according to the Cohn equation

[Equation (8.2.7)]. The most important consideration in salting out is the type of salt that is used. Salts with multiply charged anions such as sulfate, phosphate, and citrate are the most effective; while for the cation, monovalent ions should be used [20]. Following the Hofmeister or lyotropic series, the salting-out ability of the common multiply charged anions is $citrate^{2-} > phosphate^{3-} > sulfate^{2-}$; for the common monovalent cations the order is $NH_4^+ > K^+ > Na^+$.

The salt that has the most desirable properties in general for precipitating proteins is ammonium sulfate. The solubility of this salt is very high (approximately 4 M in pure water) and varies very little in the range of 0 to $30°C$. The density of a saturated solution of ammonium sulfate is 1.235 g cm^{-3}, which is enough below the density of protein aggregates (approximately 1.29 g cm^{-3}) to allow centrifugation. Another advantage of using ammonium sulfate is that protein precipitates are often very stable for years in 2 to 3 M salt; in fact, many commercial enzymes are normally sold in ammonium sulfate solution at high molarity. Furthermore, proteolysis and bacterial action are prevented in concentrated ammonium sulfate solutions. The only disadvantage of ammonium sulfate is that it cannot be used above pH 8 because of the buffering action of ammonia. Sodium citrate is very soluble and is a good alternative to ammonium sulfate when the precipitation must be performed above pH 8 [20].

Several nonionic polymers have been used to precipitate proteins, including dextran, poly(vinyl pyrrolidone), poly(propylene glycol), and poly(ethylene glycol) (PEG). Of these polymers, by far the most extensively studied is PEG. Several guidelines have been developed for the use of PEG as a protein precipitant. Solutions of PEG up to 20% w/v can be used without viscosity becoming a problem. PEG's with molecular weights above 4000 have been found to be the most effective [20]. Protein destabilization in PEG solutions does not occur until the temperature is significantly higher than room temperature ($>40°C$) [7].

8.6 Design of Precipitation Systems

Once a precipitating agent has been chosen for the protein of interest, it may be necessary to design a large-scale precipitation process. The safest procedure is to base the design on a laboratory or pilot plant system that has given acceptable results. Important considerations in obtaining the best possible plant design are the following: the type of precipitation reactor, processing conditions (flow rates, concentrations, etc.), and assumptions used to scale up to the plant scale. There are three basic types of precipitation reactor: the batch reactor, the continuous stirred tank reactor (CSTR), and the tubular reactor. These types are discussed and compared.

The batch reactor is the simplest of the three types and is often the one that is tried first at small scale. Batch precipitation is carried out by slowly adding the precipitating agent to a protein solution that is being mixed. Addition of the precipitating agent continues until the desired level of supersaturation is reached with respect to the protein being precipitated. At this point nucleation begins, and precipitation proceeds through the steps of particle growth and aggregation. Mixing continues until the precipitation is complete. The mixing in a batch reactor is generally turbulent. Protein particles precipitated in a batch reactor tend to be more compact and regular in shape than those precipitated in a tubular reactor, apparently because of the different shear profiles existing in the two reactors and the length of time the particles are exposed to this shear [21]. The shear field in a tubular reactor is essentially homogeneous; by contrast, in the batch reactor the precipitate particles are exposed to a very wide range of shears and to much longer times of exposure than in the tubular reactor, resulting in improved precipitate mechanical stability.

In a tubular reactor, precipitation takes place in volume elements that approach plug flow as they move through the tube. Thus, the distance–particle size distribution history of the particles in a volume element moving through a tubular reactor is comparable to the time–particle size distribution history of a stationary volume element in a batch reactor. In the tubular reactor the feed protein solution and the precipitating agent are contacted in a zone of efficient mixing at the reactor inlet. Good mixing can be accomplished, for example, by flowing the protein solution through a convergent nozzle to a biplanar grid and then introducing the precipitating agent just downstream of the biplanar grid [22]. The flow pattern in the reactor can be turbulent, a property that can be promoted by wire meshes at intervals along the reactor. In comparison to either the batch reactor or the CSTR, the tubular reactor has the advantages of short fluid residence times, an absence of moving mechanical parts, uniformity of flow conditions throughout the reactor, a simple and inexpensive design, and a relatively small holdup of fluid. For particles that grow relatively slowly, however, the length of the tubular reactor can be excessive.

In the CSTR, fresh protein feed contacts a mixed slurry containing precipitate aggregates. The mixing conditions in a CSTR are similar to those in a batch reactor. Upon entering the CSTR, fresh protein feed nucleates, the nucleate particles grow by diffusion, and the submicrometer-sized "primary" particles collide with and adhere to growing aggregates. The degree of supersaturation can be more easily controlled than in the batch or tubular reactor, which means that the formation of precipitates with undesirable properties is less likely.

A few general statements can be made regarding the processing conditions in precipitation systems. Flows are normally turbulent; flow must be high enough to avoid inadequate mixing and high supersaturation but low enough to avoid excessive particle breakage leading to particles that are smaller than desirable. For both the batch and tubular reactors, the flow regime can be changed from turbulent to laminar during the particle growth phase to avoid excessive particle breakage. The rate of addition of precipitant is especially important. This rate should be kept low enough to avoid high supersaturations that lead to colloidal, highly solvated precipitates. The concentration of the precipitant being added is also important, with lower concentrations leading to lower supersaturations [7].

The key parameter for scaleup of precipitation is mixing. One recommended approach to scale up mixing is to first consider using geometric similarity and constant power per unit volume (P/V) [23]. For geometric similarity all important dimensions are similar and have a common constant ratio. As seen in Figure 8.11, this method gave reasonable agreement for the d_{90} particle size (the particle size for which all particles that are larger account for 90 wt% of the total) as a function of P/V for the batch isoelectric precipitation of soy protein for vessels ranging in size from 0.27 to 200 liters. If the precipitate is susceptible to shear breakage, however, the assumption of constant P/V for scaleup may not be satisfactory. The impeller tip speed, which determines the maximum shear rate, rises when P/V is held constant upon scaleup of the reactor volume, as seen in Table 8.1. These results assume turbulent flow, where the power number is constant, so that

$$P \propto N_i^3 d_i^5 \qquad (8.6.1)$$

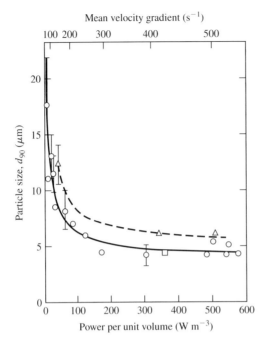

Figure 8.11 The relationship between aggregate size and power per unit volume for the preparation of isoelectric soy protein precipitate in batch stirred tanks: d_{90} is the particle size for which all particles that are larger account for 90 wt% of the total. Protein concentration is 30 kg m^{-3}. Vessel volumes: △, 0.27 liters; ○, 0.67 liters; □, 200 liters. (Data from D. J. Bell, M. Hoare, and P. Dunnill, "The formation of protein precipitates and their centrifugal recovery," *Adv. Biochem. Eng.*, vol. 26, p. 40, Springer-Verlag, Berlin, 1983.)

TABLE 8.1
Scaleup of Turbulent Agitation, Assuming Constant P/V

Volume Scaleup Factor	$\dfrac{\text{(Tip speed)}_{\text{large}}}{\text{(Tip speed)}_{\text{small}}}$
10	1.3
100	1.7
1000	2.2
10,000	2.8

Here, N_i is the impeller rotation rate, and d_i is the impeller diameter. It is also assumed that the maximum shear rate is proportional to the impeller tip velocity ($= \pi N_i d_i$) and that the volume is proportional to d_i^3. From Table 8.1 we see that even for a volume scaleup of 10,000 for constant P/V, the tip speed increases by less than a factor of 3. A volumetric scaleup factor of 10,000 is typical in scaling up from the lab to the plant.

8.7 Summary

Precipitation is the process of coming out of solution as a solid. Often, the goal of precipitation is concentration to reduce volume, although significant purification can sometimes be achieved, and in certain instances precipitation has been used as the final step in the process.

- Precipitation is based on protein solubility, which depends, in turn, on the molecular properties of the protein and on the properties of the solvent. Proteins are usually least soluble near their isoelectric pH.

- The Cohn equation is an empirical relationship between protein solubility S and ionic strength I:

$$\ln S = \beta - K_s' I$$

where β is the hypothetical solubility at zero ionic strength and depends on temperature and pH (protein charge) for a given protein, while K_s' is the "salting-out" coefficient, which also accounts for "salting in" at low I. Similar correlations apply to nonionic precipitants.

- Precipitation occurs in distinct steps that can overlap in time as a precipitate develops: (1) initial mixing to achieve homogeneity, (2) nucleation, the generation of ultramicroscopic particles, (3) growth governed by diffusion of dissolved solute molecules to the particle surface, and (4) growth governed by fluid motion, in which particles grow by colliding and sticking together. Particle breakage imposes limits on the final particle size that can be attained.

- The initial mixing time required to distribute all molecules throughout a given volume depends on the diffusivity of the protein and the distance it must diffuse within mixing eddies, which is the Kolmogoroff length. This length, l_e, is calculated by using the theory of homogeneous isotropic turbulence from the equation

$$l_e = \left(\frac{\rho v^3}{P/V} \right)^{1/4}$$

where ρ is the liquid density, v is the liquid kinematic viscosity, and P/V is the agitator power input per unit volume of liquid. The initial mixing time can be calculated from the Einstein diffusion equation by using the Kolmogoroff length and the diffusivity of the molecule being mixed.

- Intermediate particle growth depends on the diffusion of protein molecules to each growing particle. The loss rate of single molecules to this process follows the second-order rate law of Smoluchowski. The mass, measured in units of molecular weight, of a growing particle can be shown to increase linearly with time t:

$$M = M_0(1 + K_A N_0 t)$$

where M_0 is the molecular weight of the solute, N_0 is the initial number concentration of dissolved solute, and K_A is a constant that depends on the solute diffusivity and molecular diameter.

- After growing particles have reached a certain size, typically $1 \, \mu$m, further precipitant growth depends on the aggregation of these particles. In growth of particles governed by fluid motion, the particle number concentration N is given by the Smoluchowski second-order rate theory as

$$N = N_0 \exp\left(\frac{-4\alpha\phi\gamma t}{\pi} \right)$$

where N_0 is the particle number concentration at time zero, when particle growth starts to be governed by fluid motion, α is the collision effectiveness factor (fraction of collisions that result in permanent aggregates), γ is the shear rate, and ϕ is the volume fraction of the particles. Aggregation depends on the balance of electrostatic repulsion and van der Waals attraction between particles and obeys the DLVO principles introduced in Chapter 3.

- Particle population balances are calculated for a CSTR using a material balance on a differential particle size interval. Because precipitated protein leaves the CSTR at the same time that the feed protein solution enters, the product stream leaving the CSTR will have a range of particle sizes that does not vary with time, since operation is at steady state.

- Methods to precipitate proteins include pH adjustment to the isoelectric pH of the protein (isoelectric precipitation), addition of organic solvents, salting out, and addition of nonionic polymers. An example of the use of organic solvents is Cohn's method for the fractionation of serum proteins.

- The key parameter for scaleup of precipitation is mixing, which suggests holding power input per volume constant. For turbulent mixing, power input increases as the third power of the impeller rotation rate and the fifth power of the impeller diameter.

NOMENCLATURE

A	constant in the equation for linear growth rate [Equation (8.4.8)] (dimensionless)
B	volumetric breakage rate of particles in the size range L to ΔL (number s^{-1} liter cm^{-1})
c_i	concentration of component i (M, or g liter^{-1})
c_{i0}	feed concentration of component i (M, or g liter^{-1})
c_p	polymer concentration (M)
d_i	impeller diameter (cm)
$\mathscr{D}$	diffusion coefficient of solute (cm^2 s^{-1})
G	particle linear growth rate (cm s^{-1})
I	ionic strength (M)
k	Boltzmann constant (1.3807×10^{-23} J K^{-1})
k	constant in the displacement model for particle breakage [Equation (8.3.15)] (cm^{1-n} s^{-1})
K	constant in the equation for the volumetric breakage rate [Equation (8.4.9)] (s^2 cm^{-2} g^{-1})

K_1	constant in the equation for the volumetric breakage rate [Equation (8.4.9)] (liter^{-1} s^{-1})
K_A	rate constant for particle growth by diffusion (liter s^{-1})
K_i	salting-in constant (M^{-1})
K_s	salting-out constant (M^{-1})
K'	constant in the Juckes equation [Equation (8.2.1)] (M^{-1}, or liter g^{-1})
K'_s	constant in the Cohn equation [Equation (8.2.7)] (M^{-1})
l_e	mean eddy length, or Kolmogoroff length (cm)
L	particle diameter (cm)
L_e	equilibrium particle diameter (cm)
$\bar{L}_m$	mass-averaged particle diameter (cm)
L_{mol}	molecular diameter (cm)
L_0	critical particle diameter at which particles start to grow by colliding and sticking together (cm)
$\bar{L}_v$	volume-averaged particle diameter (cm)
M	molecular weight (Daltons)

M_0	molecular weight of solute (Daltons)
M_T	total mass concentration of particles (g liter^{-1})
n	population density distribution function (number liter^{-1} cm^{-1})
n	order of the displacement model for particle breakage [Equation (8.3.15)] (dimensionless)]
n_0	population density distribution function at $L = L_0$ (number liter^{-1} cm^{-1})
N	particle number concentration (number liter^{-1})
N_0	particle number concentration at time zero (number liter^{-1})
N_i	impeller rotation rate (revolutions s^{-1})
P/V	power input per volume of liquid (W liter^{-1})
Q	volumetric flow rate (liters s^{-1})
r_r, r_s	radius of polymer rod, radius of protein solute (cm)
S	protein solubility (M, or g liter^{-1})
S_0	solubility of dipolar ion at zero ionic strength (M, or g liter^{-1})
S_p	solubility of dipolar ion (M, or g liter^{-1})
t	time (s)
T	temperature (K)
u	dipole moment (C cm)
V	precipitation reactor volume (liters)
$\bar{V}$	partial specific volume of polymer (liters mol^{-1})
z_i	charge of component i (dimensionless)
Z	average protonic charge of protein

Greek Letters

α	particle collision effectiveness factor (dimensionless)
β	constant in the Cohn equation [Equation (8.2.7)] (dimensionless)
β'	constant in the Juckes equation [Equation (8.2.1)] (dimensionless)
δ	diffusion distance (cm)

$\gamma, \bar{\gamma}$	shear rate, average shear rate (s^{-1})
ε	dielectric constant of solvent (dimensionless)
μ	liquid viscosity (g cm^{-1} s^{-1})
ν	liquid kinematic viscosity (cm^2 s^{-1})
ρ	liquid density (g cm^{-3})
ρ_p	density of precipitate particle (g cm^{-3})
τ	mean residence time (s)
ϕ	volume fraction of total particles (dimensionless)
ϕ_1	volume fraction of submicrometer-sized "primary particles" (dimensionless)

PROBLEMS

8.1 **Solubility of Proteins**   Based on the Kirkwood theory of the interactions between ions and dipoles, answer the following:

  (a) When comparing the solubilities of proteins at the same ionic strength, would you expect the solubility of a protein with a high dipole moment to be high or low?

  (b) In highly polar solvents, what balance would you expect between salting in and salting out of a protein?

  Use appropriate equations to justify your answers.

8.2 **Precipitation of Albumin and γ-Globulin from Blood Plasma**   Explain the solubility principles involved in the precipitation of albumin and γ-globulin from blood plasma by the Cohn process.

8.3 **Fractionation of a Protein Solution**   Four proteins are dissolved in solution. The constants in the Cohn equation (see Example 8.1) for the proteins are given in Table P8.3 along with the concentration of each protein in solution. Assume that the Cohn

TABLE P8.3

Protein	β	K_s'' (M^{-1})	Concentration (g/liter)
1	6.30	2.84	0.3
2	10.20	4.34	0.4
3	8.60	2.48	5.0
4	8.00	3.97	10.0

equation constants are valid when other proteins are present.

(a) Determine the maximum percentage recovery of protein 4 that can be obtained if it is desired to obtain a precipitate containing 100% protein 4.

(b) Determine the maximum percentage recovery of protein 3 that can be obtained if it is desired to obtain a mother liquor containing 99% protein 3.

8.4   Mixer Power to Achieve Complete Mixing   It is desired to precipitate a protein in a solution with a density of 1.02 g/cm^3 and kinematic viscosity of 0.012 cm^2/s. The diffusion coefficient of the protein has been estimated to be 1.2×10^{-7} cm^2/s. You need to select a mixer for this precipitation. To aid in this selection, plot as a straight line the dependence of the mixer power per solution volume (kW/liter) as a function of the time (s) to completely mix the solution for mixing times up to 100 s.

8.5   Kinetics of Precipitation of a Protein   The data in Table P8.5 for particle number concentration N were obtained for the precipitation of a protein. Nucleation started at time zero.

You also know the following information:
Protein concentration = 3.0 g/liter
Power input = 1.5 kW (2.0 hp)
Liquid volume = 1000 liters
Liquid density = 1.05 g/cm^3
Liquid viscosity = 0.012 g cm^{-1} s^{-1}
Molecular weight of protein = 200,000
Density of protein particles = 1.3 g/cm^3
The flow is turbulent.

From this information, determine:

(a) The time and particle diameter at which particle growth started to be governed by particles colliding and sticking together,

(b) The constant K_A in the equation by Smoluchowski describing particle growth by diffusion,

(c) The collision effectiveness factor α for particle growth by fluid motion

8.6   Protein Precipitation in a CSTR   For a protein precipitation in a CSTR, determine the functional dependence of the mass-averaged particle size $\bar{L}_m$ (for particles above the critical aggregate diameter L_0) on the residence time τ, parameter K_0, and L_0. Assume that particle breakage is negligible. For a value of K_0 of 0.02 min^{-1} (determined for soy protein precipitating in a CSTR by A. M. Petenate and C. E. Glatz, "Isoelectric precipitation of soy protein. II. Kinetics of protein aggregate growth and breakage," *Biotech. Bioeng.*, vol. 25, p. 3059, 1983) and L_0 of 1.0 μm, plot the dependence of $\bar{L}_m$ on τ.

8.7   Scaleup with Constant Power/Volume or Maximum Shear Rate   A protein precipitation reactor with a mechanical agitator is being scaled up from volume V_1 to volume V_2 by two different methods: keeping power per unit volume (P/V) constant, or keeping the maximum shear rate (γ_{max}) constant.

(a) For the scaleup keeping P/V constant, determine the ratio of γ_{max} at volume V_2 to γ_{max} at volume V_1 as a function of V_1 and V_2:

$$\frac{(\gamma_{max})_2}{(\gamma_{max})_1} = f(V_1, V_2)$$

(b) If constant γ_{max} is being used for scaleup, determine the ratio of P/V at volume V_2 to P/V at volume V_1 as a function of V_1 and V_2:

$$\frac{(P/V)_2}{(P/V)_1} = g(V_1, V_2)$$

8.8   Design of a Tubular Reactor to Precipitate a Protein   A protein at a concentration of 2.0 g/liter is to be precipitated in a tubular reactor at 20°C and a rate of 1.0 kg/h. It is desired that the protein precipitate particles leaving the reactor have a diameter of 10 μm. The properties of the protein are as follows: molecular weight of 480,000, diffusion coefficient of 3.5×10^{-7} cm^2/s at 20°C, and precipitate particle density of 1.29 g/cm^3. Design a tubular reactor (i.e., specify diameter and length) to carry out the precipitation in turbulent flow at a Reynolds number of 10,000. It can be assumed that the particle collision effectiveness factor (α) is 0.05 for particle growth governed by fluid motion.

TABLE P8.5

t (min)	N (particles/cm^3)	t (min)	N (particles/cm^3)
0.5	3.72×10^9	3.5	4.51×10^8
1.0	1.86×10^9	4.0	2.29×10^8
1.5	1.24×10^9	4.5	8.49×10^7
2.0	9.28×10^8	5.0	3.14×10^7
2.5	7.42×10^8	5.5	1.16×10^7
3.0	6.19×10^8	6.0	4.31×10^6

References

1. Rashin, A., and Honig, B. (1984). On the environment of ionizable groups in globular proteins. *J. Mol. Biol.*, vol. 173, p. 515.
2. Miller, S., Janin, J., Lesk, A. M., and Chothia, C. (1987). Interior and surface of monomeric proteins. *J. Mol. Biol.*, vol. 196, p. 641.
3. Janin, J., Miller, S., and Chothia, C. (1988). Surface, subunit interfaces, and interior of oligomeric proteins. *J. Mol. Biol.*, vol. 204, p. 155.
4. Privalov, P. L. (1989). Thermodynamic problems of protein structure. *Annu. Rev. Biophys. Biophys. Chem.*, vol. 18, p. 47.
5. Van Oss, C. J., and Good, R. J. (1988). Orientation of the water molecules of hydration of human serum albumin. *J. Protein Chem.*, vol. 7, p. 179.
6. Juckes, I. R. M. (1971). Fractionation of proteins and viruses with polyethylene glycol. *Biochim. Biophys. Acta*, vol. 229, p. 535.
7. Rothstein, F. (1994). Differential precipitation of proteins. In *Protein Purification Process Engineering*, R. G. Harrison, ed., Dekker, New York, p. 115.
8. Bull, H. B., and Breese, K. (1978). Interaction of alcohols with proteins. *Biopolymers*, vol. 17, p. 2121.
9. Oncley, J. L. (1943). The electric moments and the relaxation times of proteins as measured from their influence upon the dielectric constants of solutions. In *Proteins, Amino Acids, and Peptides,* E. J. Cohn and J. T. Edsall, eds., Reinhold, New York, Chap. 22.
10. Kirkwood, J. G. (1943). The theoretical interpretation of the properties of solutions of dipolar ions. In *Proteins, Amino Acids, and Peptides,* E. J. Cohn and J. T. Edsall, eds., Reinhold, New York, Chap. 12.
11. Cohn, E. J. (1925). The physical chemistry of proteins. *Physiol. Rev.*, vol. 5, p. 349.
12. Bell, D. J., Hoare, M., and Dunnill, P. (1983). The formation of protein precipitates and their centrifugal recovery. In *Advances in Biochemical Engineering/Biotechnology,* vol. 26, A. Fiechter, ed., Springer-Verlag, Berlin, p. 1.
13. Ives, K. J. (1978). Rate theories. In *The Scientific Basis of Flocculation,* K. J. Ives, ed., Sijthoff and Noordhoff, Alphen aan den Rijn, The Netherlands, p. 37.
14. Smoluchowski, M. (1917). Mathematical theory of the kinetics of coagulation of colloidal solutions. *Z. Phys. Chem.*, vol. 92, p. 129.
15. Camp, T. R., and Stein, P. C. (1943). Velocity gradients and internal work in fluid motion. *Boston Soc. Civ. Engs.*, vol. 30, p. 219.
16. Brown, D. L., and Glatz, C. E. (1987). Aggregate breakage in protein precipitation. *Chem. Eng. Sci.*, vol. 42, p. 1831.
17. Twineham, M., Hoare, M., and Bell, D. J. (1984). The effects of protein concentration on the break-up of protein precipitate by exposure to shear. *Chem. Eng. Sci.*, vol. 39, p. 509.
18. Randolph, A. D., and Larson, M. A. (1971). *Theory of Particulate Processes*, Academic Press, New York.
19. Petenate, A., and Glatz, C. E. (1983). Isoelectric precipitation of soy protein. II. Kinetics of protein aggregate growth and breakage. *Biotechnol. Bioeng.*, vol. 25, p. 3059.
20. Scopes, R. (1994). *Protein Purification*, 3rd ed., Springer-Verlag, New York.
21. Bell, D. J., and Dunnill, P. (1982). The influence of precipitation reactor configuration on the centrifugal recovery of isoelectric soya protein precipitate. *Biotechnol. Bioeng.*, vol. 24, p. 2319.
22. Virkar, P. D., Hoare, M., Chan, M. Y. Y., and Dunnill, P. (1982). Kinetics of the acid precipitation of soya protein in a continuous-flow tubular reactor. *Biotechnol. Bioeng.*, vol. 24, p. 871.
23. Oldshue, J. Y. (1983). *Fluid Mixing Technology*. McGraw-Hill, New York.

Crystallization

Crystallization is the process of producing crystals from a homogeneous phase. For biochemicals, the homogeneous phase from which crystals are obtained is always a solution. Crystallization is similar to precipitation in that solid particles are obtained from a solution. However, precipitates have poorly defined morphology, while in crystals the constituent molecules are arranged in three-dimensional arrays called space lattices. In comparison to crystallization, precipitation occurs at much higher levels of supersaturation and rates of nucleation but lower solubilities. These and other differences between crystallization and precipitation are highlighted in Table 9.1. Because of these differences and because the theory of crystallization that has been developed is different from that for precipitation, crystallization is considered separately from precipitation.

Crystallization is capable of producing bioproducts at very high purity (say, 99.9%) and is considered to be both a polishing step and a purification step. Polishing refers to a process needed to put the bioproduct in its final form for use. For some bioproducts, such as antibiotics, this final form must be crystalline, and sometimes it is even necessary that a specific crystal form be obtained. In some instances, the purification that can be achieved by crystallization is so significant that other more expensive purification steps such as chromatography can be avoided.

There are actually two very different applications of crystallization in biotechnology and bioproduct engineering: crystallization for polishing and purification, and crystallization for crystallography. In the latter case, the goal is a small number of crystals with good size (0.2–0.9 mm) and internal quality. Although it has become common to crystallize proteins for characterization of their three-dimensional structure by x-ray diffraction, this is performed only at small scale in the laboratory, and the knowledge about how to crystallize proteins at large scale in a production process is less developed. However, many antibiotics and other small biomolecules are routinely crystallized in production scale processes.

This chapter is oriented toward the use of crystallization in processes that can be scaled up. The principles of crystallization are discussed first, including information about crystals, nucleation, and crystal growth. This is followed by discussions of batch crystallizers, which are widely used in the pharmaceutical industry. Strategies for the crystallization of proteins from solution are outlined. The chapter concludes with a discussion of crystallizer scaleup and design.

TABLE 9.1
Comparison of Crystallization and Precipitation

Description	Crystallization	Precipitation
Solubility	Wide range, usually medium to high	Sparingly soluble
Relative supersaturation	Low	High
Product morphology	Well-defined	Ill-defined
Product crystal size	Large	Small
Nucleation mechanism	Secondary	Primary
Nucleation rate	Low	High
Growth rate	Wide range	Low
Controllability	Controllable	Difficult to control

Adopted from Tavare [1].

9.1 Instructional Objectives

After completing this chapter, the reader should be able to do the following:

- Explain the differences between crystallization and precipitation.
- Utilize power law kinetics in the primary and secondary nucleation of crystals.
- Calculate crystal nucleation and growth rates from crystal size distribution data taken from batch experiments.
- Perform engineering analysis of a batch crystallizer.
- Outline strategies for crystallizing proteins.
- Scale up a crystallization process.

9.2 Crystallization Principles

To understand crystallization, it is important to know what crystals are and how they form and grow. This understanding can then be used as a basis for the appropriate design and operation of crystallizers.

9.2.1 CRYSTALS

When crystals are allowed to form freely, they appear as polyhedrons, or solids formed by plane faces. Although the relative sizes of the faces of the same material may differ considerably, the *angles* made by the corresponding faces of the same material do not vary—they are characteristic of that substance. Therefore, crystal forms are classified on the basis of these angles, which has resulted in the identification of seven crystal systems.

The relative sizes of the faces of a crystal in a particular system can vary considerably, resulting in a variety of crystal shapes. This variation is called a modification of *habit*. For example, if the crystals grow much more rapidly in one direction, needle-shaped crystals

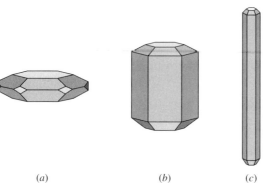

Figure 9.1 Crystal habit illustrated for hexagonal crystals: (*a*) tabular, (*b*) prismatic, and (*c*) acicular.

(*a*)   (*b*)   (*c*)

can result. This variation is illustrated in Figure 9.1 for hexagonal crystals. Crystal habit is influenced by the conditions of crystallization, particularly by the impurities present and by the particular solvent or solvents used. Impurities have been found to stunt the growth of a crystal in certain directions.

9.2.2 NUCLEATION

The generation of ultramicroscopic particles in the process of nucleation is the sum of contributions by *primary nucleation* and *secondary nucleation*. Primary nucleation occurs in the absence of crystals, while secondary nucleation is attributed to the influence of existing crystals.

Primary nucleation can be either homogeneous or heterogeneous. For homogeneous nucleation, no foreign particles are present, while there are foreign particles present during heterogeneous nucleation. Primary nucleation is usually heterogeneous in actual practice. The rate of primary nucleation, either homogeneous or heterogeneous, has been modeled by the following power law expression [1]:

$$B = \frac{dN}{dt} = k_n(c - c^*)^n \tag{9.2.1}$$

where B is the number of nuclei formed per unit volume per unit time, N is the number of nuclei per unit volume, k_n is a rate constant, c is the instantaneous solute concentration, and c^* is the solute concentration at saturation. The $(c - c^*)$ term is called the *supersaturation*. The exponent n can range up to 10 but typically is in the range of 3 to 4.

Experimental data from industrial crystallizers have indicated that secondary nucleation usually predominates. Two types of secondary nucleation are shear nucleation and contact nucleation [1]. Shear nucleation occurs as a result of fluid shear on growing crystal faces, and contact nucleation happens because of crystals colliding with each other and with the impeller and other vessel internal surfaces. The most widely used relation for the rate of secondary nucleation in crystallization is the following [1, 2]:

$$B = \frac{dN}{dt} = k_1 M_T^j (c - c^*)^b \tag{9.2.2}$$

where k_1 is a rate constant, and M_T is the suspension density. The exponent b can range up to 5 but has a most probable value of 2. The exponent j ranges up to 1.5, with 1 being the most probable value.

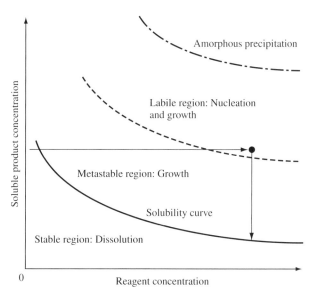

Figure 9.2 Typical phase diagram. The components in solution consist of the product (ordinate) and the precipitating reagent (abscissa). The lines with arrows outline one possible way of performing the crystallization.

As in precipitation, the solution must be supersaturated in order for particles to form in crystallization, which is reflected in Equations (9.2.1) and (9.2.2). However, as indicated in the phase diagram in Figure 9.2, the supersaturation must be above a certain value before nucleation will begin. In a part of the phase diagram called the metastable region, the supersaturation is so low that nucleation will not start. Once the supersaturation has been raised enough to be in the labile region of the phase diagram, nucleation can begin. At this point, crystals begin to grow, and the supersaturation decreases. A possible way of carrying out a crystallization is indicated by the lines with arrows in Figure 9.2. If the supersaturation becomes too high, the nucleation rate will be too great, and an amorphous precipitate will result.

9.2.3 CRYSTAL GROWTH

Crystal growth is the postnucleation process in which molecules in solution are added to the surface of existing crystals. For engineering purposes in crystallizer design, the most useful relationship describing the rate of mass deposition R during crystal growth is [1, 2]

$$R = \frac{1}{A}\frac{dW}{dt} = k_G(c - c^*)^g \tag{9.2.3}$$

where W is the mass of crystals per volume of solvent, A is the surface area of crystals per volume of solvent, and k_G is an overall mass transfer coefficient. The order, g, of the overall growth process is usually between 0 and 2.5, with an order near unity being the most common. The overall mass transfer coefficient k_G in general depends on temperature, crystal size, hydrodynamic conditions, and the presence of impurities.

It is also sometimes convenient to express the overall linear growth rate as

$$G \equiv \frac{dL}{dt} = k_g(c - c^*)^g \tag{9.2.4}$$

where L is a characteristic single dimension of the crystal, such as length. It has been shown that geometrically similar crystals of the same material grow at the rate described by Equation (9.2.4), independent of crystal size [3]. This is known as the *delta L law*. Although a few exceptions to this law have been found (e.g., when the crystals are very large), it is reasonably accurate in numerous crystallizations of industrial importance [4].

Crystal growth is actually a process that consists of two steps in series. Solute molecules must reach the crystal surface by means of diffusion. At the surface, the solute must be integrated into the crystal lattice. These two processes may be described as [1]

$$\text{Diffusion:} \quad R = k_d (c - c_i)^d \tag{9.2.5}$$

$$\text{Surface integration:} \quad R = k_r (c_i - c^*)^r \tag{9.2.6}$$

where c_i is the concentration at the interface between the liquid and solid phase, and k_d and k_r are mass transfer coefficients. When the exponents are unity, combining Equations (9.2.3), (9.2.5), and (9.2.6) gives

$$\frac{1}{k_G} = \frac{1}{k_d} + \frac{1}{k_r} \tag{9.2.7}$$

Thus, if surface integration is very fast compared with bulk diffusion, then $k_r \gg k_d$, and $k_G \cong k_d$.

9.2.4 CRYSTALLIZATION KINETICS FROM BATCH EXPERIMENTS

Estimates of nucleation and growth rates during crystallization can be obtained from batch experiments by a method known as moments analysis [1]. By making a mass balance on the number of crystalline particles within a given size range between L and $L + \Delta L$, as was done for precipitate particles (Chapter 8), the population balance equation for size-independent growth in a perfectly mixed, constant volume batch crystallizer with negligible attrition and agglomeration becomes

$$\frac{\partial n(L)}{\partial t} + G \frac{\partial n(L)}{\partial L} = 0 \tag{9.2.8}$$

The kth moment of the population density distribution in L about the origin, obtained by moment transformation of Equation (9.2.8) with respect to size, is defined as

$$\mu_k \equiv \int_0^\infty n(L) L^k \, dL \tag{9.2.9}$$

If the moments of the experimental population density function can be determined at two times that differ by a small time interval Δt over which the linearity of the model can be assumed, then the average values of the nucleation and growth rates can be expressed as

$$\bar{B} = \frac{\Delta \mu_0}{\Delta t} \tag{9.2.10}$$

and

$$\bar{G} = \frac{\Delta \mu_1}{\bar{\mu}_0 \Delta t} \tag{9.2.11}$$

TABLE 9.2
Moment Analysis of Crystal Size Distribution

Moment	k of moment [Equation (9.2.9)]	Meaning of moment	Mathematical definition	Average value for spherical crystals
μ_0	0	Number of crystals	$\int_0^\infty n(L)\,dL$	
μ_1	1	Size of crystals	$\int_0^\infty n(L)L\,dL$	$\dfrac{\mu_1}{\mu_0}$
μ_2	2	Area of crystals[a]	$\int_0^\infty n(L)L^2\,dL$	$\pi\left(\dfrac{\mu_2}{\mu_0}\right)$
μ_3	3	Mass of crystals[b]	$\int_0^\infty n(L)L^3\,dL$	$\dfrac{\pi\rho}{6}\left(\dfrac{\mu_3}{\mu_0}\right)$

[a]Multiply moment by area shape factor relating area to L^2 to obtain area of crystals per unit volume (area shape factor $= \pi$ for spheres).
[b]Multiply moment by crystal density ρ and by a volume shape factor relating volume to L^3 to obtain mass of crystals per unit volume (volume shape factor $= \pi/6$ for spheres).

Here, Δ represents the difference in values of a quantity at two different times and the bar an arithmetic average of the quantities. These equations are derived by moment transformation of the population balance equation with respect to size [Equation (9.2.8)]. Therefore, the average nucleation rate $\bar{B}$ and the average overall linear growth rate $\bar{G}$ may be calculated from two experimental population density plots obtained from a batch crystallization experiment at times t and $t + \Delta t$. In addition, the average size $\bar{L}$ at each time can be obtained as follows from the moments:

$$\bar{L} = \frac{\int_0^\infty n(L)L\,dL}{\int_0^\infty n(L)\,dL} = \frac{\mu_1}{\mu_0} \tag{9.2.12}$$

The meaning of each kth moment in the crystal population is given in Table 9.2.

Fractional moments can also be useful. For example,

$$\text{Fraction of total number of crystals between size 0 and size } L = \frac{\int_0^L n(L)\,dL}{\int_0^\infty n(L)\,dL} \tag{9.2.13}$$

Fractional moments can also be applied to the number, area, and mass of crystals.

9.3 Batch Crystallizers

Batch crystallizers are extensively used in the pharmaceutical industry to produce bio-chemicals such as antibiotics. Often, the final step of purification of an antibiotic is crystallization. Batch crystallizers have the advantages of being flexible to operate and re-quiring less capital investment than continuous crystallizers. A widely employed technique for achieving supersaturation in batch crystallizers is the method of dilution, also called the "salting-out" method if the diluent contains a salt. The addition of a diluent reduces the solubility of the solute that is desired to be crystallized. For example, antibiotics have been crystallized from aqueous solution by the slow addition of a miscible organic solvent such as ethanol. Another way of generating supersaturation in batch crystallizers is by chemical reaction to form a less soluble product; the reaction can vary from the addition or removal of a proton to the coupling of two molecules. In this section, the dilution method of batch crystallization is analyzed.

9.3.1 ANALYSIS OF DILUTION BATCH CRYSTALLIZATION

In dilution batch crystallizers, operation can be by several different modes, which include a constant diluent addition rate, a constant rate of diluent concentration change, and a constant level of supersaturation. Since the process includes the addition of a third component (the diluent), the crystallizer volume varies with time, although in some cases the diluent volume required may be relatively small, and thus the volume change may be neglected. The addition of a diluent causes the solubility of the solute being crystallized to be lowered. A commonly used expression for the dependence of the solubility of the solute upon the concentration of the diluent being added is the following [5]:

$$c^* = c_0^* \exp(-k_p c_d) \qquad (9.3.1)$$

where c_d is the diluent concentration and k_p is a constant.

Since the working volume of a batch crystallizer may vary with time, it is convenient to define the population density function, $\hat{n}$, based on the total solvent volume at any time as

$$\hat{n}(L) = n(L)V \qquad (9.3.2)$$

The resulting population balance equation for a perfectly mixed batch crystallizer with negligible attrition and agglomeration is [6]

$$\frac{\partial \hat{n}(L)}{\partial t} + \frac{\partial \hat{n}(L)G}{\partial L} = 0 \qquad (9.3.3)$$

where G is the overall linear growth rate, previously defined by Equation (9.2.4). To simplify the solution of this equation, a new variable, y, is defined such that

$$y \equiv \int_0^t G(t)\, dt \qquad (9.3.4)$$

so that

$$t = \int_0^y \frac{dy}{G(y)} \qquad (9.3.5)$$

The variable y represents the size of a crystal at any time t, that originally nucleated at time $t = 0$. With y as a variable, the population balance for a crystal system where the growth rate is independent of size becomes

$$\frac{\partial \hat{n}}{\partial y} + \frac{\partial \hat{n}}{\partial L} = 0 \qquad (9.3.6)$$

The boundary conditions ($y = 0$) for the nuclei population density can be represented by the equation [1]

$$\hat{n}(t, 0) = \hat{n}^0 = \frac{\hat{B}}{G} \qquad (9.3.7)$$

To average the distribution with respect to the internal coordinate properties, a transformation of the population balance equation [Equation (9.3.6)], leads to the following moment equations [6], where the hat symbol (^) represents quantities based on total solvent volume:

$$\frac{d\hat{N}}{dy} = \hat{n}(0, y) \tag{9.3.8}$$

$$\frac{d\hat{L}}{dy} = \hat{N} \tag{9.3.9}$$

$$\frac{d\hat{A}}{dy} = 2k_a \hat{L} \tag{9.3.10}$$

$$\frac{d\hat{W}}{dy} = \frac{3k_v \rho}{k_a} \hat{A} \tag{9.3.11}$$

where $\hat{N}$ = number of crystals
$\hat{L}$ = crystal size (length)
$\hat{A}$ = crystal area
$\hat{W}$ = mass of crystals
k_a, k_v = surface and volume shape factors
ρ = crystal density

These relationships can be combined to predict the time dependence of crystal size in a batch crystallization process, as in the following example.

EXAMPLE 9.1

Batch Crystallization with Constant Rate of Change of Diluent Concentration   Derive the relationship between y and t for an unseeded batch dilution crystallizer with a constant rate of change of diluent concentration, given the following information [6, 7]:

System kinetics: $G = 10^{-6} \Delta c$ m/s; $B = 10^6 \Delta c^2$ number (liter of solvent)$^{-1}$ s^{-1}, where $\Delta c = c - c^*$

Physical parameters: $\rho = 2 \times 10^3$ kg/m^3; $k_v = 0.5$; $k_a = 3.5$; $c_0^* = 0.25$ kg of solute/liter solvent; $k_p = 10$ liters (solvent and diluent)/liter of diluent; V_0 = initial volume of solvent = 1.0 liter; $\Delta c_0 = 0.01$ kg of solute per liter of solvent; and $k_m = 10^{-6}$ liter of diluent (liter of solvent)$^{-1}$ s^{-1}, where

$$\frac{dc_d}{dt} = k_m = \text{rate of change of diluent concentration}$$

SOLUTION

We need to determine how y depends on t in this dilution crystallizer. Assume that the nuclei population density based on the initial supersaturation level is constant throughout the volume of the batch. From Equation (9.3.7) at $t = 0$,

$$\hat{n}(0, 0) = \hat{n}_0^0 = \frac{\hat{B}}{G} = \frac{10^6 \Delta c_0^2 V_0}{10^{-6} \Delta c_0} = 10^{12} \left(0.01 \, \frac{\text{number}}{\text{m liter}}\right)(1.0 \text{ liter}) = 10^{10} \, \frac{\text{number}}{\text{m}}$$

Assuming the nuclei population density to be constant at this value, after integration of Equations (9.3.8), (9.3.9), and (9.3.10) and substitution into Equation (9.3.11), we obtain

$$\frac{d\hat{W}}{dy} = k_v \rho \hat{n}_0^0 y^3 = 0.5(2 \times 10^3)(10^{10})y^3 = 10^{13} y^3 \frac{\text{kg}}{\text{m}}$$

Since the change in solute concentration leads to crystal formation, we can write the mass balance for the crystallization as

$$d\hat{c} + d\hat{W} = 0$$

where $\hat{c} = Vc$ and $\hat{W} = VW$. From the definition of supersaturation ($\Delta\hat{c} = \hat{c} - Vc^*$), the mass balance can be rewritten as

$$\frac{d\Delta\hat{c}}{dy} + \frac{dVc^*}{dy} + \frac{d\hat{W}}{dy} = 0$$

which is called the "supersaturation balance equation."

We now evaluate the first two terms of the foregoing mass balance equation, with the objective of obtaining a term involving dt/dy that would allow us to determine y as a function of t. To evaluate the term dVc^*/dy, we first differentiate the solubility c^*, as given by Equation (9.3.1), with respect to t:

$$\frac{dc^*}{dt} = -c_0^* k_p \frac{dc_d}{dt} \exp(-k_p c_d) = -k_p c^* \frac{dc_d}{dt}$$

Substituting for dc_d/dt gives

$$\frac{dc^*}{dt} = -k_p k_m c^*$$

Assuming the initial concentration of diluent to be zero, the solution volume increases according to

$$V = \frac{V_0}{1 - k_m t}$$

Differentiating V with respect to time gives

$$\frac{dV}{dt} = \frac{V_0 k_m}{(1 - k_m t)^2} = \frac{V^2}{V_0} k_m$$

The term dVc^*/dt can be expanded to give

$$\frac{dVc^*}{dt} = V\frac{dc^*}{dt} + c^*\frac{dV}{dt}$$

Substituting for dc^*/dt and dV/dt yields

$$\frac{dVc^*}{dt} = c^* k_m V\left(\frac{V}{V_0} - k_p\right)$$

or,

$$dVc^* = c^* k_m V\left(\frac{V}{V_0} - k_p\right) dt$$

Next, we assume that volume and solubility changes are negligible (i.e., $V \cong V_0$ and $c^* \cong c_0^*$), and that changes in the supersaturation are negligible with respect to y (i.e., $d\Delta\hat{c}/dy = 0$). This last assumption is realistic because of the inherent self-regulating character of batch crystallizers [6]. Substituting for dVc^* in the supersaturation balance equation gives

$$[c_0^* k_m V_0 (k_p - 1)] \frac{dt}{dy} = \frac{d\hat{W}}{dy}$$

Substituting the result obtained previously for $d\hat{W}/dy$ and rearranging leads to

$$\frac{dt}{dy} = \frac{10^{13} y^3}{(0.25 \times 9)10^{-6}} = 4.44 \times 10^{18} y^3 \, \frac{s}{m}$$

Integrating with $y = 0$ at $t = 0$ gives

$$y = 3.08 \times 10^{-5} t^{0.25} \, m$$

This is a useful result that gives an estimate of how long the crystallization must be carried out to obtain crystals of a desired size. Note the modest time dependence of $y(t)$. Also note that this derivation is for a specific case and must be modified when different cases have different kinetic power functions.

9.4 Process Crystallization of Proteins

Because crystals of bioproducts tend to be quite pure, the use of crystallization in a purification process often results in great simplification of the process. For the purification of proteins, there are some strong reasons for utilizing crystallization in the process: (1) one or more expensive chromatography steps possibly can be eliminated; (2) in comparison to chromatography, crystallization is relatively inexpensive to carry out, since costly adsorbents are not required; and (3) proteins crystals often can be stored for long periods at low temperatures without being degraded or denatured after the addition of stabilizing agents such as ammonium sulfate, glycerol, or sucrose.

The methods that have been used to crystallize proteins include those used for precipitation: adjustment of the pH to the isoelectric point (isoelectric precipitation), the addition of organic solvents, the addition of salts (salting out), and the addition of nonionic polymers. Another crystallization method that has proven useful for proteins is the reduction of ionic strength by dialysis or diafiltration, which relies on the limited solubility of many proteins at low ionic strength (the reverse of the "salting-in" effect described in Chapter 8, Section 8.2.3). A key strategy for crystallization is to move the system *very slowly* to a state of minimum solubility of the desired protein until a limited degree of supersaturation is reached. High protein concentrations, in the range of 10 to 100 mg/ml, are recommended for crystallization, and seeding with small crystals of the desired protein is often beneficial [8].

Three studies of protein crystallization that have been reported illustrate the different approaches that can be taken:

- Alcohol oxidase, originally contained in *Pichia pastoris* yeast broth grown in a 100-liter pilot plant fermentor, was crystallized by lowering the ionic strength by diafiltration with deionized water; crystallization was preceded by lysis of the yeast cells, diafiltration with a microfilter to obtain the alcohol oxidase in the permeate, and concentration and then diafiltration with an ultrafiltrater [9]. The yield was 208 g of pure enzyme.

- Ovalbumin was crystallized in the presence of conalbumin and lysozyme by the addition of 2.5 μm seed crystals to a solution (600 ml) that had been made supersaturated by slowly adding ammonium sulfate solution [10]. The supersaturation was kept in the metastable region to avoid nucleation (see Figure 9.2). The growth rate of ovalbumin crystals was found to have a second-order dependence on the ovalbumin supersaturation, and the presence of the other two proteins did not affect the growth rate constant [k_g in Equation (9.2.4)].

- A fungal lipase was crystallized at low conductivity by slowly lowering the pH with formic acid [11]. The solution for crystallization was prepared by removing the cells by centrifugation (the lipase had been secreted from the cells), concentration by ultrafiltration, and diafiltration to lower the conductivity. An analysis of the crystal size distribution data gave a power dependency of the nucleation rate and growth rate on the lipase supersaturation of 10.8 and 6.4, respectively. These high powers were attributed to possible underestimation of the supersaturation.

9.5 Crystallizer Scaleup and Design

Since the behavior of each biochemical to be crystallized is unique, the development of a large-scale crystallization process must be based on experimental data. Thus, data are taken first in the laboratory and are followed by data from pilot plant equipment, which form the basis of the scaleup calculations and plant process design.

9.5.1 EXPERIMENTAL CRYSTALLIZATION STUDIES AS A BASIS FOR SCALEUP

Experimental work to serve as a basis for scaleup can be divided into three stages: (1) gathering basic data on solubility and metastable zone width, (2) developing the conditions in a laboratory scale batch crystallizer to obtain a product that meets required specifications at a satisfactory yield, and (3) pilot plant crystallizations that use optimal laboratory conditions as a starting point.

Data on the solubility of the solute to be crystallized should be obtained in the starting solvent as a function of temperature. These solubility data are the upper limit of the concentration of the solute at the start of the crystallization. The metastable zone width is the increase in concentration above the equilibrium solubility before primary nucleation occurs (see Figure 9.2). The metastable zone width can be determined by slowly increasing the supersaturation in a solution free of crystals until primary nucleation occurs, then noting the equilibrium solubility level at nucleation conditions [12].

A convenient size of a batch crystallizer for carrying out initial tests of crystallization is around 1 liter, assuming that enough material is available [12]. Protein equilibrium data, however, have been obtained using as little as 10 μl. A 1-liter volume is generally enough to allow for sampling during the experiment, generating of enough product slurry to obtain filtration rate data, and determining the effect of washing on crystal purity. Typically, the experiment starts with a crystal-free undersaturated solution, and for the proposed end point of the crystallization to be reached in about 2 h, supersaturation is generated by using the chosen method at an approximately linear rate. The point at which primary nucleation occurs is noted and compared with the earlier previous measurements of the metastable zone width. When crystallization has started, it is useful to measure the solution concentration as a function of time. At the end of the generation of supersaturation, the crystal slurry should be agitated for an additional hour or two, continuing to measure the solution concentration to determine the approach to equilibrium. After this holding period, the crystal slurry can be filtered to obtain filtration rate data. If there is sufficient slurry, the effect of different types of washing on crystal purity can be determined. After drying, the crystals should be characterized for particle size distribution and examined under an optical microscope, which often gives important information about crystal structure.

Problems with the final crystals not meeting specifications or with low yield should be addressed at the laboratory scale before the scale of testing is increased. One such problem that can be difficult to solve occurs when the particle size distribution is out of range, resulting in low filtration rates of the crystal slurry. In order to change the crystal size distribution, the relative rates of nucleation and growth must be changed, which can be accomplished by adjusting the supersaturation history. In general, the nucleation rate increases more than the growth rate with increasing supersaturation, as can be seen from the exponent of the supersaturation in Equations (9.2.1) and (9.2.3). Therefore, operating at low supersaturation favors growth of large crystals, while high supersaturation close to the metastable limit tends to give many more nuclei and smaller crystals. One method that may increase crystal size is the addition of seed crystals to a just-supersaturated solution.

The production of crystals with a lower than required purity is another problem that can occur in crystallization. An impurity that is the cause of the problem can be in the following locations in the crystal sample [12]:

1. Deposited on crystal surfaces due to incomplete removal of impure mother liquor
2. Trapped within voids between separate crystals in materials that agglomerate
3. Contained in inclusions of mother liquor within individual crystals
4. Distributed throughout the crystals by molecular substitution at the lattice sites

An obvious approach to eliminating the impurity from the crystal surface is to improve the solid–liquid separation and washing procedures. For crystals that agglomerate, it may be possible to reduce agglomeration by changing the crystallization conditions (such as using a different crystallization solvent). Reducing the supersaturation level to reduce both the nucleation and growth rates should help to minimize the problem of impurities in inclusions of mother liquor. The problem of the impurity being distributed throughout the crystals is difficult to solve for an individual crystallization and many require that the crystals be *recrystallized*. Recrystallization, which is widely used, is done by dissolving the crystals in fresh solvent and performing the crystallization again. Purer crystals can result if the impurity is more soluble in the solvent than the main product. Complex fractional recrystallization schemes, requiring more than one recrystallization [5], have been developed but are not used frequently at the plant scale.

Once satisfactory operating conditions have been obtained at the laboratory scale, it is often advisable to test the crystallization at the pilot plant scale. Scaleup ratios of 100 to

1000 from laboratory to pilot plant appear reasonable [1]. Common scaleup rules are outlined in the next section. The same types of measurement made in the laboratory crystallizations should also be made in the pilot plant crystallizations.

9.5.2 SCALEUP AND DESIGN CALCULATIONS

The challenge in scaleup is to generate conditions at the large scale that are similar to those at the small scale. As with precipitation (Chapter 8), the key problem is in the scaleup of mixing to control the supersaturation level both locally and globally throughout the crystallizer and to ensure that the crystals are carried throughout the volume of the crystallizer under controlled conditions of frequency and intensity of crystal–crystal interactions [12]. Crystallizer scaleup ratios of 50 to 500 from pilot plant to commercial scale are reasonable [1].

As with precipitation, it is often recommended that geometric similarity and constant power per volume be used in scaleup of crystallizers [1, 13]. For turbulent flow in vessels (Reynolds number $> 10,000$), constant power per volume means that [14]

$$\frac{P}{V} \propto N_i^3 d_i^2 \tag{9.5.1}$$

where N_i is the impeller rotation rate and d_i is the impeller diameter. For an agitated tank, the Reynolds number is given by

$$\text{Re} = \left(\frac{d_i^2 N_i \rho}{\mu} \right) \tag{9.5.2}$$

where N_i is in revolutions per unit time, ρ is the fluid density, and μ is the fluid viscosity.

Two additional strategies are also sometimes used in scaleup of crystallizers [13]:

1. Maintaining constant impeller tip speed, which implies that

$$N_i d_i = \text{constant} \tag{9.5.3}$$

2. Scaleup at the minimum speed required for particle suspension, which has been shown to require [15]

$$N_i d_i^{0.85} = \text{constant} \tag{9.5.4}$$

This last criterion should give both secondary nucleation and crystal growth rates that are more or less independent of the scale of operation.

From an analysis of the data for the operation of the crystallization at the laboratory and pilot plant scales, the scaleup equation or equations that prove to be the most accurate in describing the laboratory to pilot plant scaleup should be used in the design of the plant crystallizer. However, flexibility should be designed into the plant unit so that agitation rates and flow rates can be varied, which may be necessary because of error in the scale up to the larger scale and because of the effect that changes in the concentration of impurities in the starting solution may have on the operation.

EXAMPLE 9.2

Scaleup of Crystallization Based on Constant Power per Volume   It is desired to scale up a batch crystallization of an antibiotic based on experiments with a 1-liter crystallizer. The use of a 3 cm diameter impeller at a speed of 800 rpm led to good crystallization results. For maintaining power per volume constant upon scaleup to 300 liters, what should be the diameter and speed of the large-scale impeller? The solvent has the same density and viscosity as water.

SOLUTION

For the scaleup, we assume geometric similarity. Thus, the volume (V) scales with the impeller diameter (d) as

$$V \propto d_i^3$$

Letting subscripts 1 and 2 denote the small and large scales, respectively,

$$\frac{V_2}{V_1} = \left(\frac{d_{i,2}}{d_{i,1}}\right)^3$$

Solving for d_2,

$$d_{i,2} = d_{i,1}\left(\frac{V_2}{V_1}\right)^{1/3} = 3\left(\frac{300}{1}\right)^{1/3} = 20.1 \text{ cm}$$

From Equation (9.5.1), for power per volume to be constant,

$$N_{i,1}^3\, d_{i,1}^2 = N_{i,2}^3\, d_{i,2}^2$$

This equation is valid for turbulent flow in the crystallizer. To verify that the flow is turbulent, we use Equation (9.5.2) to calculate the Reynolds number for the lab crystallizer:

$$Re = \left(\frac{d_i^2 N_i \rho}{\mu}\right)_1 = \frac{(3 \text{ cm})^2 \times 800\, \dfrac{\text{rev}}{\text{min}} \times 1.00\, \dfrac{\text{g}}{\text{cm}^3} \times \dfrac{1 \text{ min}}{60 \text{ s}}}{0.01\, \dfrac{\text{g}}{\text{cm s}}} = 12{,}000$$

This Reynolds number indicates that the flow is turbulent. Solving for $N_{i,2}$,

$$N_{i,2} = N_{i,1}\left(\frac{d_{i,1}}{d_{i,2}}\right)^{2/3} = 800\left(\frac{3}{20.1}\right)^{2/3} = 225 \text{ rpm}$$

Thus, the rotation rate is reduced considerably compared with the laboratory scale.

9.6 Summary

The process of producing crystals from a homogeneous phase is called crystallization. Bioproducts can be obtained at very high purity (say, 99.9%) by crystallization. This process is considered to be both a polishing step and a purification step, and, in some instances, the purification that is achieved by crystallization is so significant that other more expensive purification steps such as chromatography can be avoided.

- Crystallization differs primarily from precipitation in that when a substance comes out of solution as a solid it also forms a regular lattice of molecules, or crystal.

- Crystals nucleate either by primary or secondary nucleation. Primary nucleation occurs in the absence of crystals, while secondary nucleation is attributed to the influence of existing crystals. Secondary nucleation dominates in industrial crystallizers. Two types of primary nucleation are possible: homogeneous and heterogeneous, where heterogeneous nucleation requires the presence of foreign particles for nucleation to occur and homogeneous nucleation does not.

- The rate of increase in the concentration of nuclei is proportional to a power of the supersaturation, $c - c^*$, where c^* is the concentration at equilibrium ("saturation"). The power has a most probable value of 3 to 4 for primary nucleation and 2 for secondary nucleation.

- The crystal growth rate is also proportional to a power of the supersaturation, with a power of 1 being the most common.

- The differential equation for a material balance on a crystal size interval can be subjected to a moment transformation. The zeroth moment is the number of nuclei (crystals), the first moment is the linear size, the second moment is proportional to the area, and the third moment is proportional to the mass (or volume). The crystal nucleation and growth rates can be determined from the moments of the experimental population density function.

- Batch crystallizers can be analyzed by the population balance equation.

- Scaleup of crystallizers, like scaleup of precipitation units, emphasizes mixing. The mixing parameters that can be held constant are power input per volume, impeller tip speed, and agitation sufficient to keep crystals in suspension. All three of these variables cannot be held absolutely constant upon scaleup at the same time, because each depends differently on the impeller diameter. Therefore, scaling priorities are also based on experimental measurements and plant-scale design for flexibility.

NOMENCLATURE

A surface area of crystals per volume of solvent (cm^2 $liter^{-1}$)

B nucleation rate during crystallization (number $liter^{-1}$ s^{-1})

c concentration of solute (M, or g $liter^{-1}$)

c_d concentration of diluent (liter $liter^{-1}$)

c_i concentration of solute at interface between liquid and solid phases (g $liter^{-1}$)

d_i impeller diameter (cm)

G overall linear growth rate of crystals (cm s^{-1})

P/V power input per volume of liquid (W $liter^{-1}$)

k_a surface shape factor (dimensionless)

k_d mass transfer coefficient for bulk diffusion [Equation (9.2.5)] [g cm^{-2} s^{-1} (g/liter)$^{-d}$]

k_g overall mass transfer coefficient for linear crystal growth [Equation (9.2.4)] [(cm s^{-1} (g/liter)$^{-g}$]

k_G overall mass transfer coefficient for crystal growth [Equation (9.2.3)] [g cm^{-2} s^{-1} (g/liter)$^{-g}$]

k_1 rate constant for secondary nucleation [Equation (9.2.2)] [number $liter^{-1}$ s^{-1} (g/liter)$^{b-j}$]

k_m rate of change of diluent concentration (liter $liter^{-1}$ s^{-1})

k_n rate constant for primary nucleation [Equation (9.2.1)] [number $liter^{-1}$ s^{-1} (g/liter)$^{-n}$]

k_p constant in equation for solubility of solute as a function of diluent concentration [Equation (9.3.1)] (liter $liter^{-1}$)

k_r mass transfer coefficient for surface integration [Equation (9.2.6)] [g cm^{-2} s^{-1} (g/liter)$^{-r}$]

k_v volume shape factor (dimensionless)

L characteristic dimension of crystal (cm, or cm liter^{-1})

M_T crystal suspension density (g liter^{-1})

n population density distribution function (number liter^{-1} cm^{-1})

N number concentration of nuclei (number liter^{-1})

N_i impeller rotation rate (revolutions s^{-1})

Re   Reynolds number ($d_i^2 N_i \rho / \mu$) (dimensionless)

R rate of mass deposition on crystals during growth (g cm^{-2} s^{-1})

t time (s)

V solvent volume (liter)

W mass of crystals per volume of solvent (g liter^{-1})

y size of crystals at time t [Equation (9.3.4)] (cm)

Greek Letters

μ_k kth moment of the population density distribution (units vary)

ρ crystal density (g liter^{-1})

Subscripts

f final

0 initial

Superscripts

$*$ saturation or equilibrium

$\wedge$ quantities based on total solvent volume

$-$ average quantities

PROBLEMS

9.1 **Modeling of Crystal Growth**   The data in Table P9.1 have been measured for the growth of crystals of an antibiotic. From this data set, obtain an expression for the relationship between dL/dt and the supersaturation $c - c^*$.

9.2 **Nucleation and Growth Rates from Crystal Size Distribution Data**   Crystal size distribution (CSD) data were obtained with a multichannel

TABLE P9.1

$c - c^*$ (g/liter)	dL/dt (μm/min)
0.20	0.21
0.35	0.45
0.67	0.90
1.25	1.80
2.05	3.30
2.75	5.00

Coulter counter during a crystallization experiment in a batch agitated vessel with isothermal conditions. The results of two CSD samples taken at different times differing by a small time interval are shown in Table P9.2 on the next page. Use the method of moments analysis to determine the average nucleation and growth rates. At each time, also determine the population average crystal size. (Data from N. S. Tavare, *Industrial Crystallization,* p. 180, Plenum Press, New York, 1995.)

9.3 **Batch Crystallizer with Constant Rate of Diluent Addition**   For a solute being crystallized by batch crystallization with a constant rate of change of diluent concentration, predict the effect on the crystal size at any given time for each of the changes in the variables given. Justify your answers based on an expression for the crystal size as a function of these variables.

(a) An increase in the nucleation rate

(b) An increase in the growth rate

(c) An increase in the supersaturation, for b [Equation (9.2.2)] greater than g [Equation (9.2.3)]

(d) An increase in k_m in the equation $dc_d/dt = k_m$, where c_d is the diluent concentration

(e) An increase in k_p [Equation (9.3.1)]

9.4 **Estimate of Particle Size for an Unseeded Batch Crystallizer**   For an unseeded batch crystallization of a biochemical using a constant rate of diluent concentration, the following is known:

$G = 10^{-7} \Delta c^2$ m/s

$B = 10^6 \Delta c^3$ number (liter of solvent)$^{-1}$ s^{-1}

$\rho = 2.5 \times 10^3$ kg/m^3

$k_v = 0.6$

$k_a = 3.0$

$c_0^* = 0.05$ kg of solute per liter of solvent

$k_p = 5$ liters (solvent and diluent) per liter of diluent

TABLE P9.2

Channel number	$\bar{L}$ (μm)	ΔL (μm)	$n(L) \times 10^{-10}$ (number m^{-1} kg^{-1})	
			At $t_1 = 3600$ s	At $t_2 = 4500$ s
4	17.4	4.0	180.9	176.3
5	21.9	5.0	43.6	55.7
6	27.6	6.4	24.5	32.1
7	34.8	8.0	21.8	23.4
8	43.8	10.1	13.8	17.6
9	55.2	12.7	13.9	18.1
10	69.6	16.0	6.4	9.3
11	87.7	20.2	3.8	4.8
12	110.5	25.4	0.7	1.3
13	139.2	32.0	0.4	0.3
14	175.4	40.3	0.06	0.12
15	221.0	50.8		0.09

V_0 = initial volume of solvent = 1.0 liter

Δc_0 = 0.005 kg of solute per liter of solvent

$k_m = 2 \times 10^{-5}$ liter of diluent (liter of solvent)$^{-1}$ s^{-1}, where

$$\frac{dc_d}{dt} = k_m$$

Estimate the average size of the crystals 20 min from the start of nucleation. State all the assumptions necessary for the calculation.

9.5 **Comparison of Three Scaleup Methods for a Crystallization** A batch crystallization of an antibiotic was performed using a volume of 750 ml in the laboratory with a 3.5 cm diameter impeller at a speed of 600 rpm, the minimum speed required to fully suspend the crystals. Estimate the size of the impeller and the impeller speed for scaleup to 250 liters for each of the following three assumptions as a basis for scaleup: (1) constant power per volume, (2) constant impeller tip speed, and (3) full suspension of crystals (at minimum speed).

9.6 **Deduction of Scaleup Equations for a Crystallization** The batch crystallization of a biochemical

resulted in virtually identical properties when conducted either at the laboratory scale (1 liter) or the plant scale (1000 liters). The lab crystallizer had an impeller of 3 cm diameter operating at 750 rpm, while the plant crystallizer had a 30 cm impeller with a speed of 77 rpm. What scaleup equations can be used to approximately characterize the scaleup of this crystallization? What property or properties are being assumed constant upon scaleup?

9.7 **Bulk Volume Too High for Crystalline Erythromycin** The bulk volume of dry, crystalline erythromycin (an antibiotic) produced in a plant batch crystallizer is out of specification (too high). Erythromycin free base is crystallized by the slow addition of dilute sodium hydroxide to a solution of an erythromycin salt dissolved in water–acetone. You have been asked to study this problem in the laboratory and recommend a solution based on your results. Outline the experiments you would conduct in the lab, the results of which would serve as a basis for recommendations for how to change the operation of the plant crystallizer.

References

1. Tavare, N. S. (1995). *Industrial Crystallization,* Plenum Press, New York.
2. Genck, W. J. (1997). Crystallization from solutions. In *Handbook of Separations Techniques for Chemical Engineers,* 3rd ed., P. Schweitzer, ed., McGraw-Hill, New York, pp. 2–141.

3. McCabe, W. L. (1929). Crystal growth in aqueous solutions. *Ind. Eng. Chem.,* vol. 21, pp. 30, 112.

4. Perry, R. H., Green, D. W., and Maloney, J. O., eds. (1997). Liquid–solids operations and equipment. Section 18 in *Perry's Chemical Engineers' Handbook,* McGraw-Hill, New York.

5. Mullin, J. W. (1972). *Crystallization,* 2nd ed., CRC Press, Cleveland.

6. Tavare, N. S., Garside, J., and Chivate, M. R. (1980). Analysis of batch crystallizers. *Ind. Eng. Chem. Process Des. Dev.,* vol. 19, p. 653.

7. Tavare, N. S., and Chivate, M. R. (1980). CSD analysis from a batch dilution crystallizer. *J. Chem. Eng. Jpn.,* vol. 13, p. 371.

8. McPherson, A. (1985). Crystallization of macromolecules: general principles. *Methods Enzymol.,* vol. 114, p. 112.

9. Harrison, R. G., and Nelles, L. P. (1990). Large scale process for the purification of alcohol oxidase, U.S. Patent 4,956,290.

10. Judge, R. A., Johns, M. R., and White, E. T. (1995). Protein purification by bulk crystallization: The recovery of ovalbumin. *Biotechnol. Bioeng.,* vol. 48, p. 316.

11. Jacobsen, C., Garside, J., and Hoare, M. (1998). Nucleation and growth of microbial lipase crystals from clarified concentration fermentation broths. *Biotechnol. Bioeng.,* vol. 57, p. 666.

12. Price, C. J. (1997). Take some solid steps to improve crystallization. *Chem. Eng. Prog.,* vol. 93, p. 34.

13. Nienow, A. W. (1976). The effect of agitation and scale-up on crystal growth rates and on secondary nucleation. *Trans. Inst. Chem. Eng.,* vol. 54, p. 205.

14. Rushton, J. H., Costich, W. W., and Everett, H. J. (1950). Power characteristics of mixing impellers. Part 2. *Chem. Eng. Prog.,* vol. 46, p. 467.

15. Nienow, A. W. (1968). Suspension of solid particles in turbine agitated baffled vessels. *Chem. Eng. Sci.,* vol. 23, p. 1453.

CHAPTER
10

Drying

The last step in the separation process for a biological product is usually drying, which is the process of thermally removing volatile substances (often water) to yield a solid. In the step preceding drying, the desired product is generally in an aqueous solution and at the desired final level of purity. The most common reason for drying a biological product is that it is susceptible to chemical (e.g., deamidation or oxidation) and/or physical (e.g., aggregation and precipitation) degradation during storage in a liquid formulation. Another common reason for drying is for convenience in the final use of the product. For example, it is often desirable that pharmaceutical drugs be in tablet form. Additionally, drying may be necessary to remove undesirable volatile substances. Also, although many bioproducts are stable when frozen, it is more economical and convenient to store them in dry form rather than frozen.

Drying is now an established unit operation in the process industries. However, because most biological products are thermally labile, only those drying processes that minimize or eliminate thermal product degradation are actually used to dry biological products. This chapter focuses on the types of dryer that have generally found the greatest use in the drying of biological products: vacuum-shelf dryers, batch vacuum rotary dryers, freeze dryers, and spray dryers [1]. The principles discussed, however, will apply to other types of dryers as well.

We begin with the fundamental principles of drying, followed by a description of the types of dryer most used for biological products. Then we present scaleup and design methods for these dryers.

10.1 Instructional Objectives

After completing this chapter, the reader should be able to do the following:

- Do drying calculations involving relative humidity using the psychrometric moisture chart and the equilibrium moisture curve for the material being dried.
- Calculate the relative amounts of bound and unbound water in wet solids before drying.

- Model heat transfer in conductive drying and calculate conductive drying times.
- Interpret drying rate curves.
- Calculate convective drying times of nonporous solids based on mass transfer.
- Distinguish and choose among vacuum-shelf dryers, vacuum rotary dryers, freeze dryers, and spray dryers.
- Scale up the four dryer types studied.

10.2 Drying Principles

A considerable amount of theory has been developed for drying, from simple approaches to complex theory involving the numerical solution of differential equations. We emphasize the basic theory that will illuminate underlying drying principles, along with theory that will assist in scaleup and design.

10.2.1 WATER IN BIOLOGICAL SOLIDS AND IN GASES

Before considering the removal of water by drying, it is helpful to understand how water is held within biological solids and also to know how to estimate the properties of gases, especially air, containing water vapor. As for all materials, the water contained within biological solids is in two forms, *unbound or free water* and *bound water*. Unbound water is free to be in equilibrium with water in the vapor phase; thus, unbound water has the same vapor pressure as bulk water. Unbound water is mainly held in the voids of the solid. On the other hand, bound water can exist in several conditions: (1) water in fine capillaries that exerts an abnormally low vapor pressure because of the highly concave curvature of the surface, (2) water containing a high level of dissolved solids, and (3) water in physical or chemical combination with the biological solids. Solids containing bound water are called *hygroscopic;* different solids can have widely varying hygroscopic properties [2].

Equilibrium data for moist solids in contact with humid air give valuable information about the water capacity of the solids. Typically, the water content of the solid is plotted as a function of the relative humidity of air, as shown in Figure 10.1 for three bioproducts. Curves of the type shown in Figure 10.1 are nearly independent of temperature. Humidity is defined as the mass of water per mass of dry air. From the ideal gas law, the concentration of water c_w in moles per volume can be related to the humidity $\mathcal{H}$ and total pressure p as follows:

$$c_w = \frac{p_w}{RT}$$

$$= \frac{\mathcal{H}(M_{air}/M_{water})}{1 + \mathcal{H}(M_{air}/M_{water})} \frac{p}{RT} = \frac{\mathcal{H}(28.9/18)}{1 + \mathcal{H}(28.9/18)} \frac{p}{RT} \qquad (10.2.1)$$

Here p_w is the partial pressure of water, M is molecular weight, T is the temperature, and R is the ideal gas constant. The relative humidity R_m can be determined from p_w and the

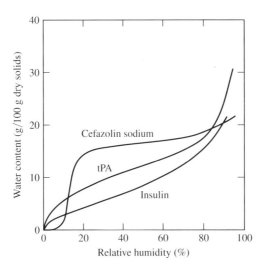

Figure 10.1 Equilibrium–moisture curves drawn to represent data for the antibiotic cefazolin sodium (crystalline) [3] and the proteins human tissue type plasminogen activator (tPA, lyophilized) [4] and human insulin (crystalline) [5].

saturation vapor pressure of water p_{ws} as follows:

$$R_m = \frac{p_w}{p_{ws}} \times 100 \tag{10.2.2}$$

The saturation vapor pressure of water is a function of temperature only and is known exactly when the temperature is known.

A convenient way of showing the properties of mixtures of air and water vapor mixtures is the humidity or psychrometric chart shown in Figure 10.2. Any point on this chart represents a specific mixture of air and water. The curved line denoted by 100% represents the humidity of air saturated with water as a function of temperature. Any point below the 100% or saturation curve represents air that is unsaturated with water, and a point on the temperature axis represents dry air. The straight lines that slant downward from the saturation curve are called adiabatic cooling lines. Thus, if an unsaturated air–water mixture at a point on one of these lines had been cooled adiabatically, the change in composition of the mixture when the temperature was lowered could be followed by moving along these lines toward the 100% saturation curve.

It can also be observed in Figure 10.2 that as the temperature is raised at constant moisture content (humidity), the relative humidity decreases. This is explained by Equation (10.2.2): while the mole fraction of water and thus the partial pressure of water (p_w) remains constant as the temperature rises, the saturation vapor pressure of water (p_{ws}) rises and thus causes R_m to decrease.

In carrying out heat calculations for dryers, the enthalpy of the drying gas must be calculated. The total enthalpy H_y at temperature T of a unit mass of gas plus water vapor is defined relative to a datum temperature T_0 as

$$H_y = C_s(T - T_0) + \mathcal{H}\lambda_0 \tag{10.2.3}$$

where C_s is the specific heat (also called the humid heat) of the gas and water vapor mixture, and λ_0 is the latent heat of the liquid at T_0 [2]. The specific heat C_s can be further

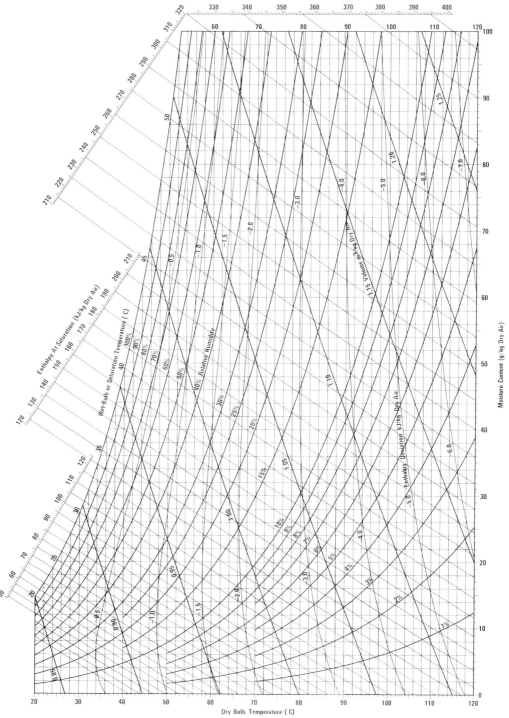

Figure 10.2 Properties of air and water vapor mixtures from 20 to 120°C. (Courtesy of Carrier Corp.)

defined as

$$C_s = C_{pg} + C_{pw} \mathcal{H} \tag{10.2.4}$$

where C_{pg} and C_{pw} are the specific heats of the gas and water vapor, respectively. For air and water vapor mixtures,

$$C_s \left(\frac{\text{kcal}}{\text{kg dry air-°C}} \right) = 0.24 + 0.46 \mathcal{H} \tag{10.2.5}$$

$$\lambda_0 = 597 \frac{\text{kcal}}{\text{kg water}} \quad \text{(at } 0° \text{ and 760 mm Hg)}$$

EXAMPLE 10.1

Drying of Antibiotic Crystals   Air at 1 atm and 25°C with a relative humidity of 50% is to be heated to 50°C and then to be used in drying wet crystals of the antibiotic cefazolin sodium. The wet crystals contain 30 g of water per 100 g of dry antibiotic. In the drying process, the air at 50°C and the crystals reach equilibrium with respect to the moisture. Determine the following: the percentages of bound and unbound water in the wet crystals before drying, the moisture content of the crystals after drying, and the water partial pressure at the drying temperature.

SOLUTION

The equilibrium moisture curve for cefazolin sodium in Figure 10.1 can be extrapolated to 100% relative humidity to give a water content of 23 g/100 g dry weight. The water corresponding to concentrations lower than 23 g/100 g dry weight is bound water, because it exerts a vapor pressure less than that of liquid water at the same temperature. The balance of water contained in the crystals is therefore unbound water. Thus,

$$\text{Bound water} = 23 \frac{\text{g}}{100 \text{ g dry weight}}$$

$$\text{Unbound water} = 30 - 23 = 7 \frac{\text{g}}{100 \text{ g dry weight}}$$

Using Figure 10.2, we move from a temperature of 25°C and relative humidity of 50% to a temperature of 50°C, keeping the moisture content constant. By interpolation, the relative humidity of the air is read to be 13%. At this relative humidity, we see from the equilibrium moisture curve for cefazolin sodium (Figure 10.1),

$$\text{Moisture content of antibiotic after drying} = 8 \text{ g water/100 g dry weight}$$

It has been assumed that the equilibrium curve in Figure 10.1 is independent of temperature for the temperature range in this example. From Equation (10.2.1),

$$p_w = \frac{\mathcal{H}(28.9/18)p}{1 + \mathcal{H}(28.9/18)}$$

From Figure 10.2 at 13% relative humidity and 50°C,

$$\mathscr{H} = 0.010 \, \frac{\text{kg}}{\text{kg dry air}}$$

so that the water partial pressure at 50°C is

$$p_w = \frac{0.010(28.9/18)(1 \text{ atm})}{1 + 0.010(28.9/18)} = 0.016 \, \text{atm}$$

10.2.2 HEAT AND MASS TRANSFER

Two fundamental processes take place during the drying of solids: the transfer of heat to evaporate liquid and the transfer of mass as a liquid or vapor within the solid and as a vapor from the surface of the solid. The rate of drying depends on the factors governing heat and mass transfer.

Heat Transfer

Because the drying of wet solids is by definition a thermal process, heat transfer is of primary importance in drying. Although diffusion can be important in some types of drying, drying rates are often limited by heat transfer. The principal heat transfer mechanisms are the following: *conduction* from a hot surface contacting the material, *convection* from a gas that contacts the material, *radiation* from a hot gas or hot surface, and *dielectric or microwave heating* in high frequency electric fields that generate heat within the wet material [6]. We will focus on conductive drying and convective drying. Conductive drying dominates in vacuum-shelf dryers, batch vacuum rotary dryers, and freeze dryers, while convective drying is predominant in spray drying.

In conductive drying, heat is supplied through the surface of the dryer and flows into the solids being dried. Either the solids are contained in trays on shelves that are heated, or they are moving freely inside the dryer and come in frequent contact with the surface of the dryer, where heat is transferred by conduction to the solid particles being dried. Fourier's law of heat conduction applies in conductive drying:

$$q = -k \frac{dT}{dy} \tag{10.2.6}$$

where q is the heat flux (flow rate of heat per unit area), T is the temperature, k is the thermal conductivity of the solid, and y is measured in the direction of heat flow. For a slab held at a different temperature on each surface, this equation states that the heat flux is constant if the thermal conductivity is constant throughout the slab.

Convective drying involves the transfer of heat from a moving gas phase, providing the heat for drying to a solid phase. For this situation, the rate of heat transfer is described in terms of a heat transfer coefficient h:

$$Q = hA(T - T_s) \tag{10.2.7}$$

where Q is the rate of heat flow into the solid, A is the surface area through which heat flows, T is the gas bulk phase temperature, and T_s is the temperature at the solid surface [7].

This can be written in terms of the heat transfer rate per unit area, or heat flux q as

$$q = h(T - T_s) \tag{10.2.8}$$

In analyzing the performance of either a conductive or a convective dryer, it is sometimes convenient to describe the system by an overall heat transfer coefficient U:

$$Q = UA\overline{\Delta T} \tag{10.2.9}$$

where A is the heat transfer area and $\overline{\Delta T}$ is the average temperature difference [2]. The average temperature difference depends on the dryer being considered and is not always the logarithmic mean temperature difference. For some types of dryer, the area available for heat transfer is not known with certainty; and it is more convenient to use a volumetric heat transfer coefficient Ua, with a being the heat transfer area per unit dryer volume:

$$Q = UaV\,\overline{\Delta T} \tag{10.2.10}$$

where V is the volume of the dryer. For this situation, a is unknown and the quantity Ua is either calculated or measured. For some dryers, the true average temperature difference for the whole dryer is difficult to define. Often, no general correlations are available for heat transfer coefficients, and these coefficients must be found experimentally.

EXAMPLE 10.2

Conductive Drying of Wet Solids in a Tray   Wet biological solids contained in trays are being dried in a vacuum-shelf dryer. The following information is known. The shelves are maintained at temperature T_0 by circulating warm water. The wet solids have thermal conductivity k. The heat of vaporization of water is λ, and mass of water per unit volume of wet solids is ρ_0. Develop an expression for the time to dry a bed of wet solids of depth d. The solids have a porous, granular character when dry.

SOLUTION

In this problem, heat flows by conduction through the bottom of the tray and into the solids, as shown in Figure E10.2. For granular solids, we can assume that water begins to evaporate at the top of the solids and then continues to evaporate at a boundary between the wet and dry solids that recedes into the bed at a distance y from the bottom of the bed. The water vapor is not bound to the dry solids lattice and is rapidly removed. This is the "receding front" or "shrinking core" model of drying [8, 9]. We assume that the temperature of the

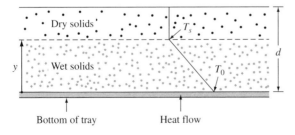

Figure E10.2 Schematic of conductive drying of biological solids in a tray, showing the receding boundary between the wet and dry solids.

moving interface stays constant at the boiling point of evaporating water at temperature T_s, which is determined by the operating vacuum pressure of the dryer. If the thermal conductivity of the tray is much greater than the thermal conductivity of the solids being dried, then we can assume that the shelf temperature is the same as the solids temperature at the bottom of the tray. Applying Fourier's law of heat conduction [Equation (10.2.6)] gives

$$q = k \frac{(T_0 - T_s)}{y}$$

Thus, the heat flux increases as the drying progresses since y decreases. Relating the rate of heat conduction to the heat to vaporize water gives

$$q A = -\lambda \frac{d}{dt}(\rho_0 y A)$$

where A is the cross-sectional area. After combining with the equation for q according to Fourier's law of conduction, this equation can be integrated with the following initial condition

$$\text{At } t = 0, \qquad y = d$$

to yield

$$\frac{d^2 - y^2}{2} = \frac{k(T_0 - T_s)}{\lambda \rho_0} t$$

The drying is complete when $y = 0$, which gives time as

$$t = \frac{d^2 \lambda \rho_0}{2k(T_0 - T_s)}$$

Mass Transfer

Drying can be limited by mass transfer in convective drying. In convective drying, the first water to evaporate is that next to the gas moving across the surface of the wet solids. After an initial warming-up period, the rate of movement of water to the surface is rapid enough that the surface remains saturated, and the drying rate remains constant for a period of time called the constant drying rate period. During this period, mass transfer is limited by a gas boundary layer at the surface of the solids. At a *critical moisture content* X_c, however, the internal rate of water movement is not fast enough to keep the surface saturated, and the drying rate begins to fall. In the falling drying rate period, the drying rate asymptotically approaches the *equilibrium moisture content* X_e. These drying periods are illustrated in Figure 10.3.

   The drying rate curve is influenced by the type of material being dried as well as by the type of mass transfer that is controlling, as indicated in the representative curves shown in Figure 10.4. Biological materials can be classified as nonhygroscopic, partially hygroscopic, or totally hygroscopic. Thus, there is no constant drying period in completely hygroscopic biological solids or in solids of any type when the mass transfer is completely controlled by internal diffusion.

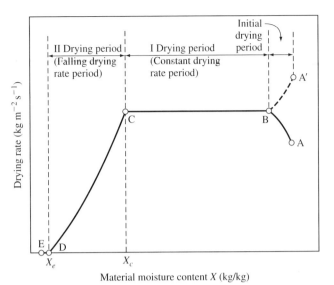

Figure 10.3 Drying rate curve, illustrating the periods of drying.

The mass transfer of water being evaporated at the solid surface by gas flowing past the surface can be defined in terms of a mass transfer coefficient k_G:

$$N_w = k_G \Delta p_w \qquad (10.2.11)$$

where N_w = molar flux of water, and Δp_w = difference in partial pressure of water between the surface and the bulk gas stream [7].

This equation can also be put in terms of concentrations by using Equation (10.2.1). Thus, when the partial pressure of water at the surface starts to drop, the critical moisture content X_c has been reached and the period of falling drying rate begins. During the constant drying rate period, the steady-state relationship between heat and mass transfer at the liquid surface is

$$q = \lambda N_w \qquad (10.2.12)$$

where q is the heat flux and λ is the heat of vaporization of water. Combining Equations (10.2.8) and (10.2.12) leads to

$$N_w = \frac{h(T - T_s)}{\lambda} \qquad (10.2.13)$$

This equation allows us to estimate the evaporation rate if we know the heat transfer coefficient and the relevant temperatures. This heat transfer coefficient has been found to vary from 10 to 100 kcal m^{-2} h^{-1} °C^{-1} for forced convection of gases [7].

The temperature at the surface of a moist solid that is undergoing drying in the constant drying rate period is usually very nearly at the *wet-bulb temperature*, which is defined as the steady state temperature of a small mass of water that is evaporating into a continuous stream of humid air [2]. The wet-bulb temperature is very nearly equal to the adiabatic

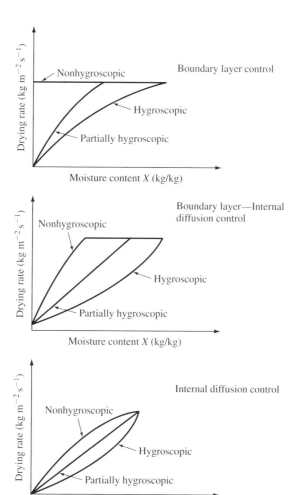

Figure 10.4 Drying rate curves for various types of materials and mass transfer conditions.

saturation temperature for air–water mixtures, which is shown in Figure 10.2 on the curve denoted 100% saturation.

During the falling rate drying period, the principal mass transfer mechanisms are *(1) liquid diffusion in continuous, homogeneous materials, (2) vapor diffusion in porous or granular materials, (3) capillary flow in porous or granular materials, (4) gravity flow in granular materials, and (5) flow caused by shrinkage-induced pressure gradients* [6]. A bed of antibiotic crystals, for example, is granular and porous, while a protein precipitate tends to be nonporous and colloidal. Vapor diffusion occurs in the dry solids when there is a shrinking interface between the dry solids and either wet solids (see Example 10.2) or frozen solids (where sublimation is occurring at the interface during freeze drying—see Section 10.3.3). One mechanism generally dominates at a given time in the drying cycle, although several mechanisms can occur together.

The liquid diffusion of water, either liquid or vapor, through a solid matrix undergoing drying can be described by Fick's second law of diffusion, written in terms of the moisture content X in mass of water per mass of dry solids:

$$\frac{\partial X}{\partial t} = \mathscr{D}\frac{\partial^2 X}{\partial y^2} \tag{10.2.14}$$

where t is time, $\mathscr{D}$ is the diffusion coefficient of water in the solid matrix, and y is the distance measured in the direction of diffusion. Because diffusion is much slower in the liquid phase than in the vapor phase, it is of interest to solve Fick's second law for the case of liquid diffusion in nonporous solids. An analytical solution of this equation has been obtained by making the following assumptions: (1) liquid diffusion is independent of moisture content (i.e., $\mathscr{D}$ is constant), (2) initial moisture distribution is uniform, and (3) material size, shape, and density are unchanging. For solids in the shape of a slab of thickness $2d$ with drying on both sides, with the following initial and boundary conditions:

$$\text{At } t = 0 \text{ and } -d < y < d, \qquad X = X_c \tag{10.2.15}$$
$$\text{At } t > 0 \text{ and } y = \pm d, \qquad X = X_e \tag{10.2.16}$$

where X_c and X_e are the critical and equilibrium moisture contents, respectively, the solution for the average moisture content is

$$\frac{X_m - X_e}{X_c - X_e} = \frac{8}{\pi^2}\left[\sum_{n=0}^{n=\infty}\frac{1}{(2n+1)^2}\exp\left[-(2n+1)^2\mathscr{D}t\left(\frac{\pi}{2d}\right)^2\right]\right] \tag{10.2.17}$$

where X_m is the average moisture content [9]. For long drying times when $\mathscr{D}t/d^2$ exceeds approximately 0.1,

$$\frac{X_m - X_e}{X_c - X_e} = \frac{8}{\pi^2}\exp\left[-\mathscr{D}t\left(\frac{\pi}{2d}\right)^2\right] \tag{10.2.18}$$

Differentiating Equation (10.2.18) with respect to time gives the following expression for the drying rate:

$$-\frac{dX_m}{dt} = \frac{\pi^2\mathscr{D}}{4d^2}(X_m - X_e) \tag{10.2.19}$$

Equation (10.2.18) can also be solved for t to obtain the time for the falling drying rate period:

$$t = \frac{4d^2}{\pi^2\mathscr{D}}\ln\left[\frac{8(X_c - X_e)}{\pi^2(X_m - X_e)}\right] \tag{10.2.20}$$

In actuality, the diffusion coefficient usually declines as the moisture content decreases, so an average value of $\mathscr{D}$ should be used over the range of moisture content during drying. Moisture diffusion coefficients have been measured for a large number of materials, especially foods. For example, the moisture diffusion coefficients of 24 different food materials were mostly in the range 10^{-8} to 10^{-4} cm^2/s for moisture contents ranging from 0.01 to 10 g/g (dry basis), respectively [10].

Another way to model the falling drying rate period is to assume that the rate of drying is proportional to the residual moisture content less the equilibrium moisture content, which approximates the falling rate period shown in Figure 10.3:

$$-\frac{dX_m}{dt} = K(X_m - X_e) \tag{10.2.21}$$

where K is a constant. Note that this equation has the same form as the analytical solution of the diffusion equation at long drying times [Equation (10.2.19)]. At the beginning of the falling rate period, X_m is at the critical moisture content X_c, which gives

$$-\frac{dX_m}{dt}\bigg|_{X_c} = K(X_c - X_e) \tag{10.2.22}$$

During the constant rate drying period, from Equation (10.2.13), we have

$$N_w = \frac{h(T - T_s)}{\lambda} = \frac{\rho_s d}{M}\left(-\frac{dX_m}{dt}\right)\bigg|_{X_c} \tag{10.2.23}$$

where ρ_s is the mass of dry solids per volume of wet solids and M is the molecular weight of water. Upon applying Equation (10.2.22), we find that Equation (10.2.23) leads to

$$K = \frac{hM(T - T_s)}{\rho_s d\lambda(X_c - X_e)} \tag{10.2.24}$$

so that during the falling drying rate period,

$$-\frac{dX_m}{dt} = \frac{hM(T - T_s)}{\rho_s d\lambda}\left(\frac{X_m - X_e}{X_c - X_e}\right) \tag{10.2.25}$$

This equation can be integrated with the following initial condition:

$$\text{At } t = 0, \qquad X_m = X_c \tag{10.2.26}$$

to give

$$t = \frac{\rho_s d\lambda(X_c - X_e)}{hM(T - T_s)}\ln\left(\frac{X_c - X_e}{X_m - X_e}\right) \tag{10.2.27}$$

For granular or porous solids, moisture tends to flow by the action of capillary forces rather than by diffusion. During the constant rate drying period, water evaporates out of the pores near the surface. The strength of the capillary forces in a pore increases as the curvature of the meniscus of the water in the pore increases; this curvature increases as the pore size decreases. Therefore, the small pores develop greater capillary forces than the larger ones and tend to pull water out of the large pores. The larger pores tend to empty first during the constant rate drying period, and the small pores are opened later during the falling rate period.

Shrinkage occurs when bound moisture is removed from colloidal nonporous solids. For the drying of thin layers, shrinkage does not usually cause much of a problem, but serious difficulties can result if the layer being dried becomes too thick. The outer layer next to the surface dries much more rapidly than the interior, and this can lead to cracking and warping of the outer layer. This problem is accentuated by the decrease in the diffusion coefficient of water as the moisture drops in the outer layer. For some cases, depending on the nature of the solids, the shrinkage and decline in the diffusion coefficient of water can be so great that a skin is formed on the surface that becomes impenetrable to water, and drying ceases as a result. This phenomenon is called case hardening [2].

EXAMPLE 10.3

Mass Flux During the Constant Rate Drying Period in Convective Drying   Wet biological solids contained in a tray are dried by blowing air with 2% relative humidity and at 60°C and atmospheric pressure across the tray. For the constant drying rate period, estimate the temperature at the surface of the solids and the maximum molar flux of water.

SOLUTION

From Figure 10.2, for air with 2% relative humidity and at 60°C,

$$T_{100\%} = 23°C$$

This temperature, found by moving up the adiabatic cooling curve to 100% relative humidity, is a good approximation of the surface temperature of the solids and of the wet-bulb temperature. At atmospheric pressure, we have

$$\lambda = 539.1 \frac{\text{kcal}}{\text{kg}}$$

From Equation (10.2.13),

$$N_w = \frac{h(T - T_s)}{\lambda}$$

The maximum molar flux would correspond to a forced convection heat transfer coefficient h of approximately 100 kcal m^{-2} h^{-1} °C^{-1} (upper limit of the range for forced convection heat transfer coefficients of gases). Thus,

$$(N_w)_{\text{max}} = \frac{100 \dfrac{\text{kcal}}{\text{m}^2\,\text{h}\,°\text{C}} \times (60 - 23)°\text{C}}{539.1 \dfrac{\text{kcal}}{\text{kg}} \times \dfrac{18\,\text{kg}}{\text{kg-mol}}} = 0.38 \frac{\text{kg-mol}}{\text{m}^2\,\text{h}}$$

EXAMPLE 10.4

Time to Dry Nonporous Biological Solids by Convective Drying   A nonporous wet cake of biological solids that is 0.5 cm thick is to be dried by blowing dry air across the top surface. The initial moisture content is 70 wt% water, and it is desired to dry the cake to a water content of 5 wt%. The diffusion coefficient of water in the cake has been estimated to be 7×10^{-6} cm^2 s^{-1}. Estimate how long the drying will take.

SOLUTION

Because the cake is nonporous, we assume that internal diffusion is controlling the drying. According to Figure 10.4 for drying with internal diffusion control, there will be no constant drying rate period, which means that the initial water content is equal to the critical moisture content X_c. Because the drying is by dry air, the water content at equilibrium (X_e) is assumed to be zero. Calculating X_c and X_m on the basis of the dry mass of solids gives

$$X_c = \frac{70}{30} = 2.333 \ \frac{\text{g water}}{\text{g dry solids}}$$

$$X_m = \frac{5}{95} = 0.053 \ \frac{\text{g water}}{\text{g dry solids}}$$

We tentatively assume that $\mathscr{D}t/d^2 > 0.1$, so that we can use Equation (10.2.20) to estimate the drying time:

$$t = \frac{4d^2}{\pi^2 \mathscr{D}} \ln\left[\frac{8(X_c - X_e)}{\pi^2(X_m - X_e)}\right]$$

$$= \frac{4(0.5)^2 \, \text{cm}^2}{\pi^2\left(7 \times 10^{-6} \, \dfrac{\text{cm}^2}{\text{s}}\right)} \ln\left[\frac{8(2.333 - 0)}{\pi^2(0.053 - 0)}\right]\frac{1 \, \text{h}}{3600 \, \text{s}} = 14.4 \, \text{h}$$

As a check, we see that

$$\frac{\mathscr{D}t}{d^2} = \frac{7 \times 10^{-6} \, \dfrac{\text{cm}^2}{\text{s}} \times 14.4(3600) \, \text{s}}{0.5^2 \, \text{cm}^2} = 1.45$$

which satisfies the assumption that $\mathscr{D}t/d^2 > 0.1$.

10.3 Dryer Description and Operation

Several types of dryers have been used to dry biological materials. However, the great majority of dryers used for these products fall into four categories: vacuum-shelf dryers, batch vacuum rotary dryers, freeze dryers, and spray dryers. Our basic description of each of these dryers will discuss the dryer's operation, as well.

10.3.1 VACUUM-SHELF DRYERS

The vacuum-shelf dryer is a relatively simple type of dryer, as shown in Figure 10.5. Trays filled with the product to be dried rest on shelves through which warm water or other suitable heat exchange medium is circulated. Heat is conducted from the shelves to the trays and into the wet solids. Vacuum is applied to the chamber containing the trays to speed up the drying and allow drying to take place at lower temperatures. The evaporating water

To vacuum

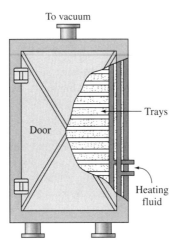

Figure 10.5 Vacuum-shelf dryer.

Trays

Door

Heating
fluid

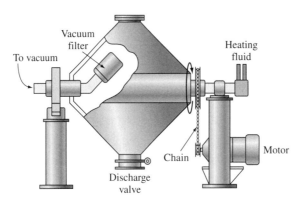

Figure 10.6 Batch vacuum rotary dryer. (Based on a design by Stokes Equipment Division, Penwalt Corporation.)

Vacuum
filter

Heating
fluid

To vacuum

Motor

Chain

Discharge
valve

vapor is drawn off in the vacuum system. Vacuum-shelf dryers up to several square meters of shelf area are available [11]. This type of dryer is used extensively for pharmaceutical products such as antibiotics, which are often in crystalline form and exhibit moderate to high heat sensitivity.

10.3.2 BATCH VACUUM ROTARY DRYERS

Another type of dryer in which the heat transfer is by conduction is the batch vacuum rotary dryer, also called a vacuum tumble dryer. Heat is supplied by warm water or other heat exchange medium circulated through a jacket on the rotating double-cone drum, shown in Figure 10.6. In this dryer, the solids are continually tumbled by rotation of the drum, so that solid particles come in contact with the walls of the jacket and with each other. Vacuum is applied to the rotating drum to be able to dry at lower temperatures and to dry more rapidly. The largest dryers of this type have volumes of up to 30 to 40 m^3 [11]. This type of dryer is generally applicable for biological products of the same type handled by the vacuum-shelf dryer, but it would not be feasible when the tumbling motion causes the particles to form larger and larger balls or when the particles stick to the metal surfaces in the drum.

10.3.3 FREEZE DRYERS

A freeze dryer is essentially a more sophisticated version of vacuum-shelf dryer. Freeze drying requires that both the temperature and pressure be controlled during the drying process. The product to be dried can be either in vials or in trays. The pharmaceutical freeze dryer shown schematically in Figure 10.7 can dry a liquid product contained in vials. When the vials are placed on the trays, the stoppers are closed only partially so that water vapor can escape. The hydraulic piston allows for the stoppers to be completely pushed into the top of the vials at the end of drying. A heat transfer fluid is circulated through the trays to provide temperature control of the vials.

The first step in the freeze-drying process is the cooling of the product to a sufficiently low temperature to allow complete solidification. The pressure in the chamber is then reduced to below the vapor pressure at the triple point of water so that drying can occur by sublimation. Ice, water, and water vapor exist in equilibrium at the triple point (0.01 °C and 4.6 mm Hg); below the triple point, water passes directly from the solid to the vapor phase by sublimation when the temperature is raised. Thus, the temperature of the shelves is increased to provide energy for sublimation. As drying occurs, a boundary between the dry solids and frozen solution can be observed in each vial (Figure 10.8), similar to the receding front in tray drying modeled in Example 10.2. Unbound water is removed in a drying phase called *primary drying*. A higher shelf temperature and additional time are required to remove the bound water in the *secondary drying phase*. Typical plots of temperature and pressure as a function of time during the freezing, primary drying, and secondary drying phases are shown in Figure 10.9. The time to complete the drying cycle is typically 24 to 48 h [12].

During the freeze-drying process, it is important that the product not exceed either the eutectic temperature or the glass transition temperature; otherwise the product can

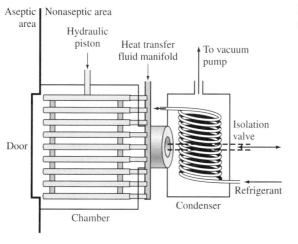

Figure 10.7 Pharmaceutical freeze dryer.

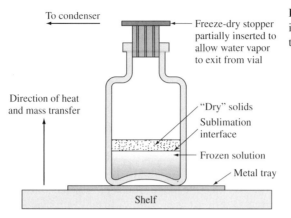

Figure 10.8 Schematic of freeze-drying in a vial, showing the boundary between the frozen solution and the dry solids.

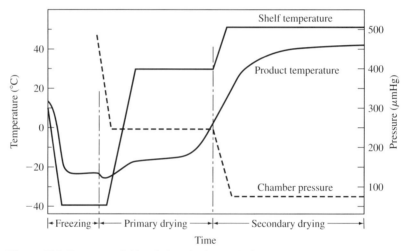

Figure 10.9 Process variables during the freeze-drying cycle.

"collapse" [13]. Collapse is the loss of either the crystalline or the amorphous structure of the product. The eutectic temperature is the temperature in crystalline systems below which no liquid exists. The glass transition temperature exists only in amorphous systems and is the temperature at which there is a change in viscosity of the system from a viscous liquid to a glass. Both the eutectic temperature and the glass transition temperature can be determined by differential thermal analysis (DTA).

Freeze drying is generally used only for biological products that have a high sensitivity to heat. Most proteins fall into this category. Because the liquid phase is not present after the product is initially frozen, reactions that can occur to the product in the liquid phase such as hydrolysis, cross-linking, oxidation, aggregation, and disulfide rearrangement are essentially eliminated [14]. Freeze drying is also advantageous in that the dry product has a very high specific surface area, which leads to a rapid and complete redissolution [13].

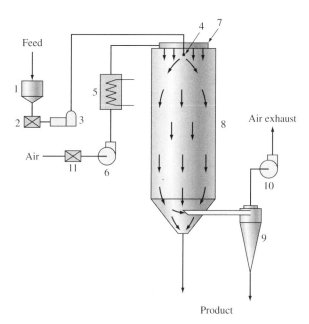

Feed

Air exhaust

Air

Product

Figure 10.10 Spray dryer with a pressure nozzle atomizer: 1, feed tank; 2, filter; 3, pump; 4, atomizer; 5, air heater; 6, fan; 7, air disperser; 8, drying chamber; 9, cyclone; 10, exhaust fan; 11, filter.

10.3.4 SPRAY DRYERS

A spray dryer transforms a feed in the liquid state into a dried particulate form by spraying the liquid into a hot gas, usually air. A schematic diagram of a spray dryer is shown in Figure 10.10. The majority of spray dryers in commercial use utilize cocurrent flow of gas and feed. Three basic unit processes are involved in spray drying: liquid atomization, gas–droplet mixing, and drying from liquid droplets. Spherical particles are produced, which are not usually obtainable by any other drying method. These particles can be either solid or hollow and range in size from 2 μm to (rarely) more than 500 μm. Drying is carried out essentially at the air wet-bulb temperature, and the time of drying is measured in seconds. Inlet gas temperatures can range from 150 to 250°C. The rapidity of drying and the relatively low temperature experienced by the solids during drying have led to the successful application of spray drying for many types of heat-sensitive products, including milk, coffee, blood, spores and various microorganisms, and antibiotics [15].

The most important operation in the spray-drying process is atomization. The type of atomizer determines the size and size distribution of the drops and their trajectory and speed. Three general types of atomizer are available. Rotary wheel atomizers (also called centrifugal disk atomizers) and pressure nozzle single-fluid atomizers are the most commonly used. Pneumatic two-fluid nozzles are used only rarely in specialized applications. For example, skim milk, milk, and whole eggs have been spray-dried with both rotary wheel and pressure nozzle atomizers [16].

The rotary wheel atomizer is generally preferred for high flow rate feed streams (i.e., >5 metric tons/h). The particles produced are relatively small (30–120 μm). This atomizer can handle a wide variety of streams with different physical properties, and its large flow ports result in a negligible tendency to clog [11, 16].

The pressure nozzle atomizer produces larger particles than the rotary wheel atomizer (e.g., 120–250 μm), and it is generally able to produce drops within a narrow range of diameters. This atomizer is not recommended for highly concentrated suspensions and abrasive materials which tend to clog and erode the orifice nozzle [11, 16].

10.4 Scaleup and Design of Drying Systems

It is not yet possible to design with confidence a large-scale dryer for a biological material based only on the physical properties of the material. Theory and empirical correlations are helpful in estimating dryer size and drying time and are also helpful in analyzing dryer performance and in predicting the effect of changes in operating conditions. Small-scale drying tests in the pilot plant and/or laboratory are an essential element of dryer scaleup and design. It is often possible to ship samples of the material to be dried to the drying equipment manufacturer for tests to be performed in the manufacturer's small-scale unit. If these tests are successful, then it is also a good idea to test the small-scale unit where the biological material is being produced. This will enable the testing of the drying for numerous batches of the material, which for biological materials can be significantly variable, and over a longer period of time [11].

10.4.1 VACUUM-SHELF DRYERS

The scaleup and sizing of large scale vacuum-shelf dryers is relatively straightforward. One good procedure is to test the drying in a bench or pilot scale dryer using the same depth of wet solids that would be used in the large-scale unit. The effect of changes in conditions upon scaleup can be estimated using the equation for the time of conductive drying determined in Example 10.2:

$$t = \frac{d^2 \lambda \rho_0}{2k(T_0 - T_s)} \tag{10.4.1}$$

The thermal efficiency of vacuum-shelf dryers is usually between 60 and 80%. The power to operate these dryers can be estimated to a good approximation based only on the power to operate the vacuum system, which is 0.06 to 0.12 kW/m^2 tray surface area for vacuums of 680 to 735 mm Hg [1].

10.4.2 BATCH VACUUM ROTARY DRYERS

Upon scaling up a batch vacuum rotary dryer, we find that the heated surface area per internal volume does not remain constant. For example, if we assume that the dryer is a perfect double cone, the heated area increases by a factor of 4.6 when the volume increases by a factor of 10 for geometrical similarity (angles constant) upon scaleup. This means that the ratio of area to volume at the large scale is 46% of the area-to-volume ratio of the small-scale unit with one-tenth the volume. By making the assumption that the time for drying is inversely proportional to the area-to-volume ratio of the dryer, the time for drying in the large-scale dryer can be estimated from the time in the small-scale one [17]:

$$\frac{t_l}{t_s} = \frac{(A/V)_s}{(A/V)_l}$$ (10.4.2)

where A is the heated area, V is the internal volume, and the subscripts s and l refer to the small- and large-scale dryers, respectively. Batch vacuum rotary dryers are usually filled to 50 to 65% of their internal volume [1].

10.4.3 FREEZE DRYERS

The design of a freeze-drying process usually requires a significant amount of development effort. For this work, it is essential to have available a pilot plant scale freeze dryer with a product shelf area of between 0.4 and 2.0 m^2 and layout and characteristics (shelf and ice condenser areas, temperatures, and pressures) similar to those of the expected large-scale production equipment [12]. The two main activities during this development work are the determination of the final product formulation and the determination of the conditions for the freeze-drying cycle.

To formulate the final product, excipients are added to the active component of the product for use as fillers or bulking agents, buffers to control pH, tonicity modifiers for osmotic pressure adjustment, structure modifiers, stabilizers to protect the active molecule, and collapse inhibitors. Commonly used excipients are mannitol, glycine, lactose, sucrose, glucose, and dextran [12].

As is indicated in Figure 10.9, the three major cycles of the freeze-drying process are freezing, primary drying, and secondary drying. The following sequence of testing is typical for determining the design conditions for freeze drying:

1. The maximum allowable temperature during primary drying is determined, typically by differential scanning calorimetry (DSC). Transition temperatures such as the eutectic temperature and the glass transition temperature can be found by means of DSC. During primary drying, the process should be designed to maintain the product temperature 3 to 4°C below the eutectic and glass transition temperatures, so that no liquid is present and drying is by sublimation only [13].

2. Chamber pressure, shelf temperature, and time for primary drying are determined. The chamber pressure should be significantly less than the vapor pressure of ice at the target product temperature; a good rule of thumb is to select a chamber pressure of about 20% of the vapor pressure of ice at the target product temperature. The shelf temperature is determined by trial and error in pilot tests, with the objective being to maintain the product temperature constant. A rough rule of thumb is to use an initial temperature difference between the shelf and the vials containing the product of 100 times the vapor pressure of ice in millibars at the target temperature (1 mm Hg = 1.33 mbar). The time for primary drying varies approximately with the initial depth of the frozen solid according to the relationship [12]

$$t = Kd^{1.5}$$ (10.4.3)

where K is a constant. Note that the exponent of d is similar, but not identical, to the exponent found in the theoretical equation for vacuum-shelf dryers [Equation (10.4.1)].

3. Chamber pressure and shelf temperature to use for secondary drying are determined. The start of secondary drying is indicated by a sharp increase in product temperature (see Figure 10.9). Drying rate is independent of pressure up to about 2.5×10^{-1} mbar, and a pressure of about 60% of this pressure (1.5×10^{-1} mbar) is often good for secondary drying. The maximum target product temperature depends on the specific product, with 24 to 30°C often used for proteins and up to 40°C for nonproteins. Some proteins and other biological materials can be damaged by overdrying, and for these cases it is recommended that the shelf temperature be maintained at about 15°C and samples tested for moisture content at various times during secondary drying [12].

The primary and secondary freeze-drying steps have been mathematically modeled [18]. The modeling is complex, however, and numerical methods are required for solutions.

10.4.4 SPRAY DRYERS

For the scaleup and design of a large-scale spray dryer, tests in small-sized equipment need to be performed to determine dryer conditions that will cause the largest particles produced to be nonsticky and at the desired moisture level. For biological materials, it is necessary to determine the degree of dryness and operating conditions that will give an active and stable product.

In the sizing of spray dryers, a key variable is the *residence time* of the air in the drying chamber. Residence time is calculated by dividing the chamber volume by the total airflow rate. The minimum residence time of the product can be assumed to be the residence time of the air. However, most of the product passing through the dryer has a residence time much higher than the air residence time. For heat-sensitive products such as biological materials, average air residence times are 35 s or more for drying coarse sprays (mean size 200–275 μm) and are in the range of 20–35 s for drying fine to semicoarse sprays (mean size < 180 μm) [15].

Once the volume of the drying chamber has been determined based on the air residence time, the dimensions of the chamber can be calculated. For spray dryers with cocurrent flow of feed and air, $H:D$, the ratio of cylindrical height to chamber diameter, falls into the following ranges: 0.6:1 to 1:1 for units with rotary wheel atomizers, and 3:1 to 4:1 for units with pressure nozzle atomizers. Because the trajectory of drops from rotary wheel atomizers is horizontal, spray dryers with these atomizers must be large enough in diameter to prevent wall encrustations caused by large drops hitting the chamber wall. The minimum chamber diameter for spray dryers with rotary atomizers should be as follows depending on the mean droplet size [15]:

1 m for 40–80 μm drops

4 m for 80–100 μm drops

5 m for 100–120 μm drops

An alternative way to scale up is to use theoretical correlations, a good example of which was provided by Gluckert [19]. The correlations developed by Gluckert rely on several key assumptions:

1. Because of the large recirculation currents in the dryer, drying conditions are uniform throughout the chamber. The entire chamber is at the gas exit temperature. This assumption has been established except near the atomizer.

2. Drying occurs from the drop surface, which consists of a liquid saturated with the solutes; experimental studies with drops have shown this to be the case. Thus in view of the prior assumption, the temperature-driving force for drying is the difference between the drying gas outlet temperature and the adiabatic saturation temperature of a saturated solution of the material being dried.

3. The convective heat transfer coefficient h can be determined by assuming that a representative drop is in a motionless fluid, with the drop moving at the same speed as the air. For this situation the Nusselt number for heat transfer (hD/k_f, where D is the drop diameter and the subscript f refers to the film around the drop) is 2. This assumption has been substantiated by considerable experimental work.

All these assumptions are helpful in understanding how spray dryers operate. For a spray dryer with pressure nozzle atomizers, for example, the chamber volume V of the large-scale unit can be calculated from the following theoretical correlation:

$$Q = \frac{10.98 k_f V^{2/3}(T - T_s)_{\text{out}}}{D_m^2} D_s \sqrt{\frac{\rho_t}{\rho_m}} \tag{10.4.4}$$

where Q = rate of heat transfer to spray (Btu/h)
k_f = thermal conductivity of gas film surrounding drop (Btu/h ft °F)
V = volume of drying chamber (ft^3)
T = gas phase bulk temperature (°F)
T_s = temperature at drop surface (°F)
D_m = maximum drop diameter, assumed to be three times the surface
per unit volume average size (ft)
D_s = diameter of pressure nozzle discharge orifice (ft)
ρ_t = density of gas at exit conditions (lb/ft^3)
ρ_m = density of spray jet at the atomizer (lb/ft^3)

This heat transfer rate is the maximum rate that can be achieved if the largest particles from the atomizer are dried as they reach the dryer wall or bottom.

EXAMPLE 10.5

Sizing of a Spray Dryer   Estimate the dimensions of a drying chamber for a spray dryer that has an output of 1000 kg/h of a heat-sensitive biological material at 60°C containing 5% moisture and having a mean particle size of 100 μm. The feed contains 40% solids by weight in an aqueous solution at 4°C. The inlet air has a humidity of 0.01 kg/kg dry air and is at 150°C, while the outlet air is at 80°C. Assume the specific heat of the dry solids is 0.3 kcal kg^{-1} °C^{-1}.

SOLUTION

We first do material balances on the overall process:

$$\text{Product rate} = 1000 \ \frac{\text{kg}}{\text{h}} \quad (5\% \text{ water})$$

$$\text{Solution feed rate} = \frac{1000(0.95)}{0.4} = 2375 \ \frac{\text{kg}}{\text{h}} \quad (40\% \text{ solids})$$

$$\text{Water evaporation rate} = 2375 - 1000 = 1375 \ \frac{\text{kg}}{\text{h}}$$

To size the dryer, we need to find the flow rate of wet gas out. Since we know the humidity of the inlet air on a dry air basis, we use the flow rate G of dry air to write a mass balance for the water:

$$G(\mathcal{H}_2 - 0.01) = 1375$$

or

$$G = \frac{1375}{\mathcal{H}_2 - 0.01} \ \frac{\text{kg dry air}}{\text{h}}$$

where G is the flow rate of dry air and $\mathcal{H}_2$ is the humidity of the outlet air. Using Equations (10.2.3), (10.2.4), and (10.2.5) and a reference temperature $T_0 = 0°\text{C}$ enables us to write

$$\text{Enthalpy of inlet air} = [0.24 + 0.46(0.01)]150 + 597(0.01) = 42.66 \ \frac{\text{kcal}}{\text{kg dry air}}$$

$$\text{Enthalpy of outlet air} = (0.24 + 0.46\mathcal{H}_2)80 + 597\mathcal{H}_2 = 19.2 + 633.8\mathcal{H}_2 \ \frac{\text{kcal}}{\text{kg dry air}}$$

Since we know the specific heat of the dry solids, we write the enthalpy of the solids and liquids based on the weight of dry solids:

$$\text{Enthalpy of entering feed} = 0.3(4) + 1.0(4)\left(\frac{60}{40}\right) = 7.2 \ \frac{\text{kcal}}{\text{kg dry solids}}$$

$$\text{Enthalpy of product leaving dryer} = 0.3(60) + 1.0(60)\left(\frac{5}{95}\right) = 21.2 \ \frac{\text{kcal}}{\text{kg dry solids}}$$

Equating the rate of enthalpy in and the enthalpy out gives the equation

$$42.66G + 7.2(950) = (19.2 + 633.8\mathcal{H}_2)G + 21.2(950)$$

Substituting into the foregoing equation for G in terms of $\mathcal{H}_2$ and solving gives

$$\mathcal{H}_2 = 0.0366 \ \frac{\text{kg water}}{\text{kg dry air}}$$

$$G = \frac{1375}{0.0366 - 0.01} = 51{,}690 \; \frac{\text{kg dry air}}{\text{h}}$$

$$\text{Rate of water vapor out} = 0.0366(51{,}690) = 1892 \; \frac{\text{kg water}}{\text{h}}$$

To calculate the flow rate of the wet air out, we need to know its density. We do this by first calculating the molar average molecular weight $\langle M \rangle$ of the wet air out to obtain

$$\langle M \rangle = 28.3$$

which assumes the molecular weight of dry air to be 28.9. From the ideal gas law at 1 atm and 80°C, the density of the wet air can be calculated, which can then be used to obtain the volumetric flow rate of wet air:

$$\rho_{\text{out}} = \frac{p\langle M \rangle}{RT} = \frac{1 \text{ atm} \times 28.3 \; \frac{\text{g}}{\text{g-mol}}}{82.06 \; \frac{\text{cm}^3 \text{ atm}}{\text{g-mol K}} \times 353 \text{ K}} = 0.000977 \; \frac{\text{g}}{\text{cm}^3} = 0.977 \; \frac{\text{kg}}{\text{m}^3}$$

$$\text{Total wet air flow rate} = \frac{(51{,}690 + 1892) \; \frac{\text{kg}}{\text{h}}}{0.977 \; \frac{\text{kg}}{\text{m}^3}} = 54{,}840 \; \frac{\text{m}^3}{\text{h}}$$

For the size of particles produced (mean size of 100 μm), the average air residence time should be in the range of 20 to 35 s. For a conservative design, we select a residence time of 35 s. This gives a spray drying chamber volume V of

$$V = 54{,}840 \; \frac{\text{m}^3}{\text{h}} \times 35 \text{ s} \times \frac{1 \text{ h}}{3600 \text{ s}} = 533 \text{ m}^3$$

A rotary wheel atomizer is appropriate for a mean particle size of 100 μm. We select a chamber cylindrical height (H) to diameter (D) ratio of 0.8:1, which falls in the 0.6:1 to 1:1 range appropriate for rotary wheel atomizers:

$$\frac{H}{D} = 0.8$$

Height and diameter are related to chamber volume by the equation

$$\frac{\pi D^2}{4} H = 533$$

Solving the two preceding equations for H and D, we obtain

$$H = 7.58 \text{ m}$$
$$D = 9.47 \text{ m}$$

This diameter is larger than the minimum diameter of 5 m recommended for particles with a mean size of 100 to 120 μm.

10.5 Summary

Drying is the process of thermally removing volatile substances (often water) to yield a solid. The drying of biological solids depends on the removal of free water and bound water and must account for the lability of biological activity as a function of temperature.

- The moisture content of a solid in equilibrium with air at a specified relative humidity is given in an equilibrium diagram for some biological solids.

- The psychrometric moisture chart relates moisture content, temperature, and relative humidity of air.

- The time to dry wet solids that are granular, such as crystalline material, by conductive drying on a heated surface is directly proportional to the square of the depth of the solids and inversely proportional to the difference between the heated surface temperature and the boiling point of the evaporating liquid.

- Drying rate curves are divided into initial, changing drying rate (warming up), constant drying rate (surface of solids is saturated), and falling drying rate periods (nonsaturated solids due to moisture content below critical value X_c).

- Drying rate curves have different shapes depending on the type of material being dried (hygroscopic, nonhygroscopic) and on the control of mass transfer (boundary layer, internal diffusion).

- The drying time for the falling rate period for a nonporous solid (continuous and homogeneous) depends on the square of the half-thickness and on the log of the ratio $(X_c - X_e)/(X_m - X_e)$, where X_m is the average moisture content at any time and X_c and X_e are the critical and equilibrium moisture contents, respectively. Moisture content data as a function of time can be used to estimate the moisture liquid diffusion coefficient and the convective heat transfer coefficient.

- *Vacuum-shelf dryers* are commonly used for trays or vials of pharmaceutical products, such as antibiotics; heat is conducted from the shelves, and water is removed by the vacuum system. *Vacuum rotary dryers* tumble the solids, which are heated by conduction of heat from the walls of the jacket of the dryer; solids aggregation sometimes occurs. In *freeze dryers* both temperature and pressure are controlled so that the sample freezes, and water sublimes into the vapor phase for removal by vacuum while avoiding eutectic and glass transition temperatures of the sample. Freeze drying is popular for proteins, which are heat sensitive. *Spray dryers* disperse liquid solution in fine droplets (atomization) into a large volume of hot air, which drives evaporation from droplets leaving solids; spray drying has successfully dried many heat-sensitive products.

- Theory, empirical correlations, and experimental testing are used to scale up drying operations. Theory and empirical correlations are helpful in estimating dryer size and drying time and are also helpful in analyzing dryer performance and in predicting the effect of changes in operating conditions. Small-scale drying tests in the pilot plant and/or laboratory are an essential element of dryer scaleup and design.

NOMENCLATURE

a heat transfer area per unit dryer volume (m^{-1})

A cross-sectional area (m^2)

c_w concentration of water vapor (mol liter^{-1})

C_{pg} specific heat of gas (kcal kg^{-1} °C^{-1})

C_{pw} specific heat of water vapor (kcal kg^{-1} °C^{-1})

C_s specific heat of gas and water vapor mixture (kcal kg^{-1} °C^{-1})

d one-half slab thickness for drying from both sides, or total thickness for drying from one side (m)

D diameter of spray dryer chamber (m)

D_m maximum drop diameter [Equation (10.4.4)] (ft)

D_s diameter of pressure nozzle discharge orifice [Equation (10.4.4)] (ft)

$\mathscr{D}$ diffusion coefficient (cm^2 s^{-1})

h convective heat transfer coefficient (kcal m^{-2} h^{-1} °C^{-1})

H height of spray drying chamber (m)

$\mathscr{H}$ humidity, mass of water per mass of dry air (dimensionless)

H_y total enthalpy (kcal kg^{-1})

k thermal conductivity (kcal m^{-1} h^{-1} K^{-1})

M molecular weight (Daltons)

n 0, 1, 2, . . ., ∞ [Equation (10.2.17)]

N_w molar flux of water (mol m^{-2} h^{-1})

p total pressure (atm)

p_w partial pressure of water (atm)

p_{ws} partial pressure of water at saturation (atm)

q heat flux (kcal m^{-2} h^{-1})

Q rate of heat flow (kcal h^{-1})

R gas law constant (0.082058 liter atm g-mol^{-1} K^{-1})

R_m relative humidity (dimensionless)

t time (s)

T temperature (K)

U overall heat transfer coefficient (kcal m^{-2} h^{-1} °C^{-1})

V dryer volume (m^3)

X moisture content, mass of water per mass of dry solids (dimensionless)

X_c critical moisture content, when the drying rate starts to fall (dimensionless)

X_e equilibrium moisture content (dimensionless)

X_m average moisture content (dimensionless)

Greek Letters

λ_0 latent heat, or heat of vaporization, of liquid (kcal kg^{-1}, or kcal mol^{-1})

ρ_m density of spray jet at atomizer [Equation (10.4.4)] (lb ft^{-3})

ρ_s mass of dry solids per volume of wet solids (g cm^{-3})

ρ_t density of gas at spray dryer exit [Equation (10.4.4)] (lb ft^{-3})

ρ_0 mass of water per volume of wet solids (g cm^{-3})

PROBLEMS

10.1   **Drying of Insulin**   Wet insulin crystals containing 32 g water per 100 g of dry insulin need to be dried in air to a moisture level of 5 g water per 100 g of dry insulin. Determine the percentage of bound and unbound water in the wet crystals before drying and the humidity of the air to accomplish the drying. For drying with air at 20°C, what should be the moisture content of the air (g moisture/g dry air)?

10.2   **Drying of Wet Antibiotic Crystals in a Vacuum-Shelf Dryer**   Small-scale drying tests were performed in a vacuum-shelf dryer with wet antibiotic crystals contained in a tray at a solids depth of 2 cm. Water at 70°C was circulating through the shelves, and the absolute pressure was maintained at 100 mm Hg. The wet crystals initially had a water content of 70%. The solids reached dryness after 10 h of drying.

It is proposed to dry these crystals in a large-scale dryer at a bed depth of 2.5 cm and using water circulating through the shelves at 70°C. The plant dryer can be operated at an absolute pressure no lower than 150 mm Hg. Estimate how long it will take to reach dryness in the large dryer. For changes in what variable is the estimated drying time the most sensitive?

TABLE P10.3

Material 1		Material 2		Material 3	
Moisture content (kg/kg)	Drying rate $\times 10^4$ (kg m^{-2} s^{-1})	Moisture content (kg/kg)	Drying rate $\times 10^4$ (kg m^{-2} s^{-1})	Moisture content (kg/kg)	Drying rate $\times 10^4$ (kg m^{-2} s^{-1})
0.06	0.3	0.03	0.5	0.02	3.0
0.13	0.9	0.07	1.6	0.06	5.5
0.21	1.5	0.12	2.7	0.11	6.7
0.26	2.4	0.18	3.8	0.16	7.7
0.31	3.3	0.22	4.8	0.20	8.3
0.35	4.4	0.26	5.7	0.23	8.6
0.38	5.5	0.29	6.4	0.27	8.6
0.40	6.8	0.32	7.0	0.34	8.6
				0.38	8.6

10.3 Classification of Drying of Biological Materials Drying rate data have been obtained for three biological solids (see Table P10.3). Based on these data, classify the drying for each solid according to whether the following apply: boundary layer control, internal diffusion control, and nonhygroscopic, partially hygroscopic, or hygroscopic solid.

10.4 Estimation of Drying Time from Drying Rate Data Drying rate data for a biological material are known (see Table P10.4). Determine the drying time for drying the material from 25 to 4 wt% moisture content. The initial material weighs 150 kg, and the material has a surface area of 1 m^2 per 40 kg of dry material.

TABLE P10.4

Moisture content (kg/kg dry weight)	Drying rate $\times 10^4$ (kg m^{-2} s^{-1})
0.020	0.26
0.060	0.75
0.122	1.49
0.173	2.17
0.222	2.75
0.255	3.16
0.295	3.50
0.345	3.51
0.391	3.49
0.450	3.51

10.5 Convective Drying of Nonporous Biological Solids A 3 mm bed of wet nonporous biological solids was dried in a tray by blowing air with 20% relative humidity across the top of the bed at 70°C and atmospheric pressure. The data in Table P10.5 were obtained for the drying of these solids:

TABLE P10.5

t (min)	X_m, mass water/ mass dry solids
0	0.190
10	0.176
20	0.153
30	0.130
40	0.107
50	0.084
60	0.064
70	0.050
80	0.039
90	0.031
100	0.026
120	0.018
140	0.014
160	0.012

From the drying data, estimate the diffusion coefficient of water in the biological material and the heat transfer coefficient for the convective heat transfer. The wet material has a bulk density of 1.1 g/cm^3.

10.6 Scaleup of a Batch Rotary Vacuum Dryer Drying tests of a biological material were carried out in a pilot scale double-cone rotary vacuum dryer having a volume of 25 liters. The drying time was 8 h. Estimate the drying time and dryer dimensions using a production scale double-cone rotary vacuum dryer with a volume of 500 liters. Assume for both the small and large dryers that each cone has the shape of an equilateral triangle.

10.7 Freeze Drying of a Pharmaceutical Product The data in Table P10.7 were taken in a pilot plant freeze dryer for the primary drying time as a function of cake depth:

TABLE P10.7

Depth (cm)	Time (h)
0.30	1.1
0.50	3.0
0.75	5.8
1.00	10.8

For operation at the same conditions (temperature and chamber pressure), it is desired to use a cake depth of 1.5 cm. Estimate the drying time that would be expected.

10.8 Spray Drying of Yeast Cells A spray dryer with pressure nozzle atomizers has a diameter of 5 m and a height of 15 m. It is desired to spray dry a feed containing yeast cells at a concentration of 30% by weight and 15°C to dry solids at 60°C with 5% moisture and average particle size of yeast cell agglomerates of 150 μm. Tests in a laboratory scale spray dryer have indicated that the air inlet and outlet temperatures should be 160 and 75°C, respectively. The inlet air has a humidity of 0.02 kg per kg of dry air. Estimate the maximum feed rate of yeast cells and the inlet flow rate of air (dry basis). Assume the specific heat of the dry solids to be 0.3 kcal kg^{-1} °C^{-1}.

References

1. Moyers, C. G., and Baldwin, G. W. (1997). Psychrometry, evaporative cooling, and solids drying. In *Perry's Chemical Engineers' Handbook,* 7th ed., R. H. Perry, D. W. Green, and J. O. Maloney, eds., McGraw-Hill, New York, p. 12-1.
2. McCabe, W. L., Smith, J. C., and Harriott, P. (1993). *Unit Operations of Chemical Engineering,* 5th ed., McGraw-Hill, New York.
3. Stephenson, G. A., and Diseroad, B. A. (2000). Structural relationship and desolvation behavior of cromolyn, cefazolin, and fenoprofen sodium hydrates. *Int. J. Pharm.,* vol. 198, p. 167.
4. Hsu, C. C., Ward, C. A., Pearlman, R., Nguyen, H. M., Yeung, D. A., and Curley, J. G. (1991). Determining the optimum residual moisture in lyophilized protein pharmaceuticals. *Dev. Biol. Standard.,* vol. 74, p. 255.
5. Pikal, M. J., and Rigsbee, D. R. (1997). The stability of insulin in crystalline and amorphous solids: Observation of greater stability for the amorphous form. *Pharm. Res.,* vol. 14, p. 1379.
6. McCormick, P. Y. (1993). Drying. In *Kirk-Othmer Encyclopedia of Chemical Technology,* vol. 8, 4th ed., M. Howe-Grant, ed., Wiley-Interscience, New York, p. 475.
7. Bird, R. B., Stewart, W. E., and Lightfoot, E. N. (2002). *Transport Phenomena,* 2nd ed., Wiley, New York.
8. Belter, P. A., Cussler, E. L., and Hu, W.-S. (1988). *Bioseparations,* Wiley, New York.
9. Pakowski, Z., and Mujumdar, A. S. (1995). Basic process calculations in drying. In *Handbook of Industrial Drying,* vol. 1, A. S. Mujumdar, ed., Dekker, New York, p. 71.

10. Marinos-Kouris, D., and Maroulis, Z. B. (1995). Transport properties in the drying of solids. In *Handbook of Industrial Drying,* vol. 1, A. S. Mujumdar, ed., Dekker, New York, p. 113.

11. Van't Land, C. M. (1991). *Industrial Drying Equipment,* Dekker, New York.

12. Snowman, J. W. (1996). Lyophilization. In *Downstream Processing of Natural Products,* M. S. Verrall, ed., Wiley, New York, p. 275.

13. Gatlin, L. A., and Nail, S. L. (1994). Freeze drying: A practical overview. In *Protein Purification Process Engineering,* R. G. Harrison, ed., Dekker, New York, p. 317.

14. Franks, F. (1998). Freeze-drying of bioproducts: Putting principles into practice. *Eur. J. Pharmaceut. Biopharmaceut.,* vol. 45, p. 221.

15. Masters, K. (1991). *Spray Drying Handbook,* 5th ed., Wiley, New York.

16. Filkova, I., and Mujumdar, A. S. (1995). Industrial spray drying systems. In *Handbook of Industrial Drying,* vol. 1, A. S. Mujumdar, ed., Dekker, New York, p. 263.

17. Fischer, J. J. (1963). Drying in vacuum tumblers. *Ind. Eng. Chem.,* vol. 55, p. 18.

18. Liapis, A. I., and Bruttini, R. (1995). Freeze drying. In *Handbook of Industrial Drying,* vol. 1, A. S. Mujumdar, ed., Dekker, New York, p. 309.

19. Gluckert, F. A. (1962). A theoretical correlation of spray-dryer performance. *AIChE. J.,* vol. 8, p. 460.

CHAPTER 11

Bioprocess Design

This chapter teaches students and practicing engineers the fundamentals of bioprocess design with emphasis on bioseparation processes. It is an attempt to combine the information presented in earlier chapters for use in the context of integrated processes. The ultimate objective is to enable the reader to efficiently synthesize and evaluate integrated bioseparation processes.

Given a product and a desired annual production rate (plant throughput), bioprocess design endeavors to answer the following questions: What are the required amounts of raw materials and utilities? What is the required size of process equipment and supporting utilities? Can the product be produced in an existing facility or is a new plant required? What is the total capital investment? What is the manufacturing cost? What is the optimum batch size? How long does a single batch take? How much product can be generated per year? During the course of a batch, what is the demand for various resources (e.g., raw materials, labor, utilities)? What is the total amount of resources consumed? Which process steps or resources constitute bottlenecks? What changes can increase throughput? What is the environmental impact of the process (i.e., amount and type of waste materials)? Which design is the "best" among several plausible alternatives?

11.1 Instructional Objectives

After completing this chapter, the reader should be able to do the following:

- Initiate a process design and choose the appropriate sequencing of processes.
- Set up a process flowsheet using the unit procedure concept.
- Choose and operate a process simulator.
- Cost capital equipment and apply cost scaling laws.
- Perform a product cost-of-step calculation.
- Perform a cost-of-process calculation.
- Perform profitability analysis.

- Construct scheduling charts.
- Perform economic evaluations and environmental impact.
- Perform process sensitivity analyses.

11.2 Definitions and Background

Process design is the conceptual work done prior to building, expanding, or retrofitting a process plant. It consists of two main activities, *process synthesis* and *process analysis*. Process synthesis is the selection and arrangement of a set of unit operations (process steps) capable of producing the desired product at an acceptable cost and quality. Process analysis is the evaluation and comparison of different process synthesis solutions. In general, a synthesis step is usually followed by an analysis step, and the results of analysis determine the subsequent synthesis step. Process design and project economic evaluation require integration of knowledge from many different scientific and engineering disciplines and are carried out at various levels of detail. Table 11.1 presents a common classification of design and cost estimates and typical engineering cost for a $50 million plant.

Figure 11.1 presents the need for design estimates of various types during the life cycle of product development and commercialization. The trapezoidal shape of the graph represents the drastic reduction in product candidates as we move from feasibility studies to commercialization. In fact, the chances of commercialization at the research stage for a new product are only about 1 to 3%, at the development stage they are about 10 to 25%, and at the pilot plant stage they are about 40 to 60% [1].

Order-of-magnitude estimates are usually practiced by experienced engineers who have worked on similar projects in the past. They take minutes or hours to complete, but the error in the estimate can be as high as 50%. Most engineers employed by operating companies usually perform level 2 and 3 studies. Such studies take days or weeks to complete using appropriate computer aids. The main objective of such studies is to evaluate

TABLE 11.1
Types of Design Estimate and Their Cost and Approximate Accuracy for a $50 Million Plant[a]

Level	Type of Estimate	Error (%)	Cost ($1000)
1	Order-of-magnitude estimate (ratio estimate) based on similar previous cost data	≤ 50	
2	Project planning estimate (budget estimation) based on knowledge of major equipment items	≤ 30	20–40
3	Preliminary engineering (scope estimate) based on sufficient data to permit the estimate to be budgeted	≤ 25	150–450
4	Detailed engineering (capital approval stage) based on almost complete process data	≤ 15	1000–2000
5	Procurement and construction (contractor's estimate) based on complete engineering drawings, specifications, and site surveys	≤ 10	3500–7000

[a]See references 1 and 2.

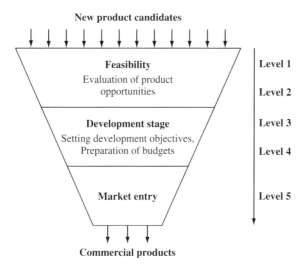

Figure 11.1 Types of design estimate during the life cycle of a product [2].

alternatives and pinpoint the most cost-sensitive areas—the economic "hot spots"—of a complex process. The results of such analyses are used to plan future research and development and to generate project budgets.

Level 4 and 5 studies are usually performed by the engineering and construction companies hired to build new plants for promising new products that are at an advanced stage of development. Such estimates are beyond the scope of this chapter. Instead, the focus of the material in the rest of this chapter will be on level 1, 2, and 3 studies. It should also be noted that opportunities for creative process design work are usually limited to preliminary studies. By the time detailed engineering work has been initiated, a process is more than 80% fixed. Furthermore, the vast majority of important decisions for capital expenditures and product commercialization are based on results of preliminary process design and cost analysis. This explains why it is so important for a new engineer to master the skills of preliminary process design and cost estimation.

Environmental impact assessment is an activity closely related to process design and cost estimation. Biochemical plants generate a wide range of liquid, solid, and gaseous waste streams that require treatment prior to discharge. The cost associated with waste treatment and disposal has skyrocketed in recent years owing to increasingly stricter environmental regulations. This cost can be reduced through minimization of waste generation at the source. However, generation of waste from a chemical or biochemical process is dependent on the process design and the manner in which the process is operated. Thus, reducing waste in an industrial process requires intimate knowledge of the process technology, in contrast to waste treatment, which essentially is an add-on at the end of the process. In addition, minimization of waste generation must be considered by process engineers at the early stages of process development. Once a process has undergone significant development, it is difficult and costly to make major changes. Furthermore, regulatory constraints that are unique to the pharmaceutical industry restrict process modifications after clinical efficacy of the drug has been established. These are only some of the reasons that process synthesis must be considered not only during, but before, the selection of unit operations for individual steps.

11.3 Synthesis of Bioseparation Processes

The development of a flowsheet for the recovery and purification of a biological product is a creative process that draws on the experience and imagination of the engineer. Attempts have been made to capture that experience on the computer in the form of expert systems [3–6] and automate to some extent the process synthesis tasks. Experienced engineers heavily rely on certain *rules of thumb,* also known as *heuristics,* for putting together the skeleton of a recovery and purification process. A few such heuristics follow:

1. Remove the most plentiful impurities first.
2. Remove the easiest-to-remove impurities first.
3. Make the most difficult and expensive separations last.
4. Select processes that make use of the greatest differences in the properties of the product and its impurities.
5. Select and sequence processes that exploit different separation driving forces.

Figure 11.2 provides a generalized structure for putting together an initial block diagram representation of a recovery process. For each product category (intracellular or extracellular) several branches exist in the main pathway. Selection among the branches and alternative unit operations is based on the properties of the product, the properties of the impurities, and the properties of the producing microorganisms, cells, or tissues. Bioprocess synthesis thus consists of sequencing steps according to the five heuristics and the structure of Figure 11.2. The majority of bioprocesses, especially those employed in the production of high value, low volume products operate in batch mode. Continuous bioseparation processes are utilized in the production of commodity biochemicals, such as organic acids and ethanol.

11.3.1 PRIMARY RECOVERY STAGES

Primary recovery comprises the first steps of downstream processing where some purification and broth volume reduction occurs. Primary recovery includes both the solids separation stage and parts of the product isolation stages discussed in Chapter 1 (Table 1.9). According to Figure 11.2, the selection of the first step depends on whether the product is intracellular (remains inside the microorganism after its expression) or extracellular (secreted into the solution). Almost all low molecular weight bioproducts are extracellular, as are many that have a high molecular weight. Recovery and purification is easier for these bioproducts than for intracellular products because of the lower amount of impurities present. Most recombinant eukaryotic proteins produced by prokaryotic microorganisms are intracellular products (see Chapter 3 for definitions of prokaryotic and eukaryotic cells). They accumulate inside the host cell in either native or denatured form; the denatured intracellular products often form insoluble inclusion bodies (IBs). A brief review of the most common primary recovery steps (described in Chapters 3, 4, 5, and 6) follows, and various rationales for unit operation selection are included.

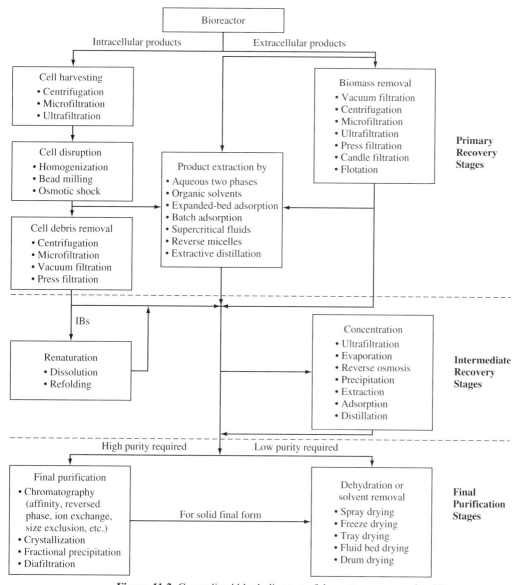

Figure 11.2 Generalized block diagram of downstream processing [5].

Intracellular Products

Cell Harvesting   The first purification step for intracellular products is cell harvesting. Removal of the extracellular liquid is in agreement with the first general heuristic: *Remove the most plentiful impurities first.*

As seen in Figure 11.2, centrifugation and membrane filtration (both micro- and ultrafiltration) are the only techniques used for large-scale cell harvesting. As explained in

Chapter 5, centrifugation has advantages for large and dense microorganisms (diameter >2 μm and density >1.03 g/cm^3). For instance, centrifugation is very efficient for harvesting yeast. For smaller microorganisms, various coagulation techniques can be used to increase the size of the settling particles (see Chapter 3). Membrane filtration has advantages for harvesting small and light cells. Another advantage of membrane filtration is in product recovery. Cell loss during centrifugation is typically 1 to 5%. However, with membrane filtration, essentially all cells are recovered unless there is cell disruption (lysis) or ripped membranes.

Cell Disruption   The second step for intracellular products is usually cell disruption, which serves to break open the host cells and release the intracellular product. The various options for cell disruption were presented in Chapter 3. Disruption of bacteria and yeast is carried out either by high pressure homogenizers or bead mills [7]. For large capacities (several cubic meters per hour) only high pressure homogenizers are practical. Osmotic shock is often used for release of periplasmic products that accumulate between the cell membrane and the cell wall.

Prior to disruption the concentrate is often diluted (by 5–10%) with a "lysis buffer" to create conditions that minimize product denaturation upon release from the cell. For hard-to-disrupt microorganisms, multiple homogenizer passes at 500 to 1000 bar are required. Multiple passes are also required if the product forms inclusion bodies, to allow the IBs to be released, and also to break the cell debris into very small particles, facilitating the separation of IBs from cell debris further downstream. Some product protein degradation occurs during cell disruption as a result of high shear at interfaces and oxidation.

Removal of Cell Debris   The cell debris that is generated by cell disruption is usually removed by centrifugation or microfiltration. Other options include rotary vacuum filtration, press filtration, depth filtration, extraction, and expanded-bed adsorption (EBA) chromatography.

Soluble Product. When the product is soluble, it is recovered during cell debris removal either in the light phase of a centrifuge or in the permeate stream of a filter. Centrifuges efficiently separate only fairly large particles of cell debris (>0.5 μm Stokes diameter). Therefore, when a centrifuge is used for cell debris removal, a polishing filtration step must follow to remove small debris particles that might otherwise cause severe problems in processes downstream such as chromatography. Filters of various types (e.g., depth, press, candle, rotary vacuum, membrane microfilters) can be used for polishing. Alternatively, these filters can be used for cell debris removal with no preceding centrifugation step. It is very difficult to predict a priori which filter will perform best for a specific product. When microfilters are used for cell debris removal, some degree of diafiltration is required to achieve an acceptable product recovery yield.

Insoluble Product. When the product is insoluble and forms inclusion bodies, it must first be separated from the cell debris particles, then dissolved and refolded (for a process of this type, see later: Section 11.6.2). Fortunately, inclusion bodies usually have a large diameter (0.3–1.0 μm) and high density (1.3–1.5 g/cm^3) [8] and can be separated from cell debris with a disk-stack centrifuge (Chapter 5). The inclusion bodies are recovered in the heavy phase of the centrifuge, while most cell debris particles remain in the light phase. The heavy phase is usually resuspended and recentrifuged two or three times to reach a high degree of inclusion body purity. Resuspension in a solution of a detergent and/or a low concentration of a chaotropic agent is often practiced to facilitate the removal of other contaminants. The pH and the ionic strength of the solution are adjusted to reduce the hydrophobicity of the cell debris particles and to enhance their removal in the light phase. Final product purity exceeding 70% is quite common.

Product Extraction/Adsorption   Product separation from cell debris can also be carried out by extraction and/or adsorption. Organic solvents are commonly used as extractants for low molecular weight products, such as various antibiotics. Aqueous two-phase systems have found applications for recovery of proteins. The criteria for extractant selection are as follows: the partition coefficient of the product should be higher than the partition coefficient of the contaminants; the extractant should not degrade the product; and the extractant should not be expensive and should be easy to recover or dispose of (see Chapter 6 for more detailed information on extraction).

In addition, product separation from debris and simultaneous concentration can be achieved by adsorptive techniques [9]. Adsorbents of various types (e.g., ion exchange, reversed phase, affinity) can be used. This type of purification requires the disrupted cells and product to be mixed in a stirred tank with an adsorbent. A washing step, where most of the cell debris particles and contaminants are washed out, follows product adsorption. More recently, expanded bed adsorption (EBA) chromatography has shown promise for separating proteins from cell debris particles [10]. The feed is pumped upward through an expanded bed. Target proteins are bound to the adsorbent while cell debris and other contaminants pass through. A washing step removes all weakly retained material. An elution step follows that releases and further purifies the product (see Chapter 7 for more detailed information on adsorption).

Extracellular Products

Biomass Removal   In agreement with the second generic heuristic (*Remove the easiest-to-remove impurities first*), biomass removal is usually the first step of downstream processing of extracellular products. This step can be accomplished by using one (or more) of the following unit operations: rotary vacuum filtration, disk-stack or decanter centrifugation, press filtration, microfiltration, ultrafiltration, and flotation. Since each unit operation has advantages and disadvantages for different products and microorganisms, the selection of the best unit operation(s) for a given system can be difficult.

Rotary Vacuum Filtration. Rotary vacuum filtration, especially with precoat, is the classical widely used method for removal of mycelial organisms [11]. Rotary vacuum filters can operate continuously for long periods of time (see Chapter 4). In addition, the filtrate flux in these units is usually higher than 200 liter m^{-2} h^{-1} and may reach 1000 liter m^{-2} h^{-1}. The most important disadvantage of this type of unit is the problem with the disposal of the mixture of filter aid and biomass. Filter aid is added in equal or higher amounts than biomass. Stringent environmental laws have made it costly to dispose of such solid materials. Therefore, if the disposal cost of filter aid is relatively high where a new plant is going to be built, alternative unit operations should be considered for biomass separation. However, if the disposal cost of filter aid is relatively low, a rotary vacuum filter is a good choice. The citric acid process, which is described later in this chapter, offers an example where rotary vacuum filtration is used for biomass removal.

Centrifugation. Disk-stack and decanter centrifuges are frequently used at large scale [12, 13]. Of the two, disk-stack centrifuges operate at higher rotational speeds and remove smaller and lighter microorganisms. However, with the use of flocculating agents, the decanter centrifuge performance improves, and choosing between the two types becomes more difficult. It appears that the only criterion being applied when disk-stack is chosen instead of decanter is the ability to remove small, light microorganisms. Centrifugation does not require filter aid, which is a significant advantage over rotary vacuum filtration. In general, the centrifuge paste contains 40 to 60% v/v extracellular liquid. To recover the product dissolved in that liquid, the paste is usually washed and recentrifuged.

Membrane Filtration. With membrane filters (micro- and ultrafilters), the extracellular product passes through the membrane while biomass and other particulate components remain in the concentrate. Concentration is usually followed by diafiltration to increase the product recovery yield (see Chapter 4 for more information on the mode of operation of membrane filters). Membrane filters are used for biomass removal mainly in recovery of low molecular weight products, such as antibiotics from mycelia. For high molecular weight products, applications are limited to cases where the amount of solids is rather small as in cell culture.

11.3.2 INTERMEDIATE RECOVERY STAGES

The primary recovery stages just described are followed by the intermediate stages, where the product is concentrated and further purified. Intermediate recovery has similarities to the product isolation stages discussed in Chapter 1 (Table 1.9). If the product is soluble, product concentration is usually the first step. If the product is denatured and insoluble, it is first dissolved and refolded and then concentrated and purified.

Product Concentration

After primary separation, the product is usually in a dilute solution. Volume reduction by concentration is in agreement with heuristics 1 and 2. Common concentration options include ultrafiltration, reverse osmosis, evaporation, adsorption, precipitation, extraction, and distillation.

Ultrafiltration   Ultrafiltration is used extensively for protein solution concentration. The molecular weight cutoff of the membrane is selected to retain the product while allowing undesirable impurities (mainly low molecular weight solutes) to pass through the membrane. The low operating temperature and the purification achieved along with concentration are some of the advantages of ultrafiltration over evaporation. The typical operating transmembrane pressure is 2 to 5 bar and the average flux is 20 to 50 liter m^{-2} h^{-1}.

Reverse Osmosis   Membranes with smaller pore sizes are used for reverse osmosis filters. The process of reverse osmosis may be used when concentrating medium to low molecular weight products (e.g., antibiotics, certain amino acids).

Evaporation   Thin-film rotating evaporators can operate at relatively low temperatures (40–50°C) under vacuum. These units compete in the market with ultrafiltration and reverse osmosis for concentrating both low and high molecular weight compounds. Unlike ultrafiltration, however, evaporation lacks the capability to provide purification during concentration. Advantages include the ability to concentrate to a higher final solids concentration and the ability to handle large throughputs [14].

Precipitation   Precipitation is often used for concentration and purification. Blood protein fractionation (see Chapter 8) and citric acid production (see later: Section 11.6.1) constitute typical applications. Addition of salts, solvents, and polymers and changes in pH, ionic strength, and temperature are commonly used to selectively precipitate compounds of interest [15]. Precipitation often follows an extraction carried out by a polymer/salt (e.g., PEG and potassium phosphate) aqueous two-phase system. When the product is recovered in the polymer-rich phase, precipitation is accomplished by addition of more polymer. It is important for economic reasons to recover and recycle the precipitating materials. Precipitation is also used to remove contaminants (i.e., nucleic acids) by adding $MnSO_4$ and streptomycin sulfate.

Distillation   The process of distillation is used for concentrating and purifying organic solvents, such as ethanol and acetic acid.

Product Renaturation

Eukaryotic proteins produced by prokaryotic microorganisms often form insoluble inclusion bodies in the host cell. Inclusion bodies can be dissolved rapidly by using solutions of strong chaotropes, such as 6 M guanidine hydrochloride or urea, in the presence of a reducing agent, such as 0.5 M 2-mercaptoethanol or 50 mM dithiothrietol [16]. The dissolved protein is then allowed to refold to its native conformation by removing the chaotropic agents through diafiltration, dilution, or chromatography, with final protein concentrations in the range of 10 to 100 mg/liter. Dilution is necessary for minimizing intermolecular interactions, which occur during product refolding and can lead to product inactivation. Addition of small amounts of thiols such as reduced glutathione (1–5 mM) and oxidized glutathione (0.01–0.5 mM) and incubation at 35 to 40°C for 5 to 10 h completes the refolding process. Thus, choosing an upstream process that forms IBs entails consideration of the large volumes, hence large waste streams, that are produced. More information on IB solubilization and protein refolding can be found in the insulin example (see later: Section 11.6.2) and also in "Solubilization and Refolding of Proteins in Inclusion Bodies" (Section 1.4.8).

11.3.3 FINAL PURIFICATION STAGES

The final purification steps are dependent on the required final product purity. Final purification includes both the purification and polishing stages discussed in Chapter 1 (Table 1.9). Pharmaceutical products require high purity, while industrial products require lower purity. For products of relatively low purity, such as laundry enzymes, the final purification step is dehydration or more generally a solvent removal step. For high purity products, the final purification stages usually involve a combination of chromatographic and filtration steps [17]. If the final product is required in solid form, then, a dehydration or solvent removal step follows.

Chromatography

Chromatography is typically done later in a process in agreement with the third generic heuristic (*Make the most difficult and expensive separations last*). With the preceding separation steps, a large fraction of contaminants is removed, thereby reducing the volume of material that needs to be treated further. In fact, a 50- to 100-fold volumetric reduction is quite common for high value biological products, resulting in a protein content of 1 to 5% w/v in the feed stream to chromatographic units.

Advances in expanded-bed adsorption chromatography promise to position chromatographic steps closer to the primary recovery stages [10]. As mentioned earlier, EBA chromatography units can be used to capture, concentrate, and purify product directly from fermentation broth that contains whole cells, cell debris and other particulate components. Consequently, EBA chromatography has the potential to eliminate some of the typical primary recovery steps (e.g., biomass and cell debris removal, product concentration).

A sequence of chromatographic steps is usually required to achieve the desired final product purity, and the fourth and fifth generic heuristics are good guides for selecting and sequencing such steps [18]. For instance, according to the fifth heuristic, an ion exchange step should not be followed by another step of the same type. Instead, it should be followed by a reversed phase, affinity, or any other chromatography step that takes advantage of a different separation driving force.

Membrane filtration steps are commonly employed between chromatographic steps to exchange buffers and concentrate the dilute product solutions. See Chapter 7 for detailed information on chromatographic separation methods and Chapter 4 for the intervening membrane filtration steps. The insulin and monoclonal antibody examples presented later in this chapter provide additional information on selection and operation of chromatographic separation units.

Crystallization and Fractional Precipitation

Crystallization and fractional precipitation can sometimes result in significant purification. Because these processes are cheaper to operate than chromatography, they should generally always be considered. The crystalline form of a bioproduct is especially advantageous, since the purity can be quite high and crystals can usually be stored for long periods of time. See Chapters 8 and 9 for a detailed discussion and analysis of precipitation and crystallization.

Dehydration or Solvent Removal

Dehydration or solvent removal is achieved with dryers. Spray, fluidized-bed, and tray dryers are used when products can withstand temperatures of 50 to 100°C. Freeze dryers are used for products that degrade at high temperatures. Freeze dryers require high capital expenditures and should be avoided if possible. See Chapter 10 for detailed information on product drying.

11.3.4 PAIRING OF UNIT OPERATIONS IN PROCESS SYNTHESIS

Besides using rules of thumb, or heuristics, for synthesizing bioseparation processes, it is often advantageous to consider how two unit operations can be paired to improve process efficiency. Some examples of operations that are logical to pair follow.

Extraction and Precipitation

The bioproduct is extracted with a solvent and then precipitated. To increase the yield, it is often desirable to concentrate the extract before the precipitation. The major hurdle to overcome for this pairing is to find a solvent that will work with both extraction and precipitation.

Precipitation and Hydrophobic Interaction Chromatography

The pairing of precipitation and hydrophobic interaction chromatography is usually accomplished for protein purification by using ammonium sulfate to precipitate impurities, leaving the desired bioproduct in the mother liquor. The ammonium sulfate is added to a concentration just below that needed to precipitate the bioproduct. After removal of precipitated impurities, the mother liquor can be applied directly to a hydrophobic interaction chromatography column, which was equilibrated to the concentration of ammonium sulfate in the mother liquor prior to the loading. The bioproduct adsorbs to the column under these conditions. The column is eluted with a reverse gradient of ammonium sulfate, and the desired bioproduct is recovered in a fraction from the elution.

Filtration and Extraction

When the bioproduct is contained in the filtrate after filtration, then it can often be extracted with an immiscible solvent. For the extraction of small molecules such as antibiotics with organic solvents, the pH must usually be adjusted to obtain the bioproduct in either its free base or free acid form so it will partition into the organic phase. For the aqueous two-phase extraction of proteins, two polymers or a salt and a polymer must be added. If the additions to the filtrate can be made on line, the filtration and extraction steps can be carried out simultaneously, resulting in a saving of processing time.

11.4 Process Analysis

The flowsheets put together during process synthesis must be analyzed and compared on the basis of capital investment, manufacturing cost, environmental impact, and other criteria to decide which ideas to consider further. Methodologies for estimating capital investment and manufacturing cost are presented in the next section of this chapter. In both cases estimation is based on the results of material and energy balances and equipment sizing. Although these basic chemical engineering calculations can be done on paper, it is highly desirable to use more sophisticated computer aids, such as spreadsheets or process simulators. Use of computer tools allows the process design team to quickly and accurately redo the entire series of calculations for a different set of assumptions and other input data.

11.4.1 SPREADSHEETS

Spreadsheet applications, such as Microsoft Excel, Lotus 1-2-3, and Corel Quattro Pro have become as easy to use as word processors and graphics packages. In its simplest form, a spreadsheet is an electronic piece of paper with empty boxes, known as cells. The user can enter data in those cells, perform calculations, and generate results. Results from spreadsheets can be easily plotted in a variety of graphs.

11.4.2 PROCESS SIMULATORS

Process simulators are software tools that enable the user to readily represent and analyze integrated processes. They have been in use in the petrochemical industries since the early 1960s. Established simulators for the petrochemical industries include Aspen Plus (from Aspen Technology, Inc.), ChemCAD (from Chemstations, Inc.), HYSYS (from Hyprotech, Ltd./AEA Engineering Software), and PRO/II (from Simulation Sciences, Inc.).

Development of simulators specific to biochemical processes began in the mid-1980s. BioProcess Simulator (BPS) (from Aspen Technology, Inc.) was the first tool of this type. For a given flowsheet, BPS was used to carry out material and energy balances, estimate the size and cost of equipment, and perform economic evaluation. BPS has had limited commercial success because it was designed as an extension of Aspen Plus, an inherently steady state simulator, and could not satisfactorily represent batch biochemical processes, which normally operate in batch mode.

BioPro Designer, the second product of this category, was initially developed at the Biotechnology Process Engineering Center (BPEC) of MIT. With a license to this technology, Intelligen, Inc. (Scotch Plains, NJ) completed the development of BioPro Designer and commercialized it, first for the Apple Macintosh and later for Microsoft Windows. SuperPro Designer, an extension of BioPro, was created to address other related industries (e.g., synthetic pharmaceuticals, agrochemicals, food processes) as well as water purification and end-of-pipe treatment processes. SuperPro handles material and energy balances, equipment sizing and costing, economic evaluation, environmental impact assessment, process scheduling, and debottlenecking of batch and continuous processes.

Biotechnology Design Simulator (BDS), the third tool of this family, was developed by Life Sciences International of Philadelphia. BDS runs on top of Gensym's G2 system and focuses on scheduling of batch operations and resource utilization as a function of time.

BATCHES from Batch Process Technologies (West Lafayette, IN) is a batch process simulator that has found applications in pharmaceuticals, biochemicals, and food processing. It is especially useful for fitting a new process into an existing facility and analyzing resource demand as a function of time. More recently, Aspen Technology and Hyprotech have introduced Batch Plus and BDK, respectively. Both mainly target synthetic pharmaceutical and specialty chemical processes, but it is expected that future versions of these tools will be able to handle biochemical processes as well.

11.4.3 USING A BIOCHEMICAL PROCESS SIMULATOR

Minimum requirements for a biochemical process simulator are the ability to handle batch as well as continuous processes and the ability to model the unit operations that are specific to bioprocessing. Because SuperPro Designer (from Intelligen, Inc.) has the ability to satisfy these requirements, we will use it to illustrate the role of such tools in bioprocess

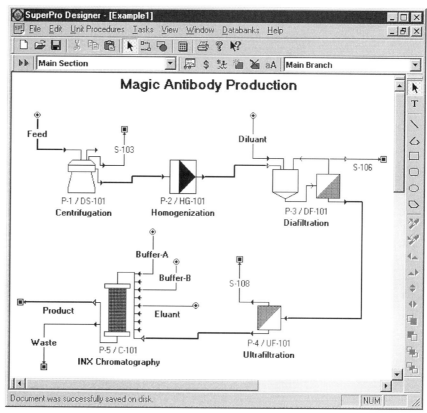

Figure 11.3 A flowsheet on the main window of SuperPro Designer.

design. A functional evaluation version of SuperPro Designer and additional information on bioprocess simulation can be obtained at the website www.intelligen.com.

To model an integrated process on the computer using a simulator, you must start by developing a flowsheet that represents the overall process. Figure 11.3, for instance, displays the flowsheet of a hypothetical process on the main window of SuperPro Designer. The flowsheet is developed by putting together the required unit operations (sometimes referred to as "unit procedures," as explained later in this section) and joining them with material flow streams. Next, the user initializes the flowsheet by registering (selecting from the component database) the various materials that are used in the process and specifying operating conditions and performance parameters for the various operations.

Most biochemical processes operate in batch or semicontinuous mode. This is in contrast to continuous operation, which is typical in the petrochemical and other industries that handle large throughputs. In continuous operations, a piece of equipment performs the same action all the time (which is consistent with the notion of unit operations). In batch processing, on the other hand, a piece of equipment goes through a cycle of operations. For instance, a typical chromatography cycle includes *equilibration, loading, washing, elution,* and *regeneration* (Chapter 7). In SuperPro Designer, the set of operations that comprise a processing step is called a "unit procedure" (as opposed to a "unit operation"). Each unit

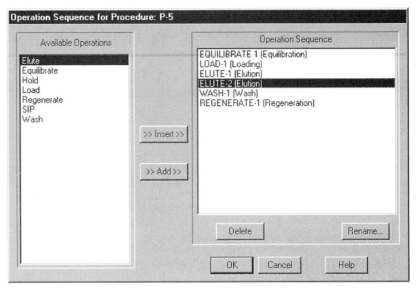

Figure 11.4 Window for adding operations to a unit procedure using SuperPro Designer.

procedure contains individual tasks (e.g., equilibration, loading) called operations. A unit procedure is represented on the screen with a single equipment icon (e.g., C-101 in Figure 11.3 represents the ion exchange chromatography procedure). In essence, a unit procedure is the recipe of a processing step that describes the sequence of actions required to complete that step. Figure 11.4 displays the dialog through which the recipe of a chromatography unit procedure is specified. On the left-hand side of that dialog, the program displays the operations that are available in a chromatography procedure; on the right-hand side, it displays the registered operations. The significance of the unit procedure is that it enables the user to describe and model the various activities of batch processing steps in detail. Later in this chapter (in the examples, Section 11.6), we will see how the execution of these activities can be visualized as a function of time.

For every operation within a unit procedure, SuperPro includes a mathematical model that performs material and energy balance calculations. Based on the material balances, SuperPro performs equipment-sizing calculations similar to some of the homework problems in this book. If multiple operations within a unit procedure dictate different sizes for a certain piece of equipment, the software reconciles the different demands and selects an equipment size that is appropriate for all operations. In other words, the equipment is sized to ensure that it will not be overfilled during any operation but is no larger than necessary (to minimize capital costs). In addition, the software checks to ensure that the vessel contents will not fall below a user-specified minimum volume (e.g., a minimum stir volume) for applicable operations.

Before any simulation calculations can be done, the user must initialize the various operations by specifying operating conditions and performance parameters through appropriate dialog windows. For instance, Figure 11.5 displays the initialization dialog of a chromatography elution operation. Through this dialog, the user specifies the elution strategy (isocratic or gradient), selects the buffer streams (two different solutions are required

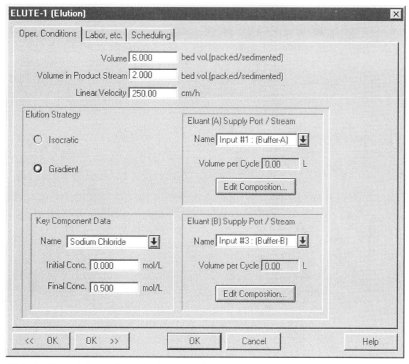

Figure 11.5 Dialog window of the elution operation.

for gradient elution), identifies the component (sodium chloride in this case) whose concentration varies during elution, and specifies its initial and final concentration, and so on. Through the Labor, etc. tab of the same dialog window, the user provides information about labor requirements during this operation. Through the Scheduling tab one specifies the start time of this operation relative to the start or end time of another operation in the same procedure, or relative to an operation in another procedure. After initialization of the operations, the simulator performs material and energy balances for the entire process and estimates the required sizes of equipment. Optionally, the simulator may be used to carry out cost analysis and economic evaluation calculations. The fundamentals of process economics are described in the next section, and pertinent examples are provided later in this chapter.

Other tasks that can be handled by process simulators include process scheduling, environmental impact assessment, debottlenecking, and throughput analysis. Issues of process scheduling and environmental impact assessment are addressed in Section 11.6. In throughput analysis and debottlenecking, the engineer analyzes the capacity and time utilization of equipment and resources (e.g., utilities, labor, raw materials) and tries to identify opportunities for increasing throughput with the minimum possible capital investment. Additional information on this subject can be found at the website www.intelligen.com.

Having developed a good model using a process simulator or a spreadsheet, the user may begin experimenting on the computer with alternative process setups and operating conditions. This has the potential of reducing the costly and time-consuming laboratory and pilot plant effort. One must be aware, however, that the GIGO (garbage in, garbage out) principle applies to all computer models. More specifically, if some assumptions and input data are incorrect, so will be the outcome of the simulation. Consequently, a certain validation of the model is necessary. In its simplest form, a review of the results by an experienced engineer can play the role of validation.

11.5 Process Economics

The preliminary economic evaluation of a project for manufacturing a biological product usually involves the estimation of capital investment, estimation of operating costs, and analysis of profitability. For biopharmaceuticals, another figure worth considering is the average cost of new drug development, which is in the range of $200 to $500 million. This number is so high because it also includes research and development (R&D) spending for all unsuccessful products. In other words, the actual average development cost per successful drug may be $20 to $50 million, but because more than 90% of new projects never reach commercialization, the average overall R&D cost skyrockets. This order-of-magnitude cost increase reinforces the need for effective process design tools and methodologies that assist engineers and scientists to efficiently evaluate and eliminate nonpromising project ideas at the very early stages of product and process development.

11.5.1 CAPITAL COST ESTIMATION

The capital investment for a new plant includes three main items: direct fixed capital (DFC), working capital, and start-up and validation cost. The DFC for small biotechnology facilities is usually in the range of $30 to $60 million, whereas for large facilities it is in the range of $100 to $250 million. For preliminary design purposes, the various items of DFC are estimated based on the total equipment purchase cost (PC) using several multipliers. Table 11.2 provides ranges and average values for the multipliers and a skeleton for the calculations. Detailed definitions of the various cost items and additional information can be found in traditional process design textbooks and the technical literature [1, 19–23].

Notice the wide range of multiplier values for estimating the cost of buildings. Plants for commodity biochemicals, such as ethanol and citric acid, fall on the low end of the range, whereas small biotech facilities that manufacture small amounts of high value products fall on the high end. The average value of 0.45 corresponds to relatively large plants that produce medium to high value products. The insulin manufacturing facility that is analyzed later in this chapter falls under this category. For more accurate estimation of building costs, it is necessary to estimate the process area required based on the footprint of the equipment and the space required around the equipment for safe and efficient operation and maintenance. Then the building cost is estimated by multiplying the area of the various sections (e.g., process, laboratory, office) of a plant by an appropriate unit cost provided in Table 11.3. This table, which was developed by Biometrics, Inc. (Waltham, MA), also provides information on air circulation rates for the various process areas, which determine the sizing and power requirements of heating, ventilation, and air conditioning (HVAC) systems.

TABLE 11.2
Fixed Capital Cost Estimation

Cost item	Average multiplier	Range of multiplier values
Total plant direct cost (TPDC)		
Equipment purchase cost (PC)		
Installation	$0.50 \times PC$	0.2–1.5
Process piping	$0.40 \times PC$	0.3–0.6
Instrumentation	$0.35 \times PC$	0.2–0.6
Insulation	$0.03 \times PC$	0.01–0.05
Electrical	$0.15 \times PC$	0.1–0.2
Buildings	$0.45 \times PC$	0.1–2.0
Yard improvement	$0.15 \times PC$	0.05–0.2
Auxiliary facilities	$0.50 \times PC$	0.2–1.0
Total plant indirect cost (TPIC)		
Engineering	$0.25 \times TPDC$	0.2–0.3
Construction	$0.35 \times TPDC$	0.3–0.4
Total plant cost (TPC)	TPDC + TPIC	
Contractor's fee	$0.05 \times TPC$	0.03–0.08
Contingency	$0.10 \times TPC$	0.07–0.15
Direct fixed capital (DFC)	TPC + Contractor's fee and contingency	

TABLE 11.3
Building Cost Estimation[a]

Space function	Unit cost ($/m²)	Air circulation rates (volume changes/h)
Process areas[b]		
Class 100,000	2000–2500	20
Class 10,000	2500–3500	35–50
Class 1,000	4500–6000	100
Class 100	6000–8000	200–600
Mechanical room (utilities)	300–600	
Laboratory	1000–2000	
Office	500–600	

[a]See reference 2.
[b]The class number refers to the maximum number of particles 0.5 μm or larger per cubic foot.

Referring again to Table 11.2, notice the wide range in the equipment installation cost multipliers. For higher accuracy, one should use multipliers that are specific to individual equipment items. In general, equipment delivered mounted on skids has a lower installation cost.

For preliminary cost estimates, Table 11.2 clearly shows that the fixed capital investment of a plant is a multiple (usually 5 to 8 times) of its equipment purchase cost. The equipment purchase cost can be estimated from vendor quotations, published data, and company data compiled from earlier projects, and by using process simulators and other computer aids. Vendor quotations are time-consuming to obtain and are therefore usually avoided for preliminary cost estimates. Instead, engineers tend to rely on the other three sources. Figures 11.6 to 11.9 provide literature data for disk-stack centrifuges, high pressure homogenizers, membrane filters, and chromatography columns. Table 11.4 gives the costs of chromatography column peripherals. The data represent average values from several vendors.

Oftentimes, cost data for one or two discrete equipment sizes are available, but the cost for a different size piece of equipment must be estimated. In such cases, the *scaling law*

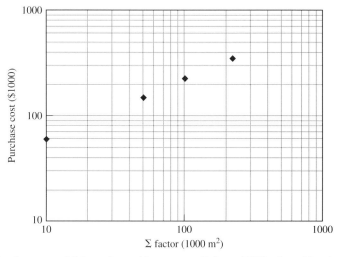

Figure 11.6 Purchase cost of disk-stack centrifuges versus Σ factor (1998 prices; Chemical Engineering Plant Cost Index $= 389.5$ for 1998).

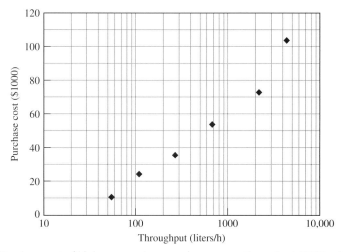

Figure 11.7 Purchase cost of high pressure homogenizers versus throughput (1998 prices).

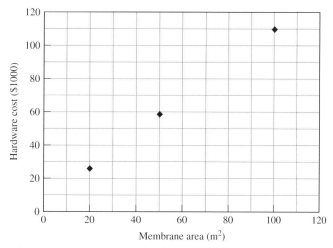

Figure 11.8 Purchase cost of micro-/ultrafiltration hardware, excluding cost of membrane (1998 prices).

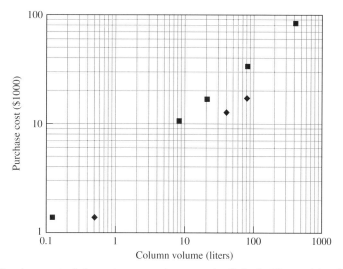

Figure 11.9 Purchase cost of chromatograpy columns made of plastic (diamonds) or 316L stainless steel (squares) for low or high pressure columns, respectively (1998 prices).

TABLE 11.4
Cost of Chromatography Column Peripherals (Pumps, Controls, etc.) (1998 Prices)

	Cost	
Throughput (liters/min)	Low pressure column	High pressure column
0.5–2.5	$95,000	$160,000
2.5–5.0	$115,000	$210,000
5.0–20.0	$125,000	$260,000

(expressed by the following equation) can be used:

$$\text{Cost}_2 = \text{cost}_1 \left(\frac{\text{size}_2}{\text{size}_1} \right)^a \qquad (11.5.1)$$

The mathematical form of the scaling law explains why cost-versus-size data graphed on logarithmic coordinates tend to fall on a straight line. The value of the exponent a in Equation (11.5.1) ranges between 0.5 and 1.0, with an average value for vessels of around 0.6 (this explains why the scaling law is also known as the "0.6 rule"). According to this rule, when the size of a vessel doubles, its cost will increase by a factor of $(2/1)^{0.6}$, or approximately 52%. This result is often referred to as the *economy of scale*. In using the scaling law, it is important to make sure that the piece of equipment whose cost is being estimated has a size that does not exceed the maximum available size for that type of equipment.

The price of equipment changes with time owing to inflation and other market conditions. That change in price is captured by the Chemical Engineering Plant Cost Index (CE Index) that is published monthly by *Chemical Engineering* magazine. The index I is used to update equipment cost data according to the following equation:

$$\text{Cost}_2 = \text{cost}_1 \frac{I_2}{I_1} \qquad (11.5.2)$$

Another factor that affects equipment purchase cost is the material of construction. As can be seen from Figure 11.9, a stainless steel chromatography column is more expensive than a plastic one of the same size. Similarly, a stainless steel tank costs 2.5 to 3 times as much as a carbon steel tank of the same size. Fortunately, in bioprocessing most of the equipment is made of stainless steel for GMP (good manufacturing practice) reasons, and selection of materials is less of a problem. Other factors that affect equipment cost include the finishing of the metal surface and the instrumentation that is provided with the equipment. This is the major cause for the wide range in prices for bioreactors.

Additional cost data for chemical processing equipment can be found in the literature [19, 20, 22]. For cost data for bioprocessing equipment, the choices are rather limited [24, 25].

In addition to direct fixed capital costs, money must be available to pay for the following items: raw materials for 1 to 2 months, labor for 2 to 3 months, utilities for a month, waste treatment/disposal for a month, and other miscellaneous expenses. "Working capital" accounts for these investments in temporary expenses and consumable materials. The required amount of working capital for a process is usually 10 to 20% of the DFC.

Start-up and validation costs can also represent a significant capital investment for a biopharmaceutical plant. A value of 10 to 20% of DFC is quite common.

11.5.2 Operating Cost Estimation

The operating cost to run a biochemical plant is the sum of all expenses associated with raw materials, labor, utilities, waste disposal, overhead, and so on. Dividing the annual operating cost by the annual production rate yields the unit production cost, in dollars per kilogram. Biotechnology is a unique industry when it comes to the range in unit production cost. There are products that cost less than $1.0/kg and others that cost more than $10,000,000/kg to make. The citric acid and therapeutic monoclonal antibody processes

TABLE 11.5
Operating Cost Items and Ranges

Cost item	Type of cost	Range of values (% of total)
Raw materials	Direct	10–80
Labor	Direct	20–50
Consumables	Direct	1–50
Lab/QC/QA[a]	Direct	2–50
Waste disposal	Direct	1–20
Utilities	Direct	1–30
Equipment-dependent cost	Indirect	10–70
Miscellaneous	Indirect	0–20

[a]By convention, costs associated with quality control (QC) and quality assurance (QA) are lumped with laboratory costs.

that are described later in this chapter (Sections 11.6.1 and 11.6.3) lie close to these two extremes. If one also considers biological wastewater treatment with a unit cost of $0.1 to $0.5/m^3 (or $0.0001/kg), then, the range in order of magnitude in the unit processing cost is 10^{11}.

Table 11.5 displays the various types of operating cost, their direct or indirect nature, and ranges for their values relative to the total operating cost. Sometimes cost items are categorized as either fixed or variable. Fixed costs are those that are incurred regardless of volume of product output. The clearest case of a fixed cost is depreciation, which is part of the equipment-dependent cost. The clearest case of a variable cost would be the cost of raw materials. Most other costs have a fixed and a variable component.

It is obvious from the wide range of values in Table 11.5 that one cannot estimate the operating cost of a product based on average values. A certain level of detailed calculations is required.

Raw Materials

Raw materials account for the cost of all fermentation media, recovery chemicals, and cleaning materials. For commodity biochemicals, such as ethanol, the cost of fermentation media is the main component. For high value products, the buffers used for product recovery and equipment cleaning can be a major part of the raw materials cost. Table 11.6 provides a list of commonly used raw materials in the biochemical industries. Note that the price of a raw material can vary widely depending on its required purity. This can be clearly seen in the case of water. Water for injection (WFI), for instance, costs 100 to 500 times as much as city water. Prices of various raw materials can be found in the *Chemical Marketing Reporter*. More recently, a number of websites have begun to offer buyers online pricing information and the opportunity to request bids from suppliers.

Labor

Labor is estimated based on the total number of operators, which in turn is calculated by summing up the operator requirements of the various operations as a function of time. As

TABLE 11.6
Common Bioprocessing Raw Materials (Year 2000 Prices), Including
Materials Used in Fermentation (Upstream Processing)

Raw material	Comments	Price ($/kg)
C Source		
Glucose	Solution 70% w/v	0.25–0.35
Corn syrup	95% Dextrose equivalent	0.35–0.45
Molasses	50% Fermentable sugars	0.08–0.12
Soybean oil	Refined	0.80–0.90
Corn oil	Refined	0.85–0.95
Ethanol	USP tax free	0.50–0.60
Methanol	Gulf Coast	0.20–0.25
n-Alkanes		0.35–0.50
N Source		
Ammonia	Anhydrous, fertilizer grade	0.20–0.25
Soybean flour	44% protein	0.25–0.30
Cottonseed flour	62% protein	0.45–0.55
Casein	13.5% w/w total N	2.40–3.00
Ammonium sulfate	Technical	0.15–0.25
Ammonium nitrate	Fertilizer grade 33.5% N, bulk	0.15–0.20
Urea	46% N, agricultural grade	0.20–0.25
Yeast	Brewers, debittered	2.60–3.20
Whey	Dried, 4.5% w/w N	0.45–0.60
Salts		
KH_2PO_4	USP, granular	1.65–1.85
K_2SO_4	Granular, purified	2.20–2.50
Na_2HPO_4		1.30–1.50
$MgSO_4 \cdot 7H_2O$		0.25–0.35
$ZnSO_4 \cdot 7H_2O$	Agricultural grade, powder	0.50–0.60
Other		
City water		0.0005
Distilled water		0.01–0.05
Water for injection		0.05–0.2
Ampicillin		250–300
Penicillin		10–20
Streptomycin		40–50

will become clear in the examples in Section 11.6, the labor requirement in a batch manufacturing facility varies with time. In a single-product facility, the number of operators in each shift must be based on maximum demand during that shift. In multiproduct facilities, each product line can employ a certain number of dedicated operators and rely on floating operators during periods of peak demand. In general, smaller facilities tend to utilize a larger number of operators per processing step because these plants are less automated. For instance, a small biotech company may utilize two or three operators to set up a fermentor, whereas in a large, highly automated fermentation facility a single operator may handle the setup of six different fermentors remotely from the control room. In general, a typical biotech company that deals with high value products will allocate at least one operator to each processing step (centrifugation, membrane filtration, chromatography, etc.) during its operation. The setup of a step may require multiple operators for a short period.

Consumables

Consumables include the cost of periodically replacing items that may be used up, fouled, or otherwise damaged during processing, such as membranes, chromatography resins, and activated carbon, etc. As the examples later in this chapter will illustrate, the high unit cost of chromatography resins and their frequent replacement can make this item a major component of the operating cost.

Laboratory/QC/QA

Laboratory, QC, and QA account for the cost of off-line analysis, quality control (QC), and quality assurance (QA) costs. Chemical and biochemical analysis and physical property characterization, from raw materials to final product, are a vital part of biochemical operations. This cost is usually 10 to 20% of the operating labor cost. However, for certain biopharmaceuticals that require a large number of very expensive assays, this cost can be as high as the operating labor. For such cases, it is important to account for the number and frequency of the various assays in detail. Changes in lot size that can reduce the frequency of analysis can have a major impact on the bottom line.

Waste Treatment and Disposal

The treatment of wastewater and the disposal of solid and hazardous materials is another important operating cost. The amount and composition of the various waste streams is derived from the material balances. Multiplying the amount by the appropriate unit cost yields the cost of treatment and disposal. Treatment of low biological oxygen demand (BOD) wastewater (<1000 mg/liter) by a municipal wastewater treatment facility usually costs $0.2 to $0.5/m^3. This is not a major expense for most biotech facilities that deal with high value products. However, disposal of contaminated solvents (generated by chromatography steps) and other regulated compounds can become a major expense, given unit disposal costs in the range of $2 to $20/kg (usually higher than the purchase price of the same chemical). Waste disposal may also become a problem if an unwanted by-product is generated as part of the recovery chemistry of a process (see the citric acid example, Section 11.6.1).

Utilities

Utilities account for heating and cooling utilities as well as electricity. The amounts are calculated as part of the material and energy balances. Aerobic fermentors are major consumers of electricity, but downstream processing equipment generally does not consume much electricity. In terms of unit cost, electricity costs around $0.1/kW h, heating steam is around $4 to $8/1000 kg, clean steam (generated utilizing purified water) is around $10 to $50/1000 kg (depending on the scale of production and level of water purity) and refrigerants around $0.05 to $0.1 per 1000 kcal of heat removed. In bioprocessing, clean steam is mainly used for sterilizing equipment as part of equipment cleaning. Another common use is for sterilizing fermentation media. Note that purified water used for buffer preparation and equipment cleaning is often classified as a utility and not as a raw material, thus increasing the cost contribution of utilities.

Equipment-Dependent Costs

Equipment-dependent costs account for the depreciation of the fixed capital investment, maintenance of equipment, insurance, and local (property) taxes, and possibly other overhead-type expenses. For preliminary cost estimates, the entire fixed capital investment is usually depreciated linearly over a 10-year period. In the real world, the government allows corporations to depreciate equipment in 5 to 7 years and buildings in 25 to 30. Land is never depreciated. The annual equipment maintenance cost can be estimated as a percentage of the equipment's purchase cost (usually 10%). Insurance rates depend to a considerable extent upon the maintenance of a safe plant in good repair condition. A value for insurance in the range of 0.5 to 1% of DFC is appropriate for most bioprocessing facilities. The processing of flammable, explosive, or dangerously toxic materials usually results in higher insurance rates. The local (property) tax is usually 2 to 5% of DFC. The factory expense represents overhead cost incurred by the operation of non-process-oriented facilities and organizations, such as accounting, payroll, fire protection, security, and cafeteria. A value of 5 to 10% of DFC is appropriate for these costs.

Miscellaneous

Included in miscellaneous costs are ongoing R&D, process validation, and other overhead-type expenses that can be ignored in preliminary cost estimates. Other general expenses of a corporation include royalties, advertising, and selling. If any part of the process or any equipment used in the process is covered by a patent not assigned to the corporation undertaking the new project, permission to use the teachings of the patent must be negotiated, and some form of royalty is usually required. Advertising and selling covers expenses associated with the activities of the sales department.

11.5.3 PROFITABILITY ANALYSIS

With estimates of capital investment, operating cost, and revenues of a project, one can proceed to assess its profitability and attractiveness from an investment point of view. There are various measures for assessing profitability. The simplest ones include gross margin, return on investment (ROI), and payback time, and they are calculated by using the following equations:

$$\text{Gross margin} = \frac{\text{gross profit}}{\text{revenues}} \tag{11.5.3}$$

$$\text{Return on investment (ROI)} = \frac{\text{net profit per year}}{\text{total investment}} \times 100\% \tag{11.5.4}$$

$$\text{Payback time (years)} = \frac{\text{total investment}}{\text{net profit per year}} \tag{11.5.5}$$

where gross profit is equal to annual revenues minus the annual operating cost, and net profit is equal to gross profit minus income taxes plus depreciation. All variables are averaged over the lifetime of a project.

Other measures that are more involved, such as the net present value (NPV) and internal rate of return (IRR) consider the cash flows of a project over its evaluation life and the value of money as a function of time. Detailed definitions for NPV and IRR can be found in the literature [19]. The examples presented next demonstrate how these measures facilitate the decision-making process.

11.6 Illustrative Examples

The use of SuperPro Designer is illustrated to analyze and evaluate the production of three biological products. The first example analyzes the production of citric acid, a commodity organic acid heavily used in the beverage industry. The second deals with the bacterial production of recombinant human insulin, the first commercial product of modern biotechnology. The third example focuses on the production of monoclonal antibodies (MAbs) from mammalian cells cultured in stirred tank bioreactors. The generation of the flowsheets for the production of all three products was based on information available in the patent and technical literature combined with our engineering judgment and experience with other biological products. We use these flowsheets to draw general conclusions on the manufacturing cost of biological products. The computer files on these examples are available as part of the evaluation version of SuperPro Designer at the website www.intelligen.com. Additional examples are given at this website.

11.6.1 Citric Acid Production

A number of organic acids are produced via fermentation. Of these, citric acid is produced in the largest amount (>400,000 metric tons per year). Citric acid is marketed as citric acid 1-hydrate or as anhydrous citric acid. The majority of citric acid (>60%) is used in the food and beverage industries to preserve and enhance flavor. In the chemical industries (25–30% of total), the uses of citric acid include the treatment of textiles, softening of water, and manufacturing of paper. In the pharmaceutical industry (10% of total), iron citrate is used as a source of iron, and citric acid is used as a preservative for stored blood, tablets, and ointments, and in cosmetic preparations [26]. Citric acid is being used more and more in the detergent industry as a replacement for polyphosphates.

Citric acid was first recovered in 1869 in England from calcium citrate, which was obtained from lemon juice. Its production by filamentous fungi has been known since 1893. The first production via surface culture fermentation was initiated in 1923. Production using stirred tank fermentors began in the 1930s, and presently this is the preferred method for large-scale manufacturing. The plant analyzed in this example produces around 10,000,000 kg of crystal citric acid per year, which represents approximately 2.5% of the current world demand.

Process Description

Upstream Section   The entire flowsheet is shown in Figure 11.10. Molasses, the carbon source of fermentation, is diluted from about 50% fermentable sugars content to 20% with water in a blending tank (V-101). Suspended particulate material is removed by filtration

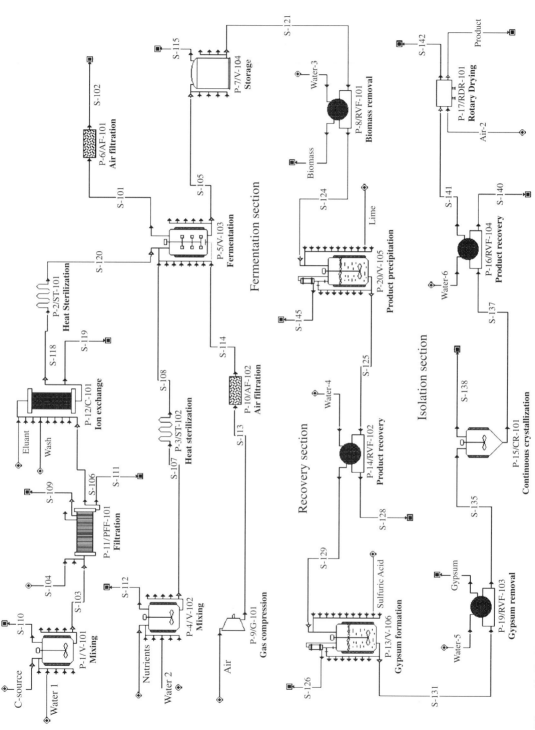

Figure 11.10 Citric acid production flowsheet.

(PFF-101). Metal ions, particularly iron, are removed by an ion exchange chromatography column (C-101). The purified raw material solution is then heat-sterilized (ST-101). Nutrients (i.e., sources of ammonium, potassium, phosphorus, magnesium, copper, and zinc) are dissolved in water (V-102) and heat-sterilized (ST-102).

The fermentation cycle is 7 days, and the production is handled by seven fermentors that operate in staggered mode. Since the plant operates around the clock, one fermentation cycle is initiated daily and another one is completed daily. Each fermentor has a vessel volume of 260 m^3 and handles broth of around 207.4 m^3. Each production fermentor (V-103) is preceded by two smaller seed fermentors, which are not shown in the flowsheet. A pure culture of the mold *Aspergillus niger* is used to inoculate the smallest seed fermentor. When optimum growth of mycelium is reached, the contents of the seed fermentor are transferred to the next stage fermentor, which is approximately 10 times larger. Similarly, this larger seed fermentor inoculates the production fermentor with about 10% volume of actively growing mycelium broth. Air is supplied by a compressor (G-101) at a rate that gradually increases from 0.15 VVM (volume of air per volume of liquid per minute) to 1.0 VVM. Cooling water removes the heat produced by the exothermic process (2990 kcal per kilogram of citric acid formed) and maintains the temperature at 28°C. The fermented broth is discharged into the holding tank (V-104), which acts as a buffer tank between the batch upstream section and the continuous downstream section.

Downstream Section Purification starts with the removal of biomass by a rotary vacuum filter (RVF-101). The clarified fermentation liquor flows to an agitated reaction vessel (V-105). Approximately 1 part of hydrated lime, $Ca(OH)_2$, for every 2 parts of liquor is slowly added to precipitate calcium citrate. The lime solution must be very low in magnesium content if losses due to the relatively soluble magnesium citrate are to be avoided. Calcium citrate is separated by a second rotary vacuum filter (RVF-102) and the citrate-free filtrate (S-128) is disposed of. The calcium citrate cake is sent to another agitated reaction vessel (V-106), where it is acidified with dilute sulfuric acid to form a precipitate of calcium sulfate (gypsum). A third filter (RVF-103) removes the precipitated gypsum and yields an impure citric acid solution. Careful control of pH and temperature of the precipitation steps is important for maximizing the yield of citric acid. The resulting solution is concentrated and crystallized using a continuous evaporator/crystallizer (CR-101). The crystals formed are separated by filtration (RVF-104) and dried in a rotary dryer (RDR-101). If the final product is required in high purity, treatment with activated carbon may precede crystallization to remove colorants. Ion exchange is sometimes used to remove metal ions and other ionic species.

Material Balances

Table 11.7 provides a summary of the overall material balances. "CA Crystal" stands for crystalline citric acid and represents the final product. Glucose represents the fermentable carbohydrates in molasses (50% w/w). Note the large amounts of $Ca(OH)_2$ and sulfuric acid consumed, and gypsum (calcium sulfate) generated. The quantities of these compounds depend on the chemistry of the recovery process and cannot be reduced without changing the recovery technology. Since this gypsum is contaminated with biomass, it has little or no commercial value. A disposal cost of $50/ton (metric) was assumed in this example. The large amount of wastewater is also worth noting.

TABLE 11.7
Overall Material Balances for Citric Acid (CA) Production (kg/year)

Component	In	Out	(Out − in)
Ammonium sulfate	156,000	13,000	−143,000
Biomass	0	1,033,000	1,033,000
CA crystal	0	10,124,000	10,124,000
Ca(OH)$_2$	6,767,000	580,000	−6,187,000
Calcium citrate	0	346,000	346,000
CO$_2$	0	1,848,000	1,848,000
Citric acid	0	365,000	365,000
Glucose	12,741,000	138,000	−12,603,000
Gypsum	0	11,087,000	11,087,000
Impurities	127,000	127,000	0
Nutrients	936,000	95,000	−841,000
Oxygen	19,062,000	15,152,000	−3,910,000
NaOH	75,000	75,000	0
Sulfuric acid	8,396,000	407,000	−7,989,000
Water	166,088,000	172,958,000	6,870,000
Total	214,348,000	214,348,000	0

Economic Evaluation

Table 11.8 provides a list of major equipment items along with their purchase costs (generated by SuperPro Designer). The total equipment cost for a plant of this capacity is around $12 million. Note that more than 50% of the equipment cost is associated with the seven production fermentors. The fermentors are made of stainless steel to minimize leaching of heavy metals that affect product formation. The final item, "cost of unlisted equipment," accounts for the cost of the seed fermentors and other secondary equipment that is not considered explicitly. Table 11.9 displays the various items of the direct fixed capital (DFC) investment. The total DFC for a plant of this capacity is around $56.5 million or approximately 4.7 times the total equipment cost.

Table 11.10 provides a summary of the operating cost. The equipment-dependent cost is the most important item, accounting for 49% of the overall operating cost. Depreciation of the fixed capital investment and maintenance of the facility are the main contributors to this cost. Raw materials account for around 18% of the overall cost. Molasses is the most expensive raw material, accounting for 82% of the raw materials cost. The following prices were assumed: $0.1/kg of molasses, $0.01/kg of 10% w/w H$_2SO_4$ solution, $0.05/kg of Ca(OH)$_2$, and $0.5/m^3 of process water. Utilities are the third largest expense, accounting for 13.5% of the overall cost. Electricity and chilled water utilized by the fermentors are the main contributors to this cost. Labor lies in the fourth position, and the environmental cost (waste treatment/disposal) is fifth. Disposal unit costs of $1/m^3 and $50/1000 kg were assumed for liquid and solid (gypsum and biomass) waste streams, respectively.

The overall unit production cost is approximately $2.2/kg, which happens to be above the current selling price of citric acid. This can be explained by noting the excess capacity

TABLE 11.8
Major Equipment Specification and Purchase Costs for Citric Acid Production
(Year 2000 Prices in U.S. Dollars)

Quantity	Name	Description	Unit cost	Cost
4	V-101	Blending tank Volume = 80 m^3	116,000	464,000
1	ST-101	Heat sterilizer Throughput = 18 m^3/h	353,000	353,000
1	ST-102	Heat sterilizer Throughput = 4 m^3/h	220,000	220,000
1	V-102	Blending tank Volume = 80 m^3	116,000	116,000
1	C-101	Ion exchange column Volume = 2.1 m^3	305,000	305,000
1	PFF-101	Plate-and-frame filter Filter area = 80 m^2	155,000	155,000
1	AF-102	Air filter Throughput = 0.13 m^3/s	4,000	4,000
1	AF-101	Air filter Throughput = 0.68 m^3/s	8,000	8,000
1	G-101	CF compressor Power = 226 kW	174,000	174,000
7	V-103	Fermentor Volume = 260 m^3	950,000	6,650,000
3	V-104	Flat-bottomed tank Volume = 300 m^3	102,000	306,000
1	RVF-101	Rotary vacuum filter Filter Area = 35.2 m^2	87,000	87,000
1	V-105	Stirred jacket vessel Volume = 3.8 m^3	99,000	99,000
1	RVF-102	Rotary vacuum filter Filter area = 80 m^2	132,000	132,000
1	V-106	Stirred jacket vessel Volume = 5 m^3	103,000	103,000
1	RVF-103	Rotary vacuum filter Filter area = 52 m^2	102,000	102,000
1	CR-101	Crystallizer Volume = 19 m^3	122,000	122,000
1	RVF-104	Rotary vacuum filter Filter area = 25.5 m^2	76,000	76,000
1	RDR-101	Rotary dryer Area = 8.5 m^2	101,000	101,000
		Cost of unlisted equipment		2,394,000
Total				11,970,000

around the world and the fact that most operating citric acid plants are rather old and partially depreciated. If depreciation is ignored, the equipment-dependent cost is reduced by more than 80% and the overall unit cost drops to around $1.3, which is slightly below the selling price of citric acid.

Based on the preliminary evaluation of this project idea, one should not recommend investing in citric acid production unless there is a combination of favorable conditions.

TABLE 11.9
Fixed Capital Estimate Summary for Citric Acid Production (Year 2000 Prices in U.S. Dollars)

Total plant direct cost (TPDC)		
Equipment purchase cost	11,970,000	
Installation	4,015,000	
Process piping	4,190,000	
Instrumentation	3,591,000	
Insulation	359,000	
Electricals	1,197,000	
Buildings	2,394,000	
Yard improvement	1,796,000	
Auxiliary facilities	1,197,000	
TPDC		30,708,000
Total plant indirect cost (TPIC)		
Engineering	7,677,000	
Construction	10,748,000	
TPIC		18,425,000
Total plant cost (TPC = TPDC + TPIC)		49,133,000
Contractor's fee	2,457,000	
Contingency	4,913,000	
	7,370,000	
Direct fixed capital (DFC) TPC + contractor's fee and contingency	56,503,000	

TABLE 11.10
Operating Cost Summary for Citric Acid Production (Year 2000 Prices in U.S. Dollars)

Cost item	Cost based on main product ($/kg citric acid crystals)	Annual cost ($/year)	Proportion of total (%)
Raw materials	0.40	3,949,000	17.79
Equipment	1.09	10,866,000	48.96
Labor	0.27	2,668,000	12.02
Consumables	0.00	19,000	0.09
Lab/QC/QA	0.04	400,000	1.80
Waste treatment and disposal	0.13	1,290,000	5.81
Utilities	0.30	3,003,000	13.53
Total	2.22	22,195,000	100.00

Obviously, availability of inexpensive equipment (e.g., by acquiring an existing facility) and raw materials (e.g., by locating the plant near a source of cheap molasses) are the most important factors. Development or adoption of a superior technology may also change the attractiveness of citric acid production. Such a technology is actually available and utilizes extraction for citric acid recovery [27]. Recovery by extraction eliminates the consumption of $Ca(OH)_2$ and H_2SO_4 and the generation of the unwanted $CaSO_4$. Butanol has been used as an extractant, as has tributyl phosphate. Ion pair extraction by means of secondary or tertiary amines dissolved in a water-immiscible solvent (e.g., octyl alcohol) provides an alternative route. With recent developments in electrodialysis membranes, the use of this technique to recover citric acid directly from the fermentation broth may become an attractive alternative [28]. The analysis of an extraction-based process is available (www.intelligen.com) as part of the examples that are distributed with the evaluation version of SuperPro Designer.

11.6.2 HUMAN INSULIN PRODUCTION

Introduction

Insulin facilitates the metabolism of carbohydrates and is essential for the supply of energy to the cells of the body. Impaired insulin production leads to the disease diabetes mellitus, which is the third largest cause of death in industrialized countries, after cardiovascular diseases and cancer [29].

Human insulin is a polypeptide consisting of 51 amino acids arranged in two chains: A with 21 amino acids, and B consisting of 30 amino acids. The A and B chains are connected by two disulfide bonds. Human insulin has a molecular weight of 5734 and an isoelectric point of 5.4. Human insulin can be produced by four different methods:

- Extraction from human pancreas
- Chemical synthesis via individual amino acids
- Conversion of pork insulin, or "semisynthesis"
- Fermentation of genetically engineered microorganisms

Extraction from the human pancreas cannot be practiced because the availability of raw material is so limited. Total synthesis, while technically feasible, is not economically viable because the yield is very low. Production based on pork insulin, also known as "semisynthesis," transforms the porcine insulin molecule into an exact replica of the human insulin molecule by substituting a single amino acid, threonine, for alanine in the G-30 position. This technology has been developed and implemented by Novo Nordisk A/S (Denmark). However, this option is also quite expensive because it requires the collection and processing of large amounts of porcine pancreases. In addition, the supply is limited by the availability of porcine pancreas.

At least three alternative technologies have been developed for producing human insulin based on fermentation and utilizing recombinant DNA technology [30].

Two-Chain Method The first successful technique of biosynthetic human insulin (BHI) production based on recombinant DNA technology was the two-chain method. This technique was developed by Genentech, Inc. (South San Francisco) and scaled up by Eli Lilly and Company (Indianapolis). Each insulin chain is produced as a β-galactosidase fusion protein in *Escherichia coli*, forming inclusion bodies. The two peptide chains are recovered from the inclusion bodies, purified, and combined to yield human insulin. Later, the β-galactosidase operon was replaced with the tryptophan (Trp) operon, resulting in a substantial yield increase.

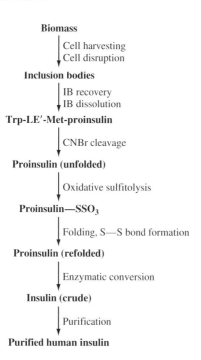

Biomass

Cell harvesting
Cell disruption

Inclusion bodies

IB recovery
IB dissolution

Trp-LE′-Met-proinsulin

CNBr cleavage

Proinsulin (unfolded)

Oxidative sulfitolysis

Proinsulin—SSO₃

Folding, S—S bond formation

Proinsulin (refolded)

Enzymatic conversion

Insulin (crude)

Purification

Purified human insulin

Figure 11.11 Human insulin from proinsulin fusion protein.

Proinsulin Method   The so-called intracellular method of making proinsulin eliminates the need for the separate fermentation and purification trains required by the two-chain method. Intact proinsulin is produced instead. The proinsulin route has been commercialized by Lilly [31]. Figure 11.11 shows the key transformation steps. The *E. coli* cells overproduce Trp-LE′-Met-proinsulin (Trp-LE′-Met is a 121 amino acid peptide signal sequence; proinsulin, with 82 amino acids, is a precursor to insulin) in the form of inclusion bodies, which are recovered and solubilized. Proinsulin is released by cleaving the methionine linker using CNBr. The proinsulin chain is subjected to a folding process to allow intermolecular disulfide bonds to form; and the C peptide, which connects the A and B chains in proinsulin, is then cleaved with enzymes to yield human insulin. A number of chromatography and membrane filtration steps are required to purify the product.

A second method of producing proinsulin was developed by Novo Nordisk A/S. It is based on yeast cells that secrete insulin as a single-chain insulin precursor [29]. Secretion simplifies product isolation and purification. The precursor contains the correct disulfide bridges and is therefore identical to those of insulin. It is converted to human insulin by transpeptidation in organic solvent in the presence of a threonine ester and trypsin followed by de-esterification. Another advantage of the secreted proinsulin technology is that by employing a continuous bioreactor–cell separator loop, it is possible to reuse the cells.

In this example, we analyze a process based on the intracellular proinsulin method.

Market Analysis and Design Basis

The annual world demand for insulin was in the range of 15,000 to 25,000 kg in the 1990s, growing at an annual rate of 5 to 6% [32, 33]. The plant analyzed in this example has a capacity of around 1800 kg of purified biosynthetic human insulin (BHI) per year. This is a relatively large plant for producing polypeptide-based biopharmaceuticals. The plant operates around the clock, 330 days a year. A new batch is initiated every 48 h, resulting in 160 batches per year. The fermentation broth volume per batch is approximately 37.5 m^3.

Process Description

The entire flowsheet for the production of BHI is shown in Figure 11.12. It is divided into four sections: Fermentation, Primary Recovery, Reactions, and Final Purification. *Note:* A "section" in SuperPro is simply a set of unit procedures (processing steps). If you open the file "insulin.spf" in SuperPro, you will see that each unit procedure in each section has its own distinctive color (blue, green, purple, and black for Fermentation, Primary Recovery, Reactions, and Final Purification, respectively).

Fermentation Section   Fermentation media are prepared in a stainless steel tank (V-101) and sterilized in a continuous heat sterilizer (ST-101). The axial compressor (G-101) and the absolute filter (AF-101) provide sterile air and ammonia to the fermentor at an average rate of 0.5 VVM. A two-step seed fermentor train (not shown in the flowsheet) is used to inoculate the 50 m^3 production fermentor (V-102) with transformed *E. coli* cells. These cells are used to produce the Trp-LE$'$-Met-proinsulin precursor of insulin, which is retained in the cellular biomass. The fermentation time in the production fermentor is about 18 h, and the fermentation temperature is 37°C. The final concentration of *E. coli* in the production fermentor is about 30 g/liter (dry cell weight). The Trp operon is turned on when the *E. coli* fermentation runs out of tryptophan. The chimeric protein Trp-LE$'$-Met-proinsulin accumulates intracellularly as insoluble aggregates (inclusion bodies), and this decreases the rate at which the protein is degraded by proteolytic enzymes. In the base case, it was assumed that the inclusion bodies (IBs) constitute 20% of total dry cell mass. At the end of fermentation, the broth is cooled down to 10°C to minimize cell lysis. After completing each processing step in the fermentation section (and subsequent sections), the equipment is washed to prepare for the next batch of product.

Primary Recovery Section   After the end of fermentation, the broth is transferred into a surge tank (V-106), which isolates the upstream from the downstream section of the plant. Three disk-stack centrifuges (DS-101) operating in parallel are used for cell harvesting. Note that a single unit procedure icon on the screen of SuperPro may represent multiple equipment items operating in parallel (to see the number of equipment items a particular icon represents, right-click on the icon, go to Equipment Data, and look at the "Number of Units" field on the Equipment tab). During centrifugation, the broth is concentrated from 37,000 liters to 9165 liters, and most of the extracellular impurities are removed. The cell recovery yield is 98%. The cell sludge is diluted with an equal volume of buffer solution (buffer composition: 96.4% w/w water for injection (WFI), 0.7% EDTA, and 2.9% Tris-base) in a blending tank (V-109). The buffer facilitates the separation of the cell debris particles from inclusion bodies. Next, a high pressure homogenizer (HG-101) is used to break the cells and release the inclusion bodies. The broth undergoes three passes under a pressure drop of 800 bar. The exit temperature is maintained at around 10°C. The same centrifuges as before (DS-101) are then used for inclusion body recovery (P-13). The reuse of these centrifuges can be seen by noting that procedures P-9 and P-13 have the

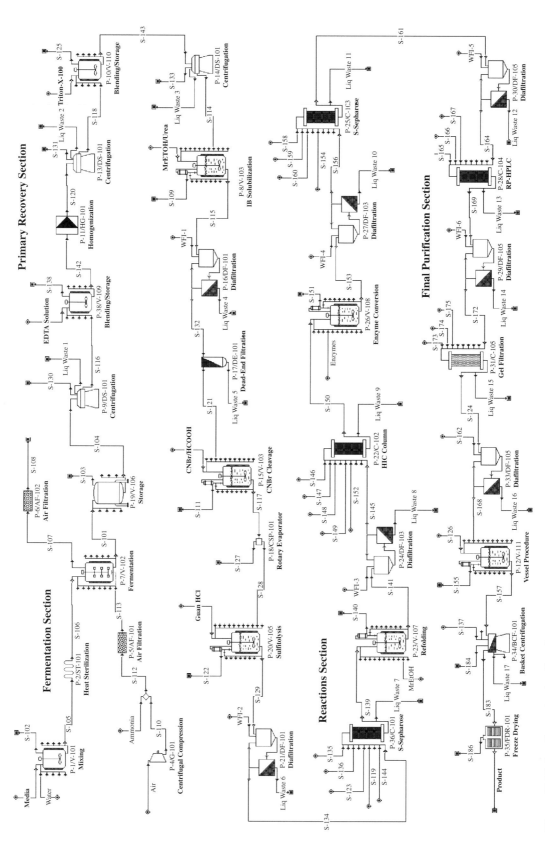

Figure 11.12 Insulin production flowsheet.

same equipment name, DS-101. The IBs are recovered in the heavy phase (with a yield of 98%) while most of the cell debris particles remain in the light phase. This is possible because the density (1.3 g/cm^3) and size (diameter about 1 μm) of the IBs are significantly greater than that of the cell debris particles. The IB sludge, which contains approximately 20% solids w/w, is washed with WFI containing 0.66% w/w Triton-X100 detergent (the volume of solution is twice the volume of inclusion body sludge) and recentrifuged (P-14) using the same centrifuges as before (DS-101). The detergent solution facilitates purification (dissociation of debris and soluble proteins from inclusion bodies). The exit temperature is maintained at 10°C. The slurry volume at the end of the primary recovery section is around 1400 liters.

Reactions Section   *Inclusion Body Solubilization.* The inclusion body suspension is transferred to a glass-lined reaction tank (V-103) and is mixed with urea and 2-mercaptoethanol to final concentrations of 300 g/liter (5 M) and 40 g/liter, respectively. Urea is a chaotropic agent that dissolves the denatured protein in the inclusion bodies, and 2-mercaptoethanol is a reductant that reduces disulfide bonds. A reaction time of 8 h is required to reach a solubilization yield of 95%. The inclusion bodies are composed of 80% w/w Trp-LE′-Met-proinsulin, with the remainder being other (contaminant) proteins. At the end of the solubilization reaction, a diafiltration unit (DF-101) is used to replace urea and 2-mercaptoethanol with WFI and to concentrate the solution. This operation is performed in 6 h with a recovery yield of 98%. All remaining fine particles (biomass, debris, and inclusion bodies) are removed by means of a polishing dead-end filter (DE-101). This polishing filter protects the chromatographic units that are used further downstream. The solution volume at this point is around 5200 liters.

CNBr Cleavage. The chimeric protein is cleaved with CNBr (cyanogen bromide) into the signal sequence Trp-LE′-Met, which contains 121 amino acids, and the denatured proinsulin (82 amino acids) in the same reactor (V-103) that was used for IB solubilization. The reaction is carried out in a 70% formic acid solution containing 30-fold molar excess CNBr (stoichiometrically, one mole of CNBr is required per mole of Trp-LE′-Met-proinsulin). The reaction takes 12 h at 20°C and reaches a yield of 95%. The mass of the released proinsulin is approximately 30% of the mass of Trp-LE′-Met-proinsulin. A small amount of cyanide gas is formed as a by-product of the cleavage reaction. Detailed information on CNBr cleavage is available in the patent literature [34]. The formic acid, unreacted CNBr, and generated cyanide gas are removed by applying vacuum and raising the temperature to around 35°C (the boiling point of CNBr). This operation is carried out in a rotary vacuum evaporator (CSP-101) and takes 1 h. Since cyanide gas is toxic, all air exhausted from the vessels is scrubbed with a solution of hypochlorite, which is prepared and maintained in situ [31].

Sulfitolysis. Sulfitolysis of the denatured proinsulin takes place in a reaction tank (V-105) under alkaline conditions (pH 9–11). This operation is designed to unfold proinsulin, break any disulfide bonds, and add SO$_3$ moieties to all sulfur residues on the cysteines. The product of interest is human proinsulin(S—SO$_3$—)$_6$ (protein–S–sulfonate). The sulfitolysis step is necessary for two reasons: (1) the proinsulin probably is not folded in the correct configuration when expressed in *E. coli* as part of a fusion protein, and (2) the cyanogen bromide treatment tends to break existing disulfide bonds. The final sulfitolysis mixture contains 50% w/w guanidine HCl (6 M), 0.35% ammonium bicarbonate

(NH$_4$HCO$_3$), 3% Na$_2$SO$_3$, and 1.5% Na$_2$S$_4$O$_6$ [35]. A reaction time of 12 h is required to reach a yield of 95%. The presence of the denaturing reagent (guanidine HCl) prevents refolding and cross-folding of the same protein molecule onto itself or two separate protein molecules onto each other. Urea may also be used as a denaturing reagent. Upon completion of the sulfitolysis reaction, the sulfitolysis solution is exchanged with WFI to a final guanidine HCl concentration of 20% w/w. This procedure, P-21, utilizes the DF-101 diafilter that also handles buffer exchange after IB solubilization. The human proinsulin(S—SO$_3$—)$_6$ is then chromatographically purified by means of three ion exchange columns (C-101) operating in parallel. Each column has a diameter of 140 cm and a bed height of 25 cm. A cation exchange resin is used (SP Sepharose Fast Flow from Amersham-Pharmacia Biotech) operating at pH 4.0. The eluant solution contains: 69.5% w/w WFI, 29% urea, and 1.5% NaCl. Urea, a denaturing agent, is used to prevent incorrect refolding and cross-folding of proinsulin(S—SO$_3$—)$_6$. The following operating assumptions were made: (1) the column is equilibrated for 30 min prior to loading, (2) the total resin binding capacity is 20 mg/ml, (3) the eluant volume is equal to 5 column volumes (CVs), (4) the total volume of the solutions for column wash, regeneration, and storage is 15 CVs, and (5) the protein of interest is recovered in 1.5 CVs of eluant buffer with a recovery yield of 90%.

Refolding. This operation catalyzes the removal of the SO$_3$ moiety and then allows disulfide bond formation and correct refolding of the proinsulin to its native form. It takes place in a reaction tank (V-107). This process step involves treatment with mercaptoethanol (MrEtOH), a reductant that facilitates the disulfide interchange reaction. It is added at a ratio of 1.5 mol of mercaptoethanol to 1 mol of SO$_3$. Dilution to a proinsulin(S—SO$_3$—)$_6$ concentration of less than 1 g/liter is required to prevent cross-folding of proinsulin molecules. The reaction is carried out at 8°C for 12 h and reaches a yield of 85%. After completion of the refolding step, the refolding reagents are replaced with WFI and the protein solution is concentrated using a diafiltration unit (DF-103), which has a product recovery yield of 95% (5% of the protein denatures). The volume of the solution at this point is around 5000 liters. Next, the human proinsulin is chromatographically purified in a hydrophobic interaction chromatography (HIC) column (C-102). The following operating assumptions were made: (1) the column is equilibrated for 30 min prior to loading, (2) the total resin binding capacity is 20 mg/ml, (3) the eluant volume is equal to 6 CVs, (4) the total volume of the solutions for column wash, regeneration, and storage is 15 CVs, (5) the protein of interest is recovered in 1 CV of eluant buffer with a recovery yield of 90%, and (6) the material of a batch is handled in three cycles.

Enzymatic Conversion. The removal of the C-peptide from human proinsulin is carried out enzymatically (using trypsin and carboxypeptidase B) in a reaction tank (V-108). Trypsin cleaves at the carboxy-terminus of internal lysine and arginine residues, and carboxypeptidase B removes terminal amino acids. The amount of trypsin used is rate limiting and allows intact human insulin to be formed. Carboxypeptidase is added to a final concentration of 4 mg/liter, while trypsin is added to a final concentration of 1 mg/liter. The reaction takes place at 30°C for 4 h and reaches a conversion yield of 95%. The volume of the solution at this point is around 4300 liters.

Final Purification Section   A purification sequence based on multimodal chromatography, which exploits differences in molecular charge, size, and hydrophobicity, is used to isolate biosynthetic human insulin. A description of all the purification steps follows.

The enzymatic conversion solution is exchanged with WFI and concentrated by a factor of 4 in a diafilter (DF-103). An ion exchange column (C-103) is used to purify the insulin solution. The following operating assumptions were made: (1) the column is equilibrated for 30 min prior to loading, (2) the total resin binding capacity is 20 mg/ml, (3) the eluant volume is equal to 8 CVs and the eluant is a 11.5% w/w solution of NaCl in WFI, (4) the total volume of the solutions for column wash, regeneration, and storage is 14 CVs, (5) the protein of interest is recovered in 1.5 CV of eluant buffer with a recovery yield of 95%, and (6) the material from each batch is handled in four cycles. The liquid volume at this point is around 1100 liters.

Next, the ion exchange eluant solution is exchanged with WFI in a diafilter (DF-105) and is concentrated by a factor of 2.0. A recovery yield of 98% was assumed for this step (2% denatures).

The purification of the insulin solution proceeds with a reversed-phase high performance liquid chromatography (RP-HPLC) step (C-104). Detailed information on the use of RP-HPLC for insulin purification is available in the literature. Analytical studies with a variety of reversed-phase systems have shown that an acidic mobile phase can provide excellent resolution of insulin from structurally similar insulin-like components. Minor modifications in the insulin molecule resulting in monodesamido formation at the 21st amino acid of the A chain, or derivatization of amines via carbamoylation or formylation, result in insulin derivatives having significantly increased retention. Derivatives of this nature are typical of the kind of insulin-like components that are found in the charge stream going into the reversed-phase purification. The use of an acidic mobile phase results in elution of all the derivatives after the insulin peak, while the use of mildly alkaline pH results in derivatives eluted on either side of the parent insulin peak. An ideal pH for insulin purification is in the region of 3.0 to 4.0, since this pH range is far enough below the isoelectric pH of 5.4 to provide for good insulin solubility. An eluant buffer with an acetic acid concentration of 0.25 M meets these operational criteria because it is compatible with the chromatography and provides good insulin solubility. A 90% insulin yield was assumed in the RP-HPLC step with the following operating conditions: (1) the column is equilibrated for 30 min prior to loading, (2) the total resin binding capacity is 15 mg/ml, (3) the column height is 25 cm, (4) the eluant volume is 6 CVs and its composition is 25% w/w acetonitrile, 1.5% w/w acetic acid, and 73.5% w/w WFI, (5) the total volume of the solutions for column wash, equilibration, regeneration, and storage is 6 CVs, and (5) the protein of interest is recovered in 1 CV of eluant buffer with a recovery yield of 90%.

The RP-HPLC buffer is exchanged with WFI and concentrated by a factor of 2.0 in a diafilter (DF-105) that has a product recovery yield of 98% (2% denatures). Purification is completed by a gel filtration chromatography column (C-105). The following operating assumptions were made: (1) the column is equilibrated for 30 min prior to loading, (2) the sample volume is equal to 5% of the column volume, (3) the eluant volume is equal to 4 CVs, (4) the total volume of the solutions for column wash, depyrogenation, stripping, and storage is 6 CVs, and (5) the protein of interest is recovered in 0.5 CV of eluant buffer with a recovery yield of 90%. The mobile phase is a solution of acetic acid.

Next, the same diafilter (DF-105) is used to concentrate the purified insulin solution by a factor of 10. The liquid volume at this point is around 500 liters, which contains approximately 12.8 kg of insulin. This material is pumped into a jacketed and agitated reaction tank (V-111). Ammonium acetate and zinc chloride are added to the protein solution until each reaches a final concentration of 0.02 M [32]. The pH is then adjusted to between 5.4 and 6.2. The crystallization is carried out at 5°C for 12 h. Insulin crystallizes with zinc with the following stoichiometry: insulin$_6$-Zn$_2$. Step recovery on insulin is around 90%.

The crystals are recovered with a basket centrifuge (BCF-101) with a yield of 95%. Finally, the crystals are freeze-dried (FDR-101). The purity of the crystallized end product is between 99.5 and 99.9% measured by analytical high pressure liquid chromatography (HPLC). Approximately 11.31 kg of product is recovered per batch. The overall recovery yield is around 32%.

Material Balances and Environmental Impact Assessment

Table 11.11 displays the raw material requirements in kilograms per year, per batch, and per kilogram of main product (MP = purified insulin crystals). Note the huge amounts of

TABLE 11.11
Raw Material Requirements for Human Insulin Production: 1 Batch = 11.31 kg Main Product (MP = Purified Insulin Crystals)

	Requirement		
Raw material	kg/year	kg/batch	kg/kg MP
Glucose	782,000	4,888	432.2
Salts	71,000	444	39.3
Air	3,648,000	22,800	2,015.9
Ammonia	76,000	475	42.0
Water	27,798,000	173,738	15,361.5
NaOH (0.5 M)	5,549,000	34,681	3,066.1
H$_3$PO$_4$ (20% w/w)	6,452,000	40,325	3,566.4
Tris base	43,000	269	23.8
WFI	61,446,000	384,038	33,955.6
EDTA	10,400	65	5.7
Triton-X100	3,000	19	1.7
CNBr	15,300	96	8.5
Formic acid	1,752,000	10,950	968.2
Urea	3,063,000	19,144	1,692.7
MrEtOH	99,000	619	54.7
NH$_4$HCO$_3$	5,500	34	3.0
Sodium sulfite	48,000	300	26.5
Na$_2$O$_6$S$_4$	24,000	150	13.3
Guanidine HCl	806,000	5,038	445.4
Sodium chloride	778,000	4,863	430.0
Sodium hydroxide	138,000	862	76.0
Acetic acid	2,435,000	15,219	1,345.6
Enzymes	3	0	0.0
Acetonitrile	767,000	4,794	423.9
Ammonium acetate	180	1	0.1
Zinc chloride	320	2	0.2
Total	115,809,703	723,814	63,998.3

WFI, water, formic acid, urea, guanidine hydrochloride, acetic acid, and acetonitrile required per kilogram of final product. All these materials end up in waste streams.

In the base case, it was assumed that this waste is treated and disposed of. However, opportunities may exist for recycling some chemicals for in-process use and recovering others for off-site use. For instance, formic acid (HCOOH), acetonitrile, and urea are good candidates for recycling and recovery. Formic acid is used in large quantities (11 tons/batch) in the CNBr cleavage step (V-103), and it is removed by means of a rotary vacuum evaporator (CSP-101), along with small quantities of CNBr, H_2O, and urea. The recovered formic acid can be readily purified by distillation and recycled in the process. Around 2 tons per batch of urea is used for the dissolution of inclusion bodies (V-103), and 17 tons per batch is used in the first chromatography step (C-101) to purify proinsulin$(S—SO3)_6$ before its refolding. Approximately 90% of the urea appears in just two waste streams (Liquid waste 4 and 7). It is unlikely that these urea-containing streams can be purified economically for in-process recycling. However, these solutions can be concentrated, neutralized, and shipped off site for further processing and utilization as a nitrogen fertilizer.

Approximately 4.8 tons per batch of acetonitrile is used in the reversed-phase HPLC column (C-104), and most of it ends up in the waste stream of the column (Liquid waste 13) along with 6.8 tons of water, 1.85 tons of acetic acid, and small amounts of NaCl and other impurities. It is unlikely that acetonitrile can be recovered economically to meet the high purity specifications for a step so close to the end of the purification train. However, there may be a market for off-site use.

Process Scheduling and Resource Tracking

Figure 11.13 displays the scheduling and equipment utilization for three consecutive batches. The plant batch time is approximately 260 h. This is the time required to go from the preparation of raw materials to final product for a single batch. However, since most of the equipment items are utilized for much shorter periods within a batch, a new batch is initiated every 48 h. Multiple bars on the same line (e.g., for DS-101, DF-101, DF-103, and DF-105) represent reuse (sharing) of equipment by multiple procedures. White space represents idle time. The equipment with the least idle time between consecutive batches is the *time (or scheduling) bottleneck* (V-103 in this case) that determines the maximum number of batches per year. Its occupancy time (approximately 43.85 h) is the minimum possible time between consecutive batches (also known as minimum effective plant batch time). This plant operates around the clock and processes 160 batches per year.

The execution in time of the various procedures and their operations can be visualized in detail through the operations Gantt chart (see Figure 11.14). Note, for instance, the operations of procedure P-8 (IB solubilization). The TRANSFER_IN operation in that procedure runs in parallel with the CENTRIFUGE_1 operation of the preceding procedure (P-14). This is the case because while the IB slurry is being centrifuged as part of P-14, the concentrate (solids stream of the centrifuge) is being pumped into the vessel (V-103) of P-8. If this detail is not captured in the model, the equipment time (scheduling) bottleneck may not be identified.

Process scheduling is closely related to the determination of the annual capacity of a batch plant. Later we will see how changes in scheduling and installation of additional equipment can be used to increase plant throughput and reduce manufacturing cost.

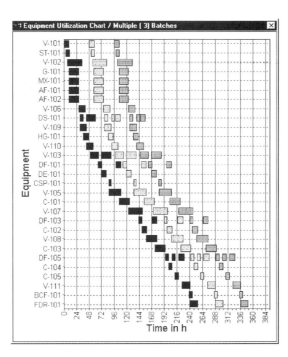

Figure 11.13 Equipment utilization as a function of time for three consecutive batches of human insulin; see Figure 11.12 for equipment icons.

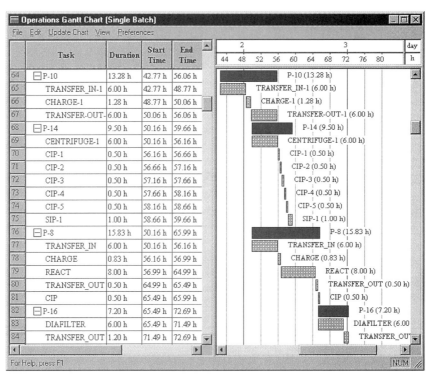

Figure 11.14 Operations Gantt chart for human insulin production.

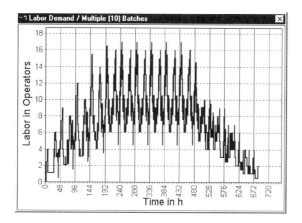

Figure 11.15 Labor demand as a function of time for 10 consecutive batches of human insulin.

TABLE 11.12
Key Economic Evaluation Results for Human Insulin Production

Direct fixed capital	$69.7 million
Total capital investment	$78.0 million
Plant throughput	1,810 kg/year
Manufacturing cost	$76.5 million/year
Unit production cost	$42.2/g
Selling price	$75/g
Revenues	$135.7 million/year
Gross profit	$59.2 million/year
Taxes (40%)	$23.7 million/year
Net profit	$42.2 million/year
IRR (after taxes)	58.2%
NPV (for 7% discount interest)	$216 million

Another characteristic of batch processing is the variable demand for resources (e.g., labor, utilities, raw materials) as a function of time. For instance, Figure 11.15 displays the labor demand (expressed in number of operators) for 10 consecutive batches. Note that for short periods there is a need for up to 17 operators to be present. If that is not possible, certain operations will need to be delayed to distribute the demand for operators more evenly. In such a case, the limited resource becomes the time bottleneck. Demand for steam and other utilities may also become a time bottleneck. The results of Figure 11.15 are also useful in staffing a facility. If the facility is dedicated to manufacturing of a single product, then the number of operators in each shift should be based on the peak demand during that shift. In multiproduct facilities, each production suite may employ a dedicated number of operators, assigning some personnel to be floating operators during periods of peak demand.

Economic Evaluation

Table 11.12 shows the results of the economic evaluation. The detailed tables for these calculations are available as part of the evaluation version of SuperPro. For a plant of this

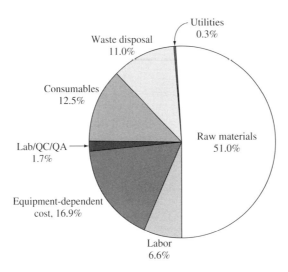

Figure 11.16 Breakdown of manufacturing cost for human insulin production.

Utilities
0.3%

Waste disposal
11.0%

Consumables
12.5%

Lab/QC/QA
1.7%

Equipment-dependent
cost, 16.9%

Labor
6.6%

Raw materials
51.0%

capacity, the total capital investment is $78 million. The unit production cost is $42.2 per gram of purified insulin crystals. Assuming a selling price of $75/g, the project yields an after-tax internal rate of return (IRR) of 58.2% and a net present value (NPV) of $216 million (assuming a discount interest of 7%). Based on these results, this project represents a very attractive investment. However, if amortization of up-front R&D costs is considered in the economic evaluation, the numbers change drastically. For instance, a modest amount of $100 million for up-front R&D cost amortized over a period of 10 years reduces the IRR to 17.7% and the NPV to $121 million.

Figure 11.16 breaks down the operating cost. The cost of raw materials is the most important, accounting for 51% of the overall manufacturing cost. The equipment-dependent cost lies in the second position, accounting for 16.9% of the overall cost. This cost item accounts for the depreciation and maintenance of the facility and other overhead expenses. Consumables, which account for 12.5% of the total cost, represent the expense of periodically replacing the resins of the chromatography columns and the membranes of the membrane filters. Treatment and disposal of waste materials account for 11% of the total cost. As mentioned in the material balance section, recycling and reuse of some of the waste materials may reduce this cost. Labor lies in the fifth position, accounting for 6.6% of the total cost. Approximately 50 operators are required to run the plant around the clock, supported by 12 scientists for QC/QA work. The utilities cost is so low because it comprises only electricity and the small amounts of heating and cooling required. The cost of purified water is treated as a raw material and not as a utility.

Figure 11.17 displays the cost distribution per flowsheet section. Only 8.0% of the overall cost is associated with fermentation. The other 92% is associated with the recovery and purification sections. This is common for high value biopharmaceuticals that are produced from recombinant *E. coli*. Most of the cost is associated with the reactions section because of the large amounts of expensive raw materials and consumables required in that section.

Finally, Table 11.13 for each raw material displays its price, annual cost, and contribution to the overall raw materials cost. H_3PO_4 (20% w/w), NaOH (0.5 M), WFI, acetic acid,

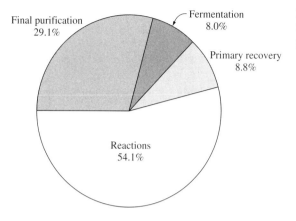

Final purification
29.1%

Fermentation
8.0%

Primary recovery
8.8%

Reactions
54.1%

Figure 11.17 Cost distribution per flowsheet section for human insulin production.

TABLE 11.13
Cost of Raw Materials Year 2000 Prices for Human Insulin Production

Raw material	Price ($/kg)	Annual cost ($)	Proportion of total (%)
Glucose	0.60	469,000	1.20
Salts	1.00	71,000	0.18
Air	0.00	0	0.00
Ammonia	0.70	53,000	0.14
Water	0.05	1,390,000	3.56
NaOH (0.5 M)	0.50	2,774,000	7.11
H_3PO_4 (20% w/w)	1.00	6,452,000	16.53
Tris base	6.00	259,000	0.66
WFI	0.10	6,145,000	15.74
EDTA	18.50	193,000	0.49
Triton-X100	1.50	4,600	0.01
CNBr	11.00	168,000	0.43
Formic acid	1.60	2,802,000	7.18
Urea	1.52	4,655,000	11.93
MrEtOH	3.00	296,000	0.76
$NH_4 HCO_3$	1.00	5,600	0.01
Sodium sulfite	0.40	19,000	0.05
$Na_2O_6S_4$	0.60	14,000	0.04
Guanidine HCl	2.15	1,732,000	4.44
Sodium chloride	1.23	957,000	2.45
Sodium hydroxide	3.50	482,000	1.23
Acetic acid	2.50	6,088,000	15.60
Enzymes	500,000.00	1,691,000	4.33
Acetonitrile	3.00	2,302,000	5.90
Ammonium acetate	15.00	2,700	0.01
Zinc chloride	12.00	3,800	0.01
Total		39,028,700	100.00

urea, formic acid, acetonitrile, guanidine HCl, and the enzymes are the major contributors to the raw materials cost. The solution of H_3PO_4 is used for equipment cleaning.

Other assumptions for the economic evaluation include the following: (1) a new manufacturing facility will be built and dedicated to production of 1800 kg/year of purified insulin; (2) the entire direct fixed capital is depreciated linearly over a period of 10 years; (3) the project life time is 15 years; (4) the unit cost of membranes is $200/$m^2$ and they are replaced every 2000 h of operation; (5) the average unit cost of chromatography resins is $300/liter; (6) the waste disposal cost is $0.005/liter for low BOD streams and $0.15/liter for streams containing significant amounts of solvents and other regulated chemicals.

Throughput Increase Options

In the base case, a new batch is initiated every 48 h. Most of the equipment items, however, are utilized for less than 24 h per batch (see Figure 11.13). If the market demand for insulin grows, this provides the opportunity for increasing plant throughput without major capital expenditures. A realistic improvement is to initiate a batch every 24 h. This will require a new fermentor of the same size, whose operation will be staggered relative to the existing unit so that one fermentor is ready for harvesting every day. Such a production change will also require additional equipment of the following types: (1) disk-stack centrifuges to reduce the occupancy of DS-101 to less than 24 h, (2) two reaction tanks to reduce the occupancy of V-103 and V-107, and (3) membrane filters to reduce the occupancy of DF-103 and DF-105.

The additional capital investment for such a change is around $20 million. This additional investment will allow the plant's capacity to be doubled, and the new unit production cost will be around $33.6/g. The reduction in the unit production cost is rather small because the majority of the cost is associated with raw materials, consumables, and waste disposal that scale approximately linearly with production.

11.6.3 THERAPEUTIC MONOCLONAL ANTIBODY PRODUCTION

Monoclonal antibodies (MAbs) are used in diagnostic tests and are being tested for therapeutic purposes. World demand for most currently approved MAbs is on the order of a few kilograms per year. However, new therapeutic MAbs are under development that require doses of several hundred milligrams to a gram over the course of therapy [36]. The annual world demand for such products will exceed 100 kg.

Current production choices for MAbs are limited to three well-established systems: ascites, stirred tank bioreactors (STR), and hollow fiber bioreactors. Alternative technologies under development include transgenic animals and genetically altered plants [37]. Currently, stirred tank bioreactors tend to be favored for production of MAbs in kilogram quantities. They are operated under batch, fed-batch, or perfusion mode.

This example analyzes the production of a typical therapeutic monoclonal antibody. In the base case, approximately 6.2 kg of purified product is produced per year in 46 batches. The manufacturing cost for producing larger quantities is estimated as part of the sensitivity analysis, presented at the end of the section.

Process Description

Upstream Section   The entire flowsheet is shown in Figure 11.18. The serum-free and low protein content media powder is dissolved in WFI in a stainless steel tank (V-101), and

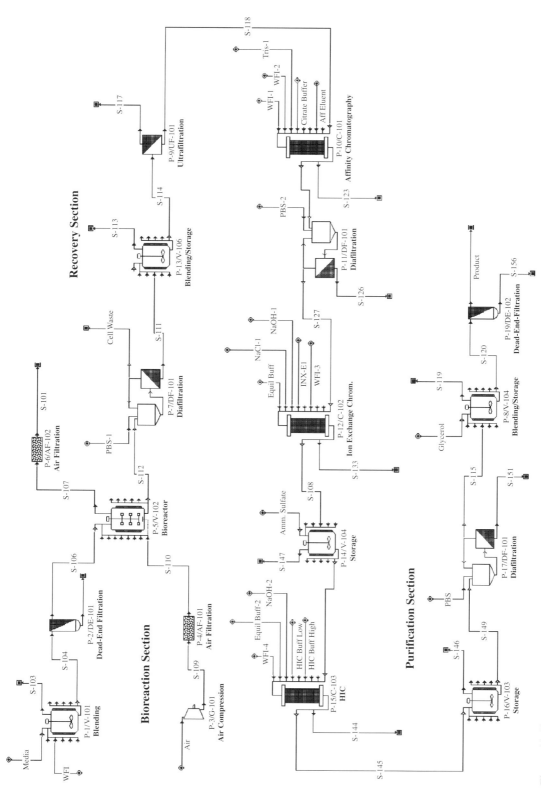

Figure 11.18 Monoclonal antibody production flowsheet.

the solution is sterilized using a 0.1 μm dead-end polishing filter (DE-101). The concentration of media powder in the feed solution is 10 g/liter. A stirred tank bioreactor (V-102) is used to grow the cells that express the therapeutic immunoglobulin G (IgG). The bioreactor operates in fed-batch mode. A cycle time of 156 h (132 h for fermentation and 24 h for turnaround) was assumed for the bioreactor. The volume of broth generated per bioreactor batch is approximately 2200 liters, containing 220 g of product (the product titer is 100 mg/liter). The total volume of the bioreactor vessel is 3000 liters.

Downstream Section   The generated biomass and other suspended compounds are removed by using a 0.65 μm membrane diafilter (DF-101). The product recovery yield of this step is 95%. This filtration step takes 5.1 h and requires a membrane area of around 30 m^2. Then a 50,000 MW cutoff ultrafilter (UF-101) is used to concentrate the clarified solution 20-fold. The recovery yield of this step is 95%. This step takes 3.6 h and requires a membrane area of 40 m^2. The bulk of the contaminant proteins are removed by using a protein A affinity chromatography column (C-101). The following operating assumptions were made: (1) resin binding capacity is 15 mg of product per milliliter of resin; (2) the eluant is a 0.1 M solution of sodium citrate, and its volume is equal to 6 CVs; (3) the product is recovered in 3 CVs of eluant buffer with a recovery yield of 95%, and the pH is maintained near neutral to ensure product stability; and (4) the total volume of the solutions for column equilibration, wash, and regeneration is 13 CVs. This step takes around 15.7 h and requires a resin volume of 24.5 liters. The protein A elution buffer is exchanged with phosphate buffer (procedure P-11), using the same diafilter (DF-101) as in P-7. The product recovery yield of this step is 95%. The purification proceeds using a cation exchange chromatography column (C-102). The following operating assumptions were made: (1) the resin's binding capacity is 20 mg of product per milliliter of resin; (2) a gradient elution step is employed with a sodium chloride concentration ranging from 0.0 to 1.0 M and a volume of 6 CVs; (3) the product is recovered in 3 CVs of eluant buffer with a recovery yield of 90%; and (4) the total volume of the solutions for column equilibration, wash, and regeneration is 17 CVs. This step takes around 13 h and requires a resin volume of 15.7 liters. Ammonium sulfate is added to a concentration of 2.0 M to increase the ionic strength of the solution and prepare it for the hydrophobic interaction chromatography (HIC) step (C-103) that follows. The following operating assumptions were made for the HIC step: (1) the resin binding capacity is 20 mg of product per milliliter of resin; (2) a gradient elution step is used in which the concentration of ammonium sulfate changes linearly from 2.0 M to 0.0 M; (3) the product is recovered in 2 CVs of eluant buffer with a recovery yield of 95%; and (4) the total volume of the solutions for column equilibration, wash, and regeneration is 22 CVs. This step takes around 13.5 h and requires a resin volume of 15.7 liters. The purified product solution is concentrated twofold and the HIC elution buffer is exchanged with phosphate buffer using the same diafilter (P-17/DF-101).

Next, glycerol is added (for product stability) to a concentration of 100 g/liter and the solution is sent to product formulation. The product concentration in the final solution is around 8 g/liter. The following additional assumptions were made: (1) WFI is used for the preparation of water solutions and buffers; and (2) to calculate the cycle time of chromatography steps, it was assumed that loading and elution operate at a linear velocity of 100 cm/h, while equilibration, washing, and regeneration operate at a linear velocity of 200 cm/h.

Material Balances

Table 11.14 provides a summary of the overall material balances per batch. The quantities are in kilograms per batch. The overall recovery yield of IgG (the product) is 64% (140 g of IgG is recovered out of the 220 g that is present in the fermentation broth). Note the large amount of process water and WFI utilized per batch. The majority of process water and WFI are utilized for equipment cleaning.

Process Scheduling

Figure 11.19 displays the scheduling and equipment utilization for two consecutive batches. The plant batch time is approximately 232 h. This is the time required to go from

TABLE 11.14
Overall Material Balances for IgG Production (kg/batch)

Component	Total inlet	Total outlet	Product
Ammonium sulfate	64.69	64.69	
Biomass	0.00	0.87	
Glycerol	1.85	1.85	
IgG	0.00	0.22	0.14
Growth media	21.76	8.41	
Na$_3$ citrate	0.80	0.80	
Phosphoric acid	1,041.00	1,041.00	
Sodium hydrophosphate	6.83	6.81	
Sodium chloride	55.18	55.19	
Sodium hydroxide	6.83	6.81	
Tris-HCl	0.69	0.69	
Water	11,459.00	11,458.00	
WFI	18,269.00	18,269.00	
Total	30,928.00	30,928.00	0.14

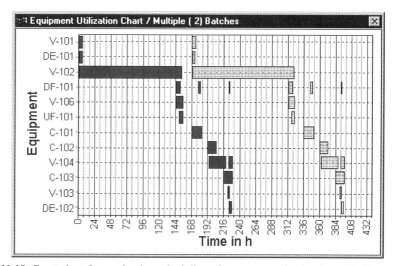

Figure 11.19 Gantt chart for production scheduling of two consecutive batches of IgG.

the preparation of raw materials to the final product in a single batch. A new batch is initiated every 7 days (168 h). The bioreactor, which is the time (scheduling) bottleneck, has a cycle time of 152 h (140 h for fermentation and 12 h for turnaround). While the bioreactor is preparing a new batch, the downstream equipment is being used to purify the product of the preceding batch. The downstream section requires two shifts per day for 4 days a week. Multiple bars on the same line (e.g., for DF-101 and V-104) represent reuse (sharing) of equipment by multiple procedures. White space represents idle time. On an annual basis, the plant processes 46 batches and produces 6.2 kg of purified IgG.

Economic Evaluation

Table 11.15 shows the key economic evaluation results generated using the built-in cost functions of SuperPro Designer. For the base case (6.2 kg/year of IgG), the total capital investment is around $16.3 million. The floor area of the production facility is around 2000 m^2. The unit production cost is around $900/g of purified IgG. Assuming a selling price of $2500/g, the project yields an after-tax internal rate of return (IRR) of 47.4% and a net present value (NPV) of $32.5 million (assuming a discount interest of 7%). However, as with the insulin example, if amortization of up-front R&D cost is considered in the economic evaluation, the numbers change drastically. For instance, a modest amount of $20 million for up-front R&D cost amortized over a period of 10 years reduces the IRR to 20% and the NPV to $12.5 million.

Figure 11.20 breaks down the operating cost. The equipment-dependent cost is the most important item, accounting for 50% of the manufacturing cost. This is common for high value products that are produced in small quantities. Labor lies in the second position, accounting for 16% of the total cost. Eight operators are required to run the plant, supported by four scientists for QC/QA work. Raw materials and consumables account for 11 and 13%, respectively. Consumables include the cost of chromatography resins and membrane filters that need to be replaced on a regular basis. In terms of cost distribution per section, 46% of the cost is associated with the upstream section and 54% with the downstream.

TABLE 11.15
Key Economic Evaluation Results for IgG Production

Direct fixed capital	$15.3 million
Total capital investment	$16.3 million
Plant throughput	6.2 kg of IgG/year
Manufacturing cost	$5.64 million/year
Unit production cost	$908/g of IgG
Selling price	$2,500/g of IgG
Revenues	$15.5 million/year
Gross profit	$9.9 million/year
Taxes (40%)	$4.0 million/year
Net profit	$7.4 million/year
IRR (after taxes)	47.4%
NPV (for 7% discount interest)	$32.5 million

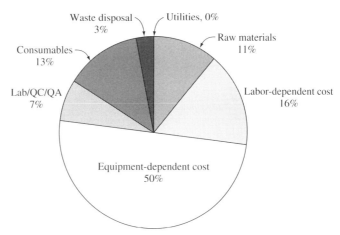

Figure 11.20 Breakdown of manufacturing cost for IgG production.

The economic evaluation relies on the following key assumptions: (1) a new manufacturing facility will be built and dedicated to production of 6.2 kg/year of IgG; (2) the entire direct fixed capital is depreciated linearly over a period of 10 years; (3) the project lifetime is 15 years; (4) the unit cost of WFI is $0.1/liter; (5) the cost of media is $5/liter (based on volume of solution fed to bioreactors); (6) all the chemicals used are of high purity grade; (7) the unit cost of membranes is $350/m^2; (8) the unit cost of chromatography resins is $6000/liter, $1600/liter, and $3200/liter for columns C-101, C-102, and C-103, respectively; (9) the chromatography resins are replaced every 20 cycles; and (10) the average waste disposal cost is $0.5/kg.

Sensitivity Analysis

After a model for the entire process has been developed on the computer, tools like SuperPro Designer can be used to ask and readily answer "what if" questions and to carry out sensitivity analysis with respect to key design variables. In this example, we looked at the impact of product titer (in the bioreactor) and production rate on unit production cost. For a product titer of 100 mg/liter, the cost drops considerably for production rates of up to 80 kg/year of purified IgG (see Figure 11.21). For higher production rates, the cost levels off and approaches a value of $260/g. Increasing the titer from 100 mg/liter to 250 mg/liter reduces the production cost by $90 to $110/g, depending on production rate. The reduction in cost is smaller (in the range of $30/g to $45/g) when the product titer is increased from 250 mg/liter to 500 mg/liter. As can be seen from Figure 11.21, the production cost reaches a minimum of $150/g as we increase throughput and product titer. For throughputs in the range of 100 kg/year and titers of 500 mg/liter, almost 80% of the manufacturing cost is associated with the downstream section. Furthermore, under such conditions the cost of purification scales approximately linearly with production rate, because most of the cost is associated with purification raw materials and consumables. Therefore, less expensive product formation options, such as transgenic animals and genetically altered plants, can have an impact only on the 20% of the total cost associated with product formation. In other words, the cost of MAbs will not drop below $120/g

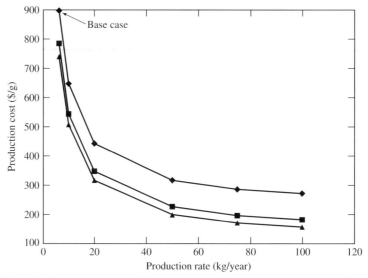

Figure 11.21 Production cost of IgG as a function of product titer and production rate: ◆, 100 mg/liter; ■, 250 mg/liter; ▲, 500 mg/liter.

(80% of \$150/g), no matter what upstream technology is used. The only way to go below the \$120/g barrier is by developing less expensive product purification technologies and deploying them in combination with inexpensive upstream technologies (such as transgenic animals).

Key assumptions for the sensitivity analysis are that the composition of fermentation media is independent of product titer and that the scheduling is independent of plant throughput (as we increase throughput, we continue to process 44 batches per year by utilizing larger and multiple pieces of the same type of equipment).

11.7 Summary

Process design consists of two main activities, process synthesis and process analysis. Process synthesis is the selection and arrangement of a set of unit operations (process steps) capable of producing the desired product at an acceptable cost and quality. Process analysis is the evaluation and comparison of different process synthesis solutions. In general, a synthesis step is usually followed by an analysis step, and the results of analysis determine the subsequent synthesis step.

- Sequencing of unit operations is based on experience and rules of thumb (heuristics), in which biomass separation (harvesting or removal) is followed by extraction (isolation) of product and then purification and polishing.
- Once a flowsheet has been established, software tools—process simulators—can be introduced. These are available from various commercial sources.

- Capital cost estimation consists of direct fixed capital, working capital, and start-up and validation costs.

- The cost of a step in a process is the sum of the costs of raw materials, labor, consumables, quality control, waste treatment/disposal, utilities, equipment depreciation, equipment maintenance, and overhead. The cost of the entire process is the sum of the costs for all the steps.

- Profitability analysis consists of the calculation of gross margin, return on investment, and payback time.

- Sensitivity analysis uses a process simulator to investigate the "what if" questions when a selected process variable is changed, such as product titer or recovery yield.

PROBLEMS

11.1 **Tissue Plasminogen Activator** Tissue plasminogen activator (tPA) was among the first products of biotechnology to be based on recombinant DNA technology. A medium molecular weight enzymatic protein, tPA finds its primary application in the treatment of myocardial infarction (heart attack) patients. First characterized in 1979, the protein was commercially developed by Genentech, with clinical trial quantities and purities being achieved in late 1984. The current market demand is around 15 kg/year of purified tPA, which generates annual revenues of around $300 million for Genentech. The typical dose of tPA is around 100 mg, and this corresponds to a price per dose of around $2000.

The management of your company believes that the annual world demand can increase to more than 200 kg if the price per dose is reduced to around $300. Before committing to the venture, your president would like you, as director of corporate planning, to evaluate a technology for using transgenic goats to produce 50 kg of tPA per year. Based on data from Genzyme Transgenics, you know that you can buy milk containing tPA at a concentration of 20 g/liter for around $200/g of tPA.

Based on information from the technical and patent literature, develop and evaluate a process that can recover and purify 50 kg of tPA per year from goat milk. More specifically, estimate the capital investment required and the unit production cost.

11.2 **Indigo** Indigo is a dye that is used by denim manufacturers (to make blue jeans). It has been tradi-

tionally produced through chemical synthesis. The chemical route, however, generates large amounts of regulated waste materials that make the process environmentally unattractive. In the late 1990s, Genencor International (a biotechnology company) commercialized a technology for producing indigo via fermentation.

Based on information from the technical and patent literature, develop and evaluate a process for producing 5,000,000 kg of indigo per year via fermentation. The product must meet the quality specifications of the denim industry. More specifically, estimate the capital investment required and the unit production cost.

11.3 **L-Lysine** L-Lysine is an amino acid that is produced in large quantities (> 100,000 metric tons/year) via fermentation. It is used as an animal feed supplement mainly for poultry and pigs.

Based on information from the technical and patent literature, develop and evaluate a process for producing 15,000,000 kg of L-lysine per year via fermentation. Your analysis should include estimation of capital and operating cost.

11.4 **Xanthan Gum** Xanthan gum is a water-soluble polysaccharide produced via fermentation. It is used in food products as a thickener, stabilizer, and an emulsifier. Xanthan gum is also used for enhanced oil recovery.

Based on information from the technical and patent literature, develop and evaluate a process for producing 10,000,000 kg of xanthan gum per year. The product should meet the specifications of the petroleum industry for enhanced oil recovery. Your analysis should include estimation of capital and operating cost. Also, perform sensitivity analysis and estimate the unit production cost for plant capacities ranging from 10 to 50 million kg of xanthan gum per year.

11.5 **Biodegradable Polymers** Because of the capacity limitations of urban landfills, biodegradable plastic packaging materials are of interest as a means to reduce the load of solid waste disposal systems.

Poly-2-hydroxybutyrate (PHB), a biodegradable polyester, is such a promising material that can be produced via fermentation. Microorganisms that synthesize PHB include Gram-positive and Gram-negative species and cyanobacteria. Some members of the *Alcaligenes* and *Azotobacter* genera are the most promising because they store high levels of PHB. PHB is synthesized and stored intracellularly as a possible future carbon and energy source. High levels of polymer are obtained under nitrogen and phosphorus limitation. At optimum conditions, PHB can reach 70 to 80% by weight of the cell mass of the organism.

Based on information from the technical and patent literature, design and evaluate a plant that produces 30,000,000 kg of PHB per year. Your analysis should include estimation of capital and operating costs.

11.6 **Laundry Enzymes** Proteolytic enzymes are used in laundering to hydrolyze and remove proteinaceous stains. The commercially important proteolytic enzymes that are used in detergents are mainly produced by *Bacillus subtilis* and *Bacillus licheniformis*. These enzymes are endo-cleaving, have broad specificity, and are active over a wide pH range; calcium improves their stability at high temperature or extremes of pH. Their molecular weight is around 30,000 and their isoelectric point is in the range of 8.5 to 9.5.

The current world demand for laundry enzymes is around 10,000 metric tons (of pure enzyme) per year, corresponding to a world market of around $300 million.

The marketing department of your company believes that the annual world demand can increase to more than 20,000 metric tons if the selling price is reduced to around $15/kg of pure enzyme. Before committing to the venture, your company would

like you to evaluate the cost structure of the current producers and find out whether it is possible (through the use of genetic engineering and modern separation technologies) to produce such enzymes for less than $10/kg.

11.7 **Therapeutic Proteins from Transgenic Tobacco** Transgenic plants (e.g., corn, tobacco, etc.) have the potential to produce complex bioactive proteins at significantly lower cost than production via transgenic animals or mammalian cell cultures. The advantages of transgenic plant production are easy and efficient introduction of stable foreign genes, cost-effective biomass production ($0.02–$0.04/kg), no possible contamination with human disease agents, and the ability to perform complex protein processing needed for many bioactive human therapeutics. Cost-effective biomass production makes this mode of production suitable for large volume recombinant proteins. The ability to perform complex protein processing is advantageous for production of therapeutic glycoproteins and bioactive peptides. Downstream processing costs are a major portion of the total unit production cost associated with transgenic plant production of large volume therapeutic proteins. Therefore, primary recovery requires significant volume reduction. Once volume reduction and biomass removal have been achieved, chromatographic purification is required to remove plant protein impurities.

Design a purification process for use in a facility manufacturing 100 metric tons per year of a recombinant human blood protein from transgenic tobacco. Assume that the expression level is 0.5 g of product protein per kilogram of tobacco. Assume that biomass production and primary recovery are performed at a separate site. The feed to the purification section is 4000 liters/day (containing 200 g/liter of product protein), and this material is purchased for $0.5/g of product protein. The process should include the appropriate filtration and chromatographic steps. Assume that a combination of affinity and ion exchange chromatography will provide a product that is more than 90% pure. (Data from Steve Griffiths, Department of Chemical Engineering, MIT.)

References

1. Douglas, J. M. (1988). *Conceptual Design of Chemical Processes,* McGraw-Hill, New York.
2. Frohlich, B. T. (1999). Considerations for developing in-house capabilities for cost of goods estimation. In *Biopharmaceutical Process Economics and Optimization* (conference proceedings), September 30–October 1, 1999, International Business Communications, Westborough, MA.
3. Asenjo, J., and Maugeri, F. (1991). An expert system for selection and synthesis of protein purification processes. In *Frontiers in Bioprocessing II,* P. Todd, S. K. Sikdar, and M. Bier, eds., American Chemical Society, Washington, DC, p. 358.
4. Leser, E. W., and Asenjo, J. A. (1992). Rational design of purification processes for recombinant proteins. *J. Chromatog.,* vol. 584, p. 43.
5. Petrides, D. P., Cooney, C. L., and Evans, L. B. (1989). An introduction to biochemical process design. In *Chemical Engineering Problems in Biotechnology,* vol. 1, M. L. Shuler, ed., American Institute of Chemical Engineers, New York, p. 351.
6. Petrides, D. P. (1994). BioPro Designer—An advanced computing environment for modeling and design of integrated biochemical processes. *Comput. Chem. Eng.,* vol. 18, p. S621.
7. Kula, M.-R., and Horst, S. (1987). Purification of proteins and the disruption of microbial cells. *Biotechnol. Prog.,* vol. 3, p. 31.
8. Taylor, G., Hoare, M., Gray, D. R., and Marston, F. A. O. (1986). Size and density of protein inclusion bodies. *Bio/Technology,* vol. 4, p. 553.
9. Palmer, D. E. (1977). High purity protein recovery. *Process Biochem.,* vol. 12, p. 24.
10. Chang, Y.-K., and Chase, H. A. (1996). Development of operating conditions for protein purification using expanded bed techniques: The effect of the degree of bed expansion on adsorption performance. *Biotechnol. Bioeng.,* vol. 49, p. 512.
11. Dlouhy, P. E., and Dahlstrom, D. A. (1968). Continuous filtration in pharmaceutical processing. *Chem. Eng. Prog.,* vol. 64, p. 116.
12. Brunner, K. H., and Hemfort, H. (1988). Centrifugal separation in biotechnological processes. In *Advances in Biotechnological Processes,* vol. 8, A. Mizrahi, ed., Liss, New York, p. 1.
13. Axelsson, H. A. C. (1985). Centrifugation. In *Comprehensive Biotechnology,* vol. 2, M. Moo-Young, ed., Pergamon Press, Oxford, p. 325.
14. Freese H. L. (1993). Evaporation. In *Fermentation and Biochemical Engineering Handbook,* H. C. Vogel, ed., Noyes, Park Ridge, NJ, p. 227.
15. Chan. M. Y. Y., Hoare, M., and Dunnill, P. (1986). The kinetics of protein precipitation by different reagents. *Biotechnol. Bioeng.,* vol. 28, p. 387.
16. Fish, W. W., Danielson, A., Nordling, K., Miller, S. H., Lam, C. F., and Bjork, I. (1985). Denaturation behavior of antithrombin in guanidinium chloride. *Biochemistry,* vol. 24, p. 1510.
17. Bjurstrom, E. (1985). Biotechnology. *Chem. Eng.,* vol. 92, p. 126.
18. Wheelwright, S. M. (1987). Designing downstream processes for large-scale protein purification. *Bio/Technology,* vol. 5, p. 789.
19. Peters, M. S., and Timmerhaus, K. D. (1991). *Plant Design and Economics for Chemical Engineers,* 4th ed., McGraw-Hill, New York.
20. Ulrich, G. D. (1984). *A Guide to Chemical Process Design and Economics,* Wiley, New York.
21. Valle-Riestra, J. F. (1983). *Project Evaluation in the Chemical Process Industries,* McGraw-Hill, New York.

22. Garrett, D. E. (1989). *Chemical Engineering Economics,* Van Nostrand Reinhold, New York.

23. Seider, W. D., Seader, J. D., and Lewin, D. R. (1999). *Process Design Principles—Synthesis, Analysis, and Evaluation,* Wiley, New York.

24. Kalk, J. P., and Langlykke, A. F. (1986). Cost estimation for biotechnology projects. In *Manual of Industrial Microbiology and Biotechnology,* A. L. Demain and N. A. Solomon, eds., American Society for Microbiology, Washington, DC, p. 363.

25. Reisman, H. B. (1988). *Economic Analysis of Fermentation Processes,* CRC Press, Boca Raton, FL.

26. Crueger, W., and Crueger, A. (1989). *Biotechnology—A Textbook of Industrial Microbiology,* 2nd ed., Sinauer, Sunderland, MA.

27. Roberts L. R. (1979). Citric acid. In *Encyclopedia of Chemical Processing and Design,* vol. 8, J. J. McKetta and W. A. Cunningham, eds., Dekker, New York, p. 324.

28. Blanch, H. W., and Clark, D. S. (1997). *Biochemical Engineering,* Dekker, New York.

29. Barfoed, H. C. (1987). Insulin production technology. *Chem. Eng. Prog.,* vol. 83, p. 49.

30. Ladisch, M. R., and Kohlmann, K. L. (1992). Recombinant human insulin. *Biotechnol. Prog.,* vol. 8, p. 469.

31. Kehoe, J. A. (1989). The story of biosynthetic human insulin. In *Frontiers in Bioprocessing,* S. K. Sikdar, M. Bier, and P. Todd, eds., CRC Press, Boca Raton, FL, p. 45.

32. Datar, R., and Rosen, C.-G. (1990). Downstream process economics. In *Separation Processes in Biotechnology,* J. A. Asenjo, ed., Dekker, New York, p. 741.

33. Petrides, D. P., Sapidou, E., and Calandranis, J. (1995). Computer-aided process analysis and economic evaluation for biosynthetic human insulin production—A case study. *Biotechnol. Bioeng.,* vol. 5, p. 529.

34. Di Marchi, R. D. (1984). Process for inhibiting undesired thiol reactions during cyanogen bromide cleavage of peptides, U.S. Patent 4,451,396.

35. Bobbitt, J. L., and Manetta, J. (1990). Purification and refolding of recombinant proteins, U.S. Patent 4,923,967.

36. Seaver, S. S. (1997). Monoclonal antibodies: Using new techniques to reduce development time. *Gene. Eng. News,* vol. 17, p. 13.

37. DeYoung, H. G. (1996). Multiple choices for monoclonal production. *Gene. Eng. News,* vol. 16, p. 12.

Laboratory Exercises in Bioseparations

Chapters 3 to 10 of this text are organized around the study of unit operations in the approximate order of their customary application to bioseparations. In this chapter, five of these operations are singled out for further exploration in the laboratory. These are flocculant screening (Chapter 3), crossflow filtration (Chapter 4), centrifugation of cells and lysate (Chapter 5), aqueous two-phase partitioning of a protein (Chapter 6), and gradient-elution ion exchange chromatography of test proteins (Chapter 7). Each section of this chapter is thus an independent laboratory exercise. The instructions can be applied flexibly to the materials and equipment available to a particular laboratory or department. The calculations, reporting, and scaleup applications are applicable to any experiment that follows the generic paradigm of each of the sets of lab instructions. The pattern to be followed consists of becoming acquainted with the equipment and describing it as a unit operation in a report, execution of a predesigned experiment, recording of appropriate data, analysis of the data in the context of this textbook, presenting reduced data in a report, critically analyzing the quality of the results, and, finally applying the actual numerical results to a scaleup to production scale. Process economics may be applied where appropriate.

12.1 Flocculant Screening

In this laboratory exercise, a flocculant will be evaluated for its ability to flocculate cells or lysate particles. Lysate particles are smaller and require, typically, higher concentration of flocculant than that required to flocculate whole cells. The flocculant concentration required will be determined by observing persistence of flocculation and clarity of supernatant measured as a function of flocculant concentration.

12.1.1 BACKGROUND

Flocculants are usually polymers with properties, such as charge, that cause them to interact with cells or lysate particles and bind them together. Their effectiveness depends on molecular weight, charge, solubility, and other properties, and their interactions with cells and particles depend, therefore, on pH, ionic strength, temperature, and dry solids concentration.

Unless a great deal is known about the suspended material of interest, or extensive experience has been published, the choice of flocculating conditions usually depends on educated trial and error. Typically, the engineer or scientist chooses a series of flocculants and tests them at different concentrations and ionic strengths, noting the time to observed flocculation, the clarity of the supernatant after centrifugation, and the quality of the flocculated precipitate. An example would be the Cysep series of cationic flocculants 4215, 4218, 4224, and 4225 (Cytec Industries, Inc.), in order of increasing positive charge; the flocculants in this series do not contain oil or surfactants, thus increasing compatibility with downstream processing involving membranes and chromatography media.

12.1.2 OBJECTIVES

The objectives of this laboratory exercise are to understand flocculation as a unit operation, monitor flocculation through visual observation, and use postcentrifugation data to evaluate the quality of flocculation used to scale a test process to an industrial process.

12.1.3 PROCEDURE

1. Obtain samples of cells or cell lysate suspended in broth or buffer and stock solutions of flocculants from the instructor.
2. Divide each group (three to six persons) according to tasks, such as pH and other preliminary measurements, addition and mixing of flocculants, benchtop centrifugation of flocs, postcentrifuge evaluations (as described later), and using Table 12.1 to keep records.
3. Record sample pH, conductivity, dry weight of solids, source, and type of cells or lysate as recommended by the instructor.
4. Make (or obtain) a polymer flocculant solution, 0.5% (w/v) if for cells, 3% (w/v) if for lysate.
5. Place 5 ml of cell suspension or lysate into each of six 15 ml centrifuge tubes.
6. Add 0, 1, 2, 3, 4, or 5 ml of polymer solution to each tube. Be sure to label tubes. Do not shake.
7. Add water to give a total liquid volume of 10 ml in each tube. Do not shake.
8. Invert tubes 20 times. Record on which inversion each flocculation is first seen and whether breakup of the flocs occurs.
9. Centrifuge the zero-flocculant control at 1000 rpm for 1 min. If it clarifies, resuspend the solids and repeat the centrifugation at lower speed. Using the highest speed at which the zero-flocculant control does not clarify, centrifuge the samples with flocculant for 1 min.
10. Qualitatively judge which supernatant is least cloudy. Score 1 for clearest and 6 for least affected. Record pellet volume and type, using the scale on the centrifuge tube. If time permits, record optical density at 600 nm of the top 2 ml of supernatant, including control.
11. If no flocculation is seen, examine higher polymer doses, 3 to 40 g/liter; if "overkill" is seen, examine lower polymer doses, especially below 10 g/liter dry weight of solids in the cell or lysate sample.
12. Record results for the group in Table 12.1.

TABLE 12.1
Tabulation of Results of Flocculation Experiments

Sample description: _____

Name and concentration of flocculant: _____

	0	1	2	3	4	5
Flocculant solution (ml)						
Flocculation at inversion number						
Still flocculated after 20 inversions?						
Clarity order postcentrifugation						
Pellet volume (ml)						
Optical density of top 2 ml of supernatant						

12.1.4 REPORT

Write a short report using the following outline.

1. Description of the experiment.
2. Record of all measurements made on sample prior to flocculation test.
3. Complete information about the flocculant solution.
4. Tabular results—use Table 12.1.
5. Identification of optimum conditions for flocculation from the results in the table.
6. A brief design of a unit operation in which you optimally flocculate 100 liters of the product used in the experiment.

12.1.5 SOME NOTES AND PRECAUTIONS

1. The procedure described here is a laboratory exercise and results in varying the ionic strength to keep volume constant for centrifugation. Real-world tests with undiluted broth must be performed to find both the optimum polymer dose and ionic strength. The optimum final dose will probably be close to that found in this procedure.
2. No pH adjustment should be required. Flocculation should be found for all biological suspensions at pH 7 (pH is recorded for posterity and as a standard laboratory practice).

3. Always dilute the as-supplied polymers to 5% or below. Store in the cool and dark for a maximum of 3 days. The Cytec cationic flocculants are stable from pH 3 to 11.
4. The flocculant dose for a given dry weight of whole cells (in any volume) should be approximately constant and less than that for a given dry weight suspension of lysate, which may vary according to the degree and mode of lysis.

12.2 Crossflow Filtration

In this laboratory exercise, permeate flux will be measured as a function of solids concentration in a crossflow filtration apparatus operated in batch concentration mode (see Chapter 4). The decline of flux as solids or solutes are concentrated is an important measurement in the monitoring of the progress of a filtration step and in the evaluation of a filtration process for scaleup. These principles apply to both microfiltration and ultrafiltration operations.

12.2.1 BACKGROUND

Common configurations ("modes") of crossflow filtration unit operations are shown in Figure 4.15. In the *total recycle mode* there is no net concentration of solids or retained solutes because both the permeate and the retentate are returned to and remixed with the feed. In the *batch concentration mode* only the retentate is returned to and remixed with the feed. Since permeate is continuously removed from the system, there is an unsteady increase in retained solids or macromolecules in the feed. Again, however, the compositions of the recirculation loop and the material in the feed tank are identical. The batch concentration configuration will be used in this laboratory exercise. *Continuous, or feed-and-bleed, configurations* (Figure 4.17) are designed for steady state operation. Permeate flows are adjusted so that the recirculation loop composition is constant and generally equal to the final desired composition. Neither the permeate nor the retentate is returned to the feed tank, although the result would be unchanged if the retentate recirculation loop returned directly to the tank.

12.2.2 OBJECTIVES

The objectives of this laboratory exercise are to understand crossflow filtration as a unit operation, monitor permeate flux through a batch concentration operation and determine its dependence on solids concentration, and use crossflow filtration data to scale a pilot process to an industrial process.

12.2.3 PROCEDURE

1. Test the crossflow filtration unit to ascertain that there are no pressure leaks in the filter cartridge. Be sure to understand the meaning of each pressure (inlet, back pressure, and permeate outlet pressure) and to adhere to the limits specified by the instructor. Be careful not to exceed the back pressure specified by the instructor.
2. Determine the volume of feed suspension, which consists of a fermentation broth containing *E. coli* cells. The cells are to be concentrated by operating the filter unit in the batch concentration mode.

3. Assign a member of the research team to each task recommended by the instructor, such as flux measurement, pressure and temperature recording, and determining residual volume in the feed tank and concentration of solids.
4. Set pressures and temperatures to desired values while operating the unit in total recycle mode (return permeate line to the feed tank).
5. Begin filtration in the batch concentration mode, noting time, feed volume, pressures, and temperatures. At specified intervals, determine the permeate flux by collecting a known volume of permeate in a graduated cylinder while timing with a stopwatch.
6. Determine solids concentration in the feed tank at the same time intervals. Note that this can be calculated by knowing the initial solids content of the whole broth and the volumetric concentration factor derived from the new volume of feed in the tank at each time point. Try to make feed volume and permeate flux determinations simultaneously. The solids concentration can be confirmed by a calibrated concentration measurement, such as turbidity, if such method is specified by the instructor. Record the area of the membrane used.

12.2.4 REPORT

Prepare an *individual* written report using the following outline.

1. One-paragraph summary.
2. Experimental procedure, including equipment sketch.
3. Graphical summary (with accompanying narrative) of experimental results, plotting permeate flux versus time, feed volume and/or solids concentration versus time, and permeate flux versus solids concentration. A log scale on the abscissa is useful.
4. Process flow diagram showing membrane filter, proper configuration, and computed feed and permeate rates.
5. Concise written discussion of results, errors, and engineering inaccuracies.
6. A scaling calculation, performed by using the experimental membrane filter area to calculate the membrane area required to achieve a $5 \times$ concentration of whole broth from a production scale fermentor. The volume of the production scale fermentor is 180,000 liters, and the fermentation time including turnaround is 4 days. The whole broth contains 2% solids by weight. Assume a *feed-and-bleed* configuration around the filtration unit. Estimate the capital investment required to do the job, using the information in Chapter 11 about the cost of crossflow filters. Discuss errors and engineering inaccuracies that affect your design.

12.3 Centrifugation of Flocculated and Unflocculated Particulates

In this laboratory exercise, effluent particle concentration will be measured as a function of flow rate in a continuous centrifugation apparatus. The escape of particles is an important measurement in the monitoring of the success of a centrifugation step and in the evaluation of a centrifugation process for scaleup. These principles apply equally to cell and lysate separations, both of which can be enhanced by flocculation prior to centrifugation. There-fore, the effects of flocculation on particle retention will be evaluated on the basis of "breakthrough curves," which establish allowable flow rates for a given combination of centrifuge and product mixture.

12.3.1 BACKGROUND

A disk-stack centrifuge (Figure 5.4) can be operated in flowing batch or continuous mode. In the latter case, solids are ejected continuously through a nozzle. In this experiment, the flowing batch mode will be used. The rate at which feed can be pumped into the centrifuge (Q) is determined by the geometry and speed of the centrifuge (Σ factor—see Sections 5.4.1 and 5.4.2) and the sedimentation properties of the particles in the feed (v_g—see Section 5.3.4). It is possible to control v_g by flocculation, the conversion of individual particles to larger hydrodynamic units.

If a disk-stack centrifuge is unavailable, this experiment can also be performed using a bowl centrifuge, such as the Beckman JCF-Z continuous flow rotor [1], the inner work-ings of which are shown in Figure 12.1. In this case, the measurements required for Σ de-termination can be found in Section 5.4.1. If it is not possible to make direct measurements on the centrifuge rotor, whether disk-stack or bowl, these dimensions will be provided by the instructor.

12.3.2 OBJECTIVES

The objectives of this laboratory exercise are to understand centrifugation as a unit operation, monitor particle escape through a flowing batch operation and determine its dependence on

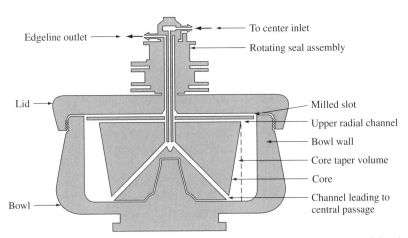

Figure 12.1 Cross section of a commercial bowl centrifuge rotor. Feed enters from the right, above the rotary seals, and passes down the central tube to a branching channel that carries the pumped feed to the bottom of the bowl. Feed flows upward in the space between the core and the bowl wall, then su-pernatant passes up the flow annulus to the outlet. (Based on a design by Beckman Coulter, Inc., Spinco Division, 1995.)

solids properties with and without flocculation, and use breakthrough centrifugation data to scale a pilot process to an industrial process.

12.3.3 PROCEDURE

1. Obtain one of four possible samples from the instructor:
   (a) Unflocculated whole cells (bacteria or yeast) in broth
   (b) Flocculated whole cells in broth
   (c) Unflocculated lysate of cells in broth
   (d) Flocculated lysate of cells in broth

   in a feed container to be connected to the feed pump for pumping into the centrifuge.

2. Assign a member of the experiment team to each task recommended by the instructor, such as flow rate measurement and/or recording, sample collection, sample dilution if dilution is necessary, reading the spectrophotometer, recording spectrophotometer readings, assisting with cleanup, obtaining the correct measurements and speed of the centrifuge required for determining Σ, and collecting data on a common sheet and submitting a copy for other groups. The bowl and core combination shown in Figure 12.1 for the Beckman JCF-Z continuous flow rotor is termed "large pellet" rotor, since the filled bowl can accommodate up to 800 ml of solids. This bowl has a top acceleration of $39,900 \times g$, the inner radius of the capture zone is 5.6 cm, and the outer radius is 8.9 cm. The rotor has a volume capacity of 1250 ml, and its internal height is 8.2 cm.

   You should determine the Σ value for the centrifuge from the dimensions and speed. Using breakthrough Q's, get values for v_g. Use these results to scale up disk-stack centrifuges (not bowl centrifuges).

3. Connect the feed container, via tubing, to the feed pump and the pump to the centrifuge. Pump the suspension through the centrifuge at a given flow rate from 500 to 2200 ml/min.

4. Collect the clear liquid coming out from the upper chute of the centrifuge.

5. Measure the optical density (OD) at 600 nm of the clear broth collected at this flow rate.

6. Change the flow rate to another value and repeat steps 3 to 5. When the changing flow rate, it is advisable to randomly increase or decrease pump speed rather than simply increase or decrease speed linearly. The pump rate should be allowed to reach steady state before the feed is switched into the centrifuge inlet. The pump rate can be measured by using a graduated cylinder and a stopwatch.

7. Record and plot OD versus flow rate for the given feed material using the data table provided (Table 12.2) and the data from other groups. The plot should look like the example breakthrough curve in Figure 12.2.

8. Collect data from other groups for other feed materials.

12.3.4 REPORT

Each person will prepare an *individual* written report using the following outline.

1. One-paragraph summary.
2. Experimental procedure, including equipment sketch.

TABLE 12.2
Form for Tabulating Effluent Concentration vs Flow Rate
for Plotting in a Graph such as Figure 12.2

Unflocculated Whole Cells

Flow rate					
Supernatant OD					

Flocculated Whole Cells

Flow rate					
Supernatant OD					

Unflocculated Lysate

Flow rate					
Supernatant OD					

Flocculated Lysate

Flow rate					
Supernatant OD					

3. Table of data and graphical summary (with accompanying narrative) of experimental results (tabulated using Table 12.2 format), plotting breakthrough OD versus flow rate for the four different suspensions. Indicate which suspension was used by your group.

4. Process flow diagram showing rotor components (symbolically), feed (and pumping) configuration, and given feed and supernatant flow rates and concentrations.

5. A concise written discussion of results, errors, and engineering inaccuracies.

6. Scaling calculation performed by using the experimental centrifuge to calculate the number of centrifuges required to achieve a 0.07 OD suspension from whole broth from a production-scale fermentor. The production scale fermentor is 180,000 liters, and the fermentation time including turnaround is 4 days. The whole broth contains 1.58 OD units of unflocculated whole cells. Using appropriate relationships, indicate the possible changes that could be made to centrifuge design to reduce the number of centrifuges required. State the centrifuge specifications required to perform this job with six mechanically feasible centrifuges. Discuss errors and engineering inaccuracies that affect your design.

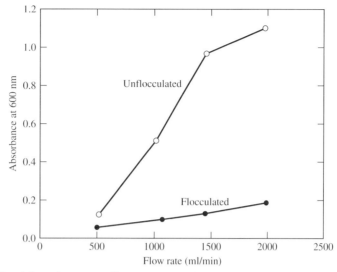

Figure 12.2 Breakthrough curves (effluent concentration versus feed flow rate) for flocculated and unflocculated cells in broth in a typical experiment. Note that the flow rate can be more than four times as great for flocculated cells as that for unflocculated cells. This set of breakthrough curves is for *E. coli* treated with Cytec E-2706 cationic flocculant at pH 5 compared with unflocculated suspension. Centrifuge: Alfa Laval at 14,000 rpm.

12.4 Aqueous Two-Phase Extraction

Aqueous two-phase extraction is a technique used in the biochemistry profession and, to a lesser extent, in the biotechnology industry, to isolate and purify a variety of proteins and biological materials including subcellular particles and whole cells. This technique was developed to its present form by Albertsson [2].

Two-phase aqueous systems are made by combining two incompatible water-soluble polymers (e.g., PEG and dextran) or a polymer and a salt in water, above a "critical" concentration. These systems separate into two immiscible liquid phases, one enriched in one polymer and the other enriched in the other polymer or salt. Soluble materials, such as proteins, or particulate separands (including living cells), added to the system, partition themselves between the two phases [3]. This is the basis for extraction. For this experiment, PEG/dextran is the phase system of choice because its physical properties make it suitable for protein extraction without any harmful effects on the protein. The partition coefficients of different separands can be manipulated in this two-polymer system by the addition of acid, base, partitioning ions, nonpartitioning ions, and surfactants.

12.4.1 PHYSICAL MEASUREMENTS

In a single-stage extraction, a dextran solution (heavy phase) along with a certain amount of protein is thoroughly mixed with a solution of PEG (light phase). The protein partitions between the two phases until equilibrium is reached. The equilibrium partition coefficient

K is defined as

$$K = y/x \qquad (12.4.1)$$

where y = concentration of protein in the extract phase
x = concentration of protein in the raffinate phase.

The objectives of this experiment are as follows:

1. Show that the PEG/dextran mixture forms two phases above certain concentrations.
2. Estimate phase volume ratios.
3. Determine the equilibrium partition coefficient of a protein.
4. Determine the product yield.

12.4.2 PROCEDURE

1. Use the following aqueous stock solutions:
   (a) 50% (w/w) PEG, MW 8000
   (b) 10 mg/ml hemoglobin, MW 64,500, in 0.1 M Na phosphate buffer, pH 7.8
   (c) 0.1 M Na phosphate buffer, pH 7.8
   (d) 25% (w/w) dextran, MW 480,000

2. Determine the mass of each solution and of dry NaCl needed to combine to form each of the four phase systems with the initial compositions specified on the phase diagram given (Figure 12.3). You will make a total of six phase systems, one each of systems 1, 2, and 4, and three of phase system 3.
   (a) Plan to add 0.1, 0.2, and 0.5 ml of the hemoglobin solution to phase system 3 and 0.2 ml to the other three systems.
   (b) Record all weights.

3. Formulation of the phase systems:
   (a) Place a 15 ml graduated conical tube in a beaker on the balance and tare it.
   (b) Carefully add the appropriate amounts of PEG, dextran, and buffer stock solutions to the tube. Be sure to wait for the pipet to fill with the appropriate mass of dextran solution, which is very viscous.
   (c) Add 0.5 g of NaCl.
   (d) Add the chosen amount of hemoglobin plus 1 ml of buffer (0.1 M Na phosphate, pH 7.8) and enough water (in grams) for a total final mass of 10 g.

   Table 12.3 summarizes the compositions of the six systems. The masses of the PEG and dextran stock solutions required for system 4 should be determined from the table in Figure 12.3.

4. Vortex each phase system. Make sure the NaCl is fully dissolved. If necessary, invert the tube several times to help in dissolving the salt.

5. Centrifuge each system at 1500 rpm for 10 min to separate the two phases. Make sure the centrifuge is balanced by checking the volumes of diametrically opposite samples.

6. Use the calibrated markings on the tube to estimate and record the volume of each phase.

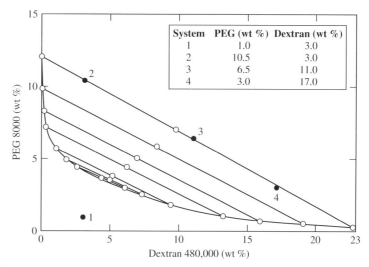

System	PEG (wt %)	Dextran (wt %)
1	1.0	3.0
2	10.5	3.0
3	6.5	11.0
4	3.0	17.0

Figure 12.3 Phase diagram of the PEG 8000/dextran 480,000 system, indicating the total compositions of systems 1, 2, 3, and 4 by the positions of the corresponding numbered closed circles. Indicated on the tie lines are the experimental data used to construct the phase diagram (open circles) showing compositions of the two phases (points on the binodial curve) and the composition of the total system (points between the points on the binodial curve). (Data from P.-Å. Albertsson, *Partition of Cell Particles and Macromolecules,* 3rd ed., Wiley, New York, 1986.)

TABLE 12.3
Compositions of the Six PEG/Dextran Two-Phase Systems

	System					
	1	2	3a	3b	3c	4
PEG stock, g	0.2	2.1	1.3	1.3	1.3	
Dextran stock, g	1.2	1.2	4.4	4.4	4.4	
Hemoglobin solution, ml	0.2	0.2	0.1	0.2	0.5	0.2
NaCl, g	0.5	0.5	0.5	0.5	0.5	0.5
Buffer, ml	1.0	1.0	1.0	1.0	1.0	1.0

7. Remove a sample from the top and bottom phases by carefully drawing up 0.6 to 0.7 ml of the top phase with a pipet, taking care not to disturb the interface. Gently perforate the tube to allow the bottom phase to drain into a tube for diluting and measuring optical absorbance in the spectrophotometer.

8. Determine the concentration of protein in each phase using a spectrophotometer and the calibration curve (Figure 12.4).

   (a) Set wavelength λ at 405 nm.

   (b) Use pure water to set blank.

   (c) Make a 1:6 dilution of each sample with pure water.

   (d) Record absorbance of each phase sample.

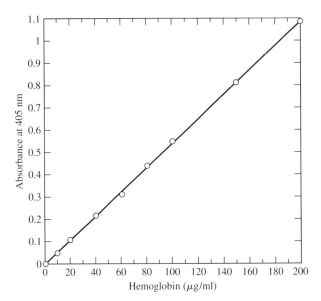

Figure 12.4 Hemoglobin calibration curve, absorbance at 405 nm versus protein concentration.

12.4.3 CALCULATIONS AND REPORT

Reports should be no more than 10 pages long and should be written in the format of a scientific paper, which should include the following:

1. *Title page* with your name, date, title of report, and a 100- to 150-word abstract that summarizes the purpose, methods, and results of the experiments.

2. *Introduction,* which gives background on biphasic aqueous extraction and describes the objectives of the experiments.

3. *Materials and methods,* describing the materials, equipment, and procedures you used. Provide sketches of equipment as appropriate. Cite volumes, concentrations, and so on that were used. This section should be written in a way that would allow the reader to use it as a recipe to repeat your experiments.

4. *Results and discussion,* including the raw data that you measured, together with the calculated results and equations used. Present calculated results in tables or figures, as appropriate. Define all symbols used, and number the equations, tables, and figures; refer to these display items by number in your text. Include error analysis (means, standard errors, confidence limits) and discussion of problems or difficulties, especially those leading to systematic errors. In particular, the following must be included:

   (a) Statement of which of the four systems formed two phases.

   (b) Report of the phase volume ratios for each system and, from the binodial curve on the phase diagram (Figure 12.3), a determination of the equilibrium composition of the upper and lower phases. Calculate the mass ratio of each phase from the phase diagram using the lever rule (see instructions in Section 12.4.4 and Figure 12.5). Explain the trends observed in the phase volume ratios and compare those with the calculated mass ratios.

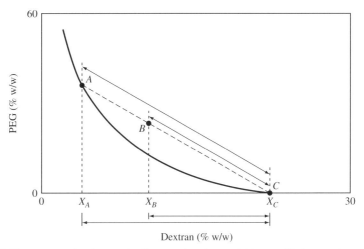

Figure 12.5 Phase separation lever rule diagram showing variables used in inverse lever rule calculations for the purpose of determining phase volume ratios.

(c) Report of the concentration of hemoglobin in both phases. Calculate the percentage recovery and the equilibrium partition coefficient K of hemoglobin in each phase system. Answer the following questions:

   (1) Did K vary as a function of volume ratio?

   (2) Did K vary as a function of protein concentration (explain)?

   (3) What is the significance of the ratio of mass of hemoglobin in the top phase to that in the bottom phase? What is the extraction factor, and what is the extracted fraction in each case? If these factors varied, why did they vary?

   (4) Why was the NaCl added to the phase system?

5. *Concluding remarks* in which findings are summarized. In particular, explain how the data obtained would be used in an extraction procedure and discuss why the PEG/dextran phase system is suitable for protein extraction.

6. *Notation* page with a table of definitions of dimensions or units of symbols used.

7. *References* utilized or relevant to this exercise, including authors, year, article title, journal or book title, volume number, and page numbers.

12.4.4 Inverse Lever Rule

The line connecting points A and C on the binodial curve in the phase diagram shown in Figure 12.5 is known as a *tie line*. The tie line connects the compositions at which the top and bottom phases are in equilibrium. Since the top and bottom phases are always of some uniform composition, the relative amount of each phase can be calculated using a material balance. If the system consists of N moles (or grams) total, then at any point:

$$N = N_T + N_B \tag{12.4.2}$$

where N_T and N_B are the number of grams in the top and bottom phases, respectively.

Thus for a system of overall composition X_B, referring to Figure 12.5,

$$NX_B = N_T X_A + N_B X_C \qquad (12.4.3)$$

Substituting Equation (12.4.2) into Equation (12.4.3), we get

$$(N_T + N_B)X_B = N_T X_A + N_B X_C \qquad (12.4.4)$$

Rearranging leads to

$$\frac{N_T}{N_B} = \frac{X_C - X_B}{X_B - X_A} = \frac{\text{length of line } BC}{\text{length of line } AB} \qquad (12.4.5)$$

This equation is known as the *inverse lever rule*. Thus, the ratio of the masses of the top and bottom phases can be calculated simply as the ratio of the length of two segments of the tie line.

12.5 Chromatography Scaleup

In this laboratory exercise, elution volume, peak width, resolution, height equivalent of a theoretical plate, and the number of theoretical plates will be determined for a small-scale column and a large-scale column and compared to determine the success of scaleup without significantly changing resolution.

12.5.1 BACKGROUND

Column chromatography is a common procedure for purifying biological compounds such as proteins, nucleic acids, and antibiotics (see Chapter 7). The basis for separating one dissolved species from another in a mixture depends on the interaction of the solute molecules with the solid support (packing) and the liquid mobile phase. The ionic charge, molecular weight, or relative hydrophobicity may be used as the basis for separation. When one wants to separate solutes based on their ionic charge, one uses an ion exchange column.

12.5.2 OBJECTIVES

The goal of this experiment is to obtain hands-on experience in running an ion exchange chromatography column and to learn how to operate the column such that a separation of two different proteins is obtained. Equipment used in this experiment will be chromatography columns and instruments commonly found in purification laboratories. A scaleup calculation will be performed using the Yamamoto (constant resolution) approach.

12.5.3 PROCEDURE

This exercise will require some prelaboratory preparation before the laboratory procedures can be implemented. The prelaboratory preparation will aid in being prepared to make all the appropriate measurements.

Prelaboratory Preparation

Review the small-scale column chromatogram (Figure 12.6) and calculate the resolution (R_s), the gradient slope (G), and the Yamamoto number (Y_m) for the separation of ribonuclease A and cytochrome c only. Use the relevant equations given shortly. Assume for S Sepharose Fast Flow (the packing that will be used) that $\varepsilon = 0.3$ and $d_p = 100\,\mu$m. Bring these calculations to the laboratory, and be prepared to review them at the time of the experiment.

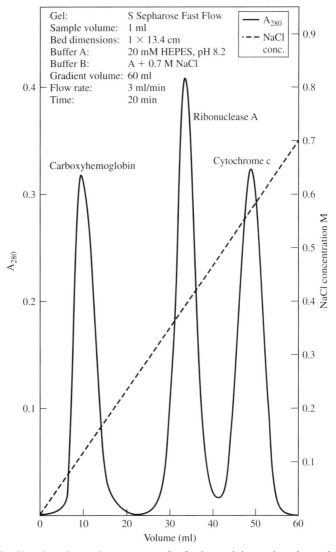

Figure 12.6 Small-scale column chromatogram for feed containing carboxyhemoglobin (2.0 mg), ribonuclease A (2.0 mg), and cytochrome C (1.8 mg). (Courtesy of Amersham Biosciences Corp.)

Constant Resolution Relationships

The scaleup principle described in Chapter 7 (Section 7.8) is based on the principle of having the same resolution at all levels of scaleup. From Equation (7.4.3) for resolution, using average elution volumes instead of average retention times, we write

$$R_s = \frac{V_2 - V_1}{\frac{1}{2}(w_2 + w_1)} \tag{12.5.1}$$

where V_2 and V_1 are the average elution volumes at which components 2 and 1 emerge from the column, respectively, and w_2 and w_1 are the widths of the respective elution peaks. Peak width w is defined in Figure 7.6, where it is seen that

$$w = 4\sigma \tag{12.5.2}$$

for peak standard deviation σ.

From constant resolution theory, we next define a number, Y_m, which, when held constant, can be used to calculate scaled operating values. For the purpose of this exercise, the following form is used [see Equation (7.8.11)]:

$$Y_m = \left(\frac{L}{G(1 - \varepsilon)u d_p^2}\right)^{1/2} \tag{12.5.3}$$

where L is bed height, u is interstitial flow velocity, d_p is resin particle diameter, ε is void fraction, and G is the normalized gradient slope, given by Equation (7.8.17). If we assume that void fraction and resin particle size are not changed with scaleup, then the relationship of interest is simply

$$\left(\frac{Y_{m1}}{Y_{m2}}\right)^2 = 1 = \frac{L_1 G_2 u_2}{L_2 G_1 u_1} \tag{12.5.4}$$

Finally, for constant resolution, the volume of gradient components to be used in the laboratory column can be determined by writing Equation (7.8.17) as

$$V_g = \frac{V(c_{hi} - c_{low})}{G} \tag{12.5.5}$$

where V is the column volume and V_g is the total gradient volume.

Laboratory Procedure

1. With the help of the instructor, determine the proper operating parameters for the particular column with which you will work. Each group may work with a different set of column parameters. Pitfalls, according to experience, could consist of improper chart connections, pumping speed, or calibrations. Assignment of tasks is important.

2. Equilibrate the column with the buffer provided for this purpose, and set up the UV detector and recorder to be sure it is functioning properly.

3. Load the protein mixture onto the column.

4. Separate the proteins from one another by eluting the column with the salt gradient. During this time, record the optical absorbance and conductivity at the outlet, and monitor the elution volume by collecting the eluant in a graduated cylinder.

5. Obtain a copy of the chromatogram (recording of optical density versus time), and record the volumetric data, column parameters, equipment sketch, and any other relevant information needed for completing the report.

6. Reequilibrate the column(s).

12.5.4 Report

Each person will prepare in *individual* written report using the following outline:

1. One-paragraph summary.

2. Experimental procedure, including equipment sketch.

3. Calculation of R_s, G, and Y_m for the small-scale column using relationships in Equations (12.5.1), (12.5.2), (12.5.3), and (7.8.17).

4. Graphical summary, including a copy of your experimental chromatogram obtained during the experiment with the salt gradient superimposed.

5. Calculation of values of V_e and w in milliliters, height equivalent of a theoretical plate (HETP), number of theoretical plates (N), and R_s for the experiment column. Discuss whether the scaleup was or was not successful. N is determined from the following form of Equation (7.4.2):

$$N = \frac{V_e^2}{(w/4)^2} \tag{12.5.6}$$

where w is the peak width and V_e is the elution volume to elute a particular separand out of the column. The corresponding "plate height" is

$$HETP = \frac{L}{N} \tag{12.5.7}$$

Each of these values may be calculated for each separand.

6. Concise written discussion of results, errors, and measurement inaccuracies.

7. Scaling calculation performed by describing how you would operate a 10-liter column to give the same resolution as the small-scale column or the large-scale column. Include column diameter, bed height, linear and volumetric flow rates, and total gradient volume calculations for the 10-liter column. The column you choose should be available from Millipore (see Table 12.4), and the flow properties of S Sepharose Fast Flow packing should be considered (Figure 12.7). Discuss why you selected your particular design, and list your assumptions and considerations.

Limit reports to seven pages, including graphs and sketches.

TABLE 12.4
Characteristics of Moduline® Acrylic Biochromatography Columns: Sizes and Pressure Ratings

Diameter × length (mm)	Adjustable length (mm)	Cross-sectional area (cm²)	Adjustable capacity (liters)	Pressure rating (bar)
70 × 500	0–500	38	0–1.9	4
70 × 1000	500–1000	38	1.9–3.8	4
90 × 500	0–500	64	0–3.2	4
90 × 1000	500–1000	64	3.2–6.4	4
140 × 500	0–500	154	0–7.7	4
140 × 1000	500–1000	154	7.7–15.4	4
180 × 500	0–500	254	0–12.6	4
180 × 1000	500–1000	254	12.6–25.4	4
250 × 500	0–500	500	0–25	2
250 × 1000	500–1000	500	25–50	2
350 × 500	0–500	962	0–48	2
350 × 500	0–500	962	0–48	3
350 × 1000	500–1000	962	48–96	2
350 × 1000	500–1000	962	48–96	3
440 × 500	0–500	1520	0–76	1
440 × 500	0–500	1520	0–76	3
440 × 1000	500–1000	1520	76–152	1
440 × 1000	500–1000	1520	76–152	3
630 × 500	0–500	3119	0–156	3
630 × 1000	500–1000	3119	156–312	3

Source: Millipore, Inc., 2002

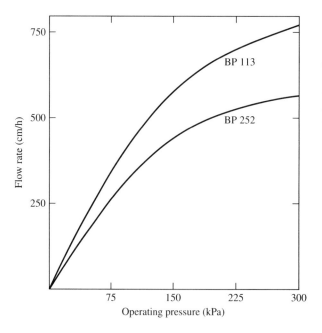

Figure 12.7 Pressure–flow characteristics for S Sepharose Fast Flow resin columns. BP 113 column: 15 cm bed height and 100 cm² cross-sectional area. BP 252 column: 15 cm bed height and 500 cm² cross-sectional area. (Courtesy of Amersham Biosciences Corp.)

References

1. Spinco Division of Beckman Coulter, Inc. (1995). *Beckman Coulter JCF-Z Instruction Manual,* Palo Alto, CA.
2. Albertsson, P.-Å. (1986). *Partition of Cell Particles and Macromolecules,* Wiley, New York.
3. Walter, H., Brooks, D. E., and Fisher, D. (1985). *Partitioning in Aqueous Two-Phase Systems,* Academic Press, Orlando, FL, 1985.

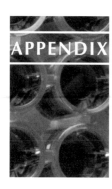

Table of Units and Constants

UNITS, MECHANICAL

Mass: 1 kg
2.2046 lb_m

Length: 1 m
3.2808 ft
39.370 in.

Volume: 1 m^3
1000 liters
35.314 ft^3
264.17 U.S. gal

Force: 1 newton (N) (kg m s^{-2})
10^5 dynes (g cm s^{-2})
0.22488 lb_f

Pressure: 1 atm
101.33 kPa (kN m^{-2})
1.0133 bars
1.0133×10^6 g cm^{-1} s^{-2} (dyne cm^{-2})
760 mm Hg (torr) at 0°C
14.696 psia (lb_f $in.^{-2}$)
29.921 in. Hg at 0°C

Energy: 1 joule (J) (N m)
10^7 g cm^2 s^{-2} (erg) (dyne cm)
0.23901 cal
3.7251×10^{-7} hp h
2.7778×10^{-7} kW h
0.73756 ft lb_f
9.4783×10^{-4} Btu

Power: 1 W (J s^{-1})
10^7 g cm^2 s^{-3} (erg s^{-1})
0.73756 ft lb_f s^{-1}

1.3410×10^{-3} hp

9.4782×10^{-4} Btu s^{-1}

Viscosity: 1 centipoise (cp) (commonly used unit)

0.01 poise (g cm^{-1} s^{-1}) (dyne s cm^{-2})

10^{-3} Pa s (N s m^{-2}) (kg m^{-1} s^{-1})

6.7197×10^{-4} lb$_m$ ft^{-1} s^{-1}

2.0886×10^{-5} lb$_f$ s ft^{-2}

UNITS, ELECTRICAL

Potential: 1 volt (V)

1 J C^{-1} (C = coulomb)

Charge: 1 electron charge

1.6022×10^{-19} C

CONSTANTS

The Gas Constant, R

The gas constant R occurs in many forms, depending on the applications. The first two values apply to thermal energy in an ideal gas (often applied to solutions). The remaining values apply to conversions involving pressure explicitly and are useful in osmotic pressure applications (van't Hoff approximation).

8.3145 J g-mol^{-1} K^{-1}

1.9872 cal g-mol^{-1} K^{-1}

8.3145×10^3 Pa m^3 g-mol^{-1} K^{-1}

0.082058 liter atm g-mol^{-1} K^{-1}

0.73020 ft^3 atm lb-mol^{-1} R^{-1}

Other Constants

Acceleration of free fall on earth	$g = 9.8066$ m s^{-2} = 980.66 cm s^{-2} = 32.173 ft s^{-2}
Avogadro's number	$\mathcal{N} = 6.0221 \times 10^{23}$ molecules g-mol^{-1}
Boltzmann's constant	$k = 1.3807 \times 10^{-23}$ J K^{-1}
Dielectric constant (relative) for water at $25°$ C	$\varepsilon_r = 78.3$
Electron charge	$e = 1.6022 \times 10^{-19}$ C
Faraday's constant	$F = 9.6485 \times 10^4$ C g-mol^{-1}
Hamaker constant (approximate)	$A = 10^{-19}$ J
Permittivity of free space (vacuum)	$\varepsilon_0 = 8.8542 \times 10^{-12}$ C^2 J^{-1} m^{-1} (farads m^{-1})

Bibliography

Bird, R. B., Stewart, W. E., and Lightfoot, E. N. (2002). *Transport Phenomena,* 2nd ed., Wiley, New York.

Dean, J. A., ed. (1992). *Lange's Handbook of Chemistry,* 14th ed., McGraw-Hill, New York.

Weast, R. C., ed. (1985). *CRC Handbook of Chemistry and Physics,* 66th ed., CRC Press, Boca Raton, FL.

Index